# Combinatorial Identities
## for Stirling Numbers

**The Unpublished Notes of H. W. Gould**

# Combinatorial Identities for Stirling Numbers

## The Unpublished Notes of H. W. Gould

Jocelyn Quaintance
University of Pennsylvania, USA

H. W. Gould
West Virginia University, USA

 World Scientific

NEW JERSEY · LONDON · SINGAPORE · BEIJING · SHANGHAI · HONG KONG · TAIPEI · CHENNAI · TOKYO

*Published by*

World Scientific Publishing Co. Pte. Ltd.
5 Toh Tuck Link, Singapore 596224
*USA office:* 27 Warren Street, Suite 401-402, Hackensack, NJ 07601
*UK office:* 57 Shelton Street, Covent Garden, London WC2H 9HE

**British Library Cataloguing-in-Publication Data**
A catalogue record for this book is available from the British Library.

**COMBINATORIAL IDENTITIES FOR STIRLING NUMBERS**
**The Unpublished Notes of H W Gould**

ISBN 978-981-4725-26-2

Printed in Singapore

To my parents for all their patience, love, and support.
To Jean, fundraiser extraordinaire.

# Foreword

The book you hold in your hands is a uniquely valuable work revealing the tastes and interests of Henry Gould. It is a mixture of Henry's own personal development of the Stirling numbers, combined with an incredible historical knowledge of the subject. Indeed, this book reminds me in many ways of Nathan Fine's book, *Basic Hypergeometric Series and Applications* (for which I wrote the preface).

Henry Gould's first paper on Stirling numbers appeared in 1960 (*Proc. American Math. Soc.*, 11 (1960), 447-451). This first paper is typical of Gould's style and approach. The question addressed concerned formulas by Schlömilch and Schäfli which present representations of Stirling numbers of the first kind by Stirling numbers of the second kind. Gould asked, quite naturally, can we reverse this so that numbers of the second kind are represented by numbers of the first kind? The beautiful answer he provided is:

$$S_2(n - k, k) = \sum_{j=0}^{k} \binom{k-n}{k+j}\binom{k+n}{k-j} S_1(k+j-1, k).$$

In the interim, he has written many further papers on this topic. Indeed this book is, in some sense, an extensive expansion and completion of one aspect of class notes he prepared a number of years ago entitled *Sums of Powers of Numbers*.

Essential for this book is the historical understanding that Henry Gould brings to the subject. This historical expertise is perhaps most clearly revealed in a letter he wrote some time ago to Richard Askey. Henry never officially got a Ph.D. While this is uncommon today especially in the U.S., it was a common practice in England early in the 19th and early 20th century; in particular, G.H. Hardy never had a Ph.D. (he would have considered it beneath him). However, Henry is actually an unofficial student of Leonard Carlitz. Here better than anything I can say are Henry's words describing his study of history and his debt to Carlitz:

> " I remember my first visit to Duke, circa 1953, before I got my B.A. degree (that came in 1954). It was at the suggestion of Leonard Carlitz. I had read some of his papers and wrote to him, sending him some formulas I had discovered. He responded that I should visit Duke and look at this book and that book and this journal and another. ... that some of my formulas were new while others were old. For example I had "rediscovered" Abel's extension of the binomial formula, etc. while I was in high school.
>
> Carlitz knew what was in most every paper in the library. I also used to have that ability early on. I used to spend whole days in the libraries at Virginia, Carolina, Duke, Library of Congress, and numerous other places, trying to find what was known. I began years ago jotting things down in notebooks, then on file cards, and now have over 30,000 3" by 5" file cards on the literature that interested me. When I first came here to WVU 38 years ago, I began going through Crelle's and Liouville's journals and the Jahrbuch ueber die Fortschritte der Mathematik, Zentralblatt, Math. Reviews and every journal we had in fact, page by page. I remember many nights when I would sleep in the library, on the hard floor sometimes when I got tired after making notes on 50 volumes of some journal.
>
> Consequently I can recall times when I would visit Carlitz or talk with him on the phone and mention some paper by (e.g.) Gegenbauer, and we would both know the paper and both remark that the author proved 300 formulas or whatever ... it was all at our mental fingertips. I often take a student to the library and just walk over and pull down a journal volume to show the student some specific reference and the student is always amazed that I remember exactly where to look!"

That, dear reader, is Henry Gould, a man deeply in love with mathematics, especially combinatorial mathematics. Not surprisingly, many of the 30,000 file cards are related to Stirling numbers. It is good to see that his insights and understanding are finally being published.

We owe Jocelyn Quaintance a debt for guiding the writing and persevering in making sure that this grand project finally made it into print.

*George E. Andrews*

# Preface

This book is the result of a life changing relationship. In 2006 I was a recently graduated Ph.D. trying to publish my first combinatorial research papers with little success. In order to become a research mathematician, I realized that I had to learn what topics led to publication-worthy research projects. I thought the best way to gain these insights was to take a year off and work with an expert in the field of enumerative combinatorics. I asked George Andrews who he recommended I contact as a potential advisor. He suggested Henry Gould. I sent Henry an email not knowing what to expect. Much to my surprise and delight, Henry quickly replied that he would be most pleased to become my mentor. Therefore, I quit my job and in August 2006 travelled to Morgantown, WV to begin my studies with Henry. Henry introduced me to the wonderful world of binomial identities, special functions, Catalan numbers, and Stirling numbers. Under his guidance, I learned how to formulate research questions, the techniques needed to solve these questions, and how to present the results for publication. During the next four years, Henry and I collaborated on over ten papers. When working on these papers, Henry often referred to his "bible", a much used copy of his book entitled *Combinatorial Identities*, a collection of tables containing over 500 binomial identities. I asked Henry if he had the proofs for all of the identities in this book. He told me that his proofs existed in an unpublished manuscript of seven handwritten volumes which he started writing before he was an undergraduate at the University of Virginia. Henry said he was loathe to show anyone his private notebooks since they were a diary of his mathematical development over the past 50 years and were not in publication ready form. But since we had developed such a rapport, he would show me a few excerpts. When I saw the excerpts, I realized that I was looking at something special that needed to be shared with the mathematical community. I told Henry this and because he trusted me, he was willing to give me a copy of the 2100+ handwritten pages to prepare for publication. Thus began a three year odyssey, the end result is this book. The first eight chapters introduce readers to the special techniques Henry uses in proving his binomial identities. We do not assume the reader has any special mathematical background and have written this material in a style which should be readily accessible by anyone who has taken a year of calculus and an undergraduate level course in discrete mathematics. The second half of the book uses the techniques espoused in the first half and is geared towards the research mathematician. It focuses on connections between various kinds of Stirling numbers, a topic on which Henry is a world-renowned expert. As far as we know, Chapters 9 through 15 are the only source which systematically records Henry's unique results interrelating Stirling numbers of the first kind, Stirling numbers of the second kind, Worpitzky numbers, Bernoulli numbers, and Nörlund polynomials. Researchers should particularly focus on Chapters 13 through 15.

My fervent hope is this book will introduce a new generation of mathematicians to the world of combinatorial identities by teaching them the skills necessary for discovering new identities, just as Henry patiently taught me over the past ten years. That would be the most fitting tribute to Henry's legacy.

*Jocelyn Quaintance*

# Acknowledgments

Henry and Jocelyn thank Dr. Addison Fischer for his three years of financial support. Jocelyn thanks Drs. Monique and Harry Gingold (and Beshy) for being gracious hosts. She also thanks João Sedoc for his help in typesetting the manuscript in the publisher's format.

# Contents

# Chapter 1

# Basic Properties of Series

The purpose of this book is to develop Professor Gould's formulas for relating Stirling numbers of the second kind to Stirling numbers of the first kind via Bernoulli numbers. Many of these relationships rely on Professor Gould's techniques for evaluating series whose summands are binomial coefficients. Therefore, the first eight chapters of this book will be a primer on these various techniques. Assume $n$ and $k$ are nonnegative integers with $k \leq n$. Define $n!$ to be the product of the first $n$ positive integers and $\binom{n}{k} := \frac{n!}{k!(n-k)!}$ with $0! := 1$. Combinatorially $\binom{n}{k}$ counts the number of subsets of size $k$ made from a set with $n$ elements. We say $\binom{n}{k}$ is a **binomial coefficient**. The binomial coefficients are often displayed in Pascal's triangle. Pascal's triangle begins with $\binom{0}{0}$, and for row $i$, with $i \geq 2$, the $j^{th}$ entry from the left is $\binom{i-1}{j-1}$. Each row of the Pascal's triangle contains only a finite number of entries since $\binom{n}{k} = 0$ whenever $k > n$.

| | | | | | 1 | | | | | |
|---|---|---|---|---|---|---|---|---|---|---|
| | | | | 1 | | 1 | | | | |
| | | | 1 | | 2 | | 1 | | | |
| | | 1 | | 3 | | 3 | | 1 | | |
| | 1 | | 4 | | 6 | | 4 | | 1 | |
| 1 | | 5 | | 10 | | 10 | | 5 | | 1 |
| 1 | 6 | | 15 | | 20 | | 15 | | 6 | 1 |
| 1 | 7 | 21 | | 35 | | 35 | | 21 | 7 | 1 |

Table 1.1: A portion of Pascal's triangle

There is an inductive way to construct the rows of Pascal's triangle which is known as Pascal's identity.

**Pascal's Identity:** Let $n$ and $k$ be nonnegative integers, with $0 \le k \le n$. Define $\binom{n}{k} = 0$ if $k$ is a negative integer. Then

$$\binom{n+1}{k} = \binom{n}{k} + \binom{n}{k-1}, \tag{1.1}$$

Here is a combinatorial proof of Pascal's identity. In this and other combinatorial arguments we show both sides of a given equation count the same quantity. Suppose $S = \{1, 2, 3, ..., n+1\}$. The left side of Equation (1.1) counts the number of subsets of $S$ with size $k$. Call such a subset a $k$-subset of $S$. We claim the right side of Equation (1.1) also counts the $k$-subsets of $S$. There are two distinct types of $k$-subsets. The first type of $k$-subset contains $n+1$. To complete such a subset, we must choose $k-1$ elements from $\{1, 2, ..., n\}$ in $\binom{n}{k-1}$ possible ways. The second type of $k$-subset does not contain $n+1$, and we must select $k$ elements from $\{1, 2, ..., n\}$ in $\binom{n}{k}$ possible. Adding the two possibilities together counts all the $k$-subsets of $S$ without repetition and produces Equation (1.1).

Mathematicians often generalize definitions, and the binomial coefficients are no exceptions. The typical way to generalize $\binom{n}{k}$ is to let $n$ be an arbitrary complex number. We will always assume, unless otherwise specified, that $k$ is a nonnegative integer, and $x$ is a complex number. Define $\binom{x}{0} := 1$ and

$$\binom{x}{k} = \frac{x(x-1)(x-2)...(x-k+1)}{k!}, \qquad \text{for complex } x. \tag{1.2}$$

We say $\binom{x}{k}$ is a **general binomial coefficient**. Whenever $x = n$, Equation (1.2) corresponds to the traditional combinatorial definition of a binomial coefficient.

It is important to note that $\binom{x}{k}$ is a polynomial in $x$ of degree $k$. This observation, along the with Fundamental theorem of algebra, proves many binomial identities.

Equation (1.1) generalizes as follows:

**Pascal's Identity for general binomial coefficients:** Let $x$ be a complex number and $k$ a nonnegative integer. Define $\binom{x}{k} = 0$ if $k$ is a negative integer. Then

$$\binom{x+1}{k} = \binom{x}{k} + \binom{x}{k-1}. \tag{1.3}$$

The proof of Equation (1.3) is a matter of applying the definition provided by Equation (1.2). In particular

$$\binom{x}{k} + \binom{x}{k-1} = \frac{x(x-1)...(x-k+1)}{k!} + \frac{x(x-1)...(x-k+2)}{(k-1)!}$$

$$= \frac{x(x-1)...(x-k+2)[(x-k+1)+k]}{k!}$$

$$= \frac{(x+1)x(x-1)...(x-k+2)}{k!} = \binom{x+1}{k}.$$

There are four other binomial identities necessary for understanding the material in the following chapters. All of these identities are proven by applications of Equation (1.2). We leave the proofs of the first three identities for the reader and provide a detailed proof of the $-\frac{1}{2}$-Transformation.

**Committee/Chair Identity:** Let $x$ be a complex number and $n$ be a nonnegative integer. Then

$$(n+1)\binom{x}{n+1} = x\binom{x-1}{n}. \tag{1.4}$$

**Cancellation Identity:** Assume $x$ is a complex number, while $n$ and $k$ are nonnegative integers such that $n \geq k$. Then

$$\binom{x}{n}\binom{n}{k} = \binom{x-k}{n-k}\binom{x}{k}. \tag{1.5}$$

$-1$-**Transformation:** Let $x$ be a complex number and $k$ be a nonnegative integer. Then

$$\binom{x}{k} = (-1)^k \binom{-x+k-1}{k}. \tag{1.6}$$

$-\frac{1}{2}$-**Transformation:** Let $n$ be a nonnegative integer. Then

$$\binom{-\frac{1}{2}}{n} = \frac{(-1)^n}{2^{2n}}\binom{2n}{n}. \tag{1.7}$$

**Proof:** Using Equation (1.2) we see that

$$\binom{-\frac{1}{2}}{n} = \frac{(-\frac{1}{2})(-\frac{3}{2})...(-\frac{1}{2}-n+1)}{n!} = \frac{(-1)^n}{2^n} \cdot \frac{1 \cdot 3 \cdot ... \cdot (2n-1)}{n!}$$

$$= \frac{(-1)^n}{2^n} \cdot \frac{1 \cdot 3 \cdot ... \cdot (2n-1)}{n!} \cdot \frac{2 \cdot 4 \cdot ... \cdot 2n}{2 \cdot 4 \cdot ... \cdot 2n}$$

$$= \frac{(-1)^n}{2^n} \cdot \frac{(2n)!}{n!} \cdot \frac{1}{2^n(1 \cdot 2 \cdot ... \cdot n)} = \frac{(-1)^n}{2^{2n}} \cdot \frac{(2n)!}{n!n!}$$

$$= \frac{(-1)^n}{2^{2n}}\binom{2n}{n}. \qquad \square$$

Throughout the first eight chapters the primary object of study will be finite series $\sum_{k=0}^{n} a_k = a_1 + a_2 + \cdots + a_n$ where each $a_k$ is either a real or complex number. The $k$ is the index of the series and $a_k$ is often a product of binomial coefficients. Occasionally we will work with infinite series $\sum_{n=0}^{\infty} a_n$. In most cases the infinite series will be of the form $\sum_{n=0}^{\infty} a_k x^k$, where $x$ is a formal parameter. We interpret $\sum_{n=0}^{\infty} a_k x^k$ as a *formal power series*, ignore questions of convergence, and instead focus on the algebraic and combinatorial manipulations of series. For more information on formal powers series, we refer the reader to [Wilf, 2014, Chap.2].

## 1.1   General Considerations of $\sum_{k=a}^{n} f(k)$

We begin our study of algebraic manipulations for finite series by stating the following collection of basic identities which are implicitly used throughout Professor Gould's work. Since many of these identities are self evident, we leave the proofs to the reader. Without loss of generality we assume $f$ and $g$ are arbitrary functions whose domains are the set of all nonnegative integers and whose ranges are the set of complex numbers. We also assume, unless otherwise specified, that $n$ and $a$ are nonnegative integers.

**Linearity Properties:** Let $c$ be a complex number not indexed by $k$. Then

$$\sum_{k=a}^{n} f(k) + \sum_{k=a}^{n} g(k) = \sum_{k=a}^{n} [f(k) + g(k)] \qquad (1.8)$$

$$\sum_{k=a}^{n} cf(k) = c\sum_{k=a}^{n} f(k). \qquad (1.9)$$

**Identity 1:** Let $0 < k < a - 1 \leq n$. Then

$$\sum_{k=0}^{n} f(k) = \sum_{k=0}^{a-1} f(k) + \sum_{k=a}^{n} f(k). \qquad (1.10)$$

**Identity 2:** Let $0 < k_1 < k_2 < \ldots < k_{m-1} < n$.

$$\sum_{j=0}^{n} f(j) = \sum_{j=0}^{k_1} f(j) + \sum_{j=k_1+1}^{k_2} f(j) + \ldots + \sum_{j=k_{m-2}+1}^{k_{m-1}} f(j) + \sum_{j=k_{m-1}+1}^{n} f(j)$$

$$= \sum_{i=0}^{m-1} \sum_{j=k_i+1}^{k_{i+1}} f(j), \qquad \text{where } k_0 + 1 := 0 \text{ and } k_m = n. \quad (1.11)$$

**Identity 3:** Let $x$ be a real number. Let $\lfloor x \rfloor$ denote the greatest integer less than or equal to $x$. Let $a$ and $n$ be nonnegative integers such that $n \geq a + 1$. Then

$$\sum_{k=a}^{n} f(k) + \sum_{k=a}^{n} (-1)^k f(k) = 2 \sum_{k=\lfloor \frac{a+1}{2} \rfloor}^{\lfloor \frac{n}{2} \rfloor} f(2k). \tag{1.12}$$

**Identity 4:** Let $x$ be a real number and let $\lfloor x \rfloor$ denote the greatest integer less than or equal to $x$. Let $a$ and $n$ be nonnegative integers such that $n \geq a + 1$. Then

$$\sum_{k=a}^{n} f(k) - \sum_{k=a}^{n} (-1)^k f(k) = 2 \sum_{k=\lfloor \frac{a+2}{2} \rfloor}^{\lfloor \frac{n+1}{2} \rfloor} f(2k - 1). \tag{1.13}$$

If we add Identity 3 to Identity 4 we obtain

**Identity 5:** (bisection of a series) Let $n \geq a + 1$. Then

$$\sum_{k=a}^{n} f(k) = \sum_{k=\lfloor \frac{a+1}{2} \rfloor}^{\lfloor \frac{n}{2} \rfloor} f(2k) + \sum_{k=\lfloor \frac{a+2}{2} \rfloor}^{\lfloor \frac{n+1}{2} \rfloor} f(2k - 1)$$

$$= \sum_{k=\lfloor \frac{a+1}{2} \rfloor}^{\lfloor \frac{n}{2} \rfloor} f(2k) + \sum_{k=\lfloor \frac{a}{2} \rfloor}^{\lfloor \frac{n-1}{2} \rfloor} f(2k + 1). \tag{1.14}$$

We now state the following generalization of Identity 5:

**Identity 6:** (multi-section of a series) Let $r$ be a positive integer. Let $n$ and $a$ be nonnegative integers such that $n - a + 1 \geq r$. Then

$$\sum_{k=a}^{n} f(k) = \sum_{j=0}^{r-1} \sum_{k=\lfloor \frac{a+r-1-j}{r} \rfloor}^{\lfloor \frac{n-j}{r} \rfloor} f(rk + j). \tag{1.15}$$

Identity 5 is the case when $r = 2$.

For a fixed $r$ Identity 6 partitions the sum modulo remainder classes. The number of remainder classes is indexed by $j$.

In practice Identity 6 often occurs with $a = 0$ and the number of summands being a multiple of $r$, in which case Equation (1.15) is equivalent to

**Identity 6a:** (multi-section of a series) Let $r$ be a positive integer. Then

$$\sum_{k=0}^{rn-1} f(k) = \sum_{j=0}^{r-1}\sum_{k=0}^{n-1} f(rk+j). \tag{1.16}$$

A useful special case of Identity 5 occurs when $f(k) \to (-1)^k f(k)$, in which case we have

**Identity 7:** (alternating bisection formula) Let $n$ and $a$ be nonnegative integers such that $n \geq a+1$. Then

$$\sum_{k=a}^{n}(-1)^k f(k) = \sum_{k=\lfloor\frac{a+1}{2}\rfloor}^{\lfloor\frac{n}{2}\rfloor} f(2k) - \sum_{k=\lfloor\frac{a+2}{2}\rfloor}^{\lfloor\frac{n+1}{2}\rfloor} f(2k-1)$$

$$= \sum_{k=\lfloor\frac{a+1}{2}\rfloor}^{\lfloor\frac{n}{2}\rfloor} f(2k) - \sum_{k=\lfloor\frac{a}{2}\rfloor}^{\lfloor\frac{n-1}{2}\rfloor} f(2k+1). \tag{1.17}$$

A similar substitution in Identity 6 provides

**Identity 8:** (alternating multi-section formula) Let $r$ be a positive integer. Let $n$ and $a$ be nonnegative integers such that $n - a + 1 \geq r$. Then

$$\sum_{k=a}^{n}(-1)^k f(k) = \sum_{j=0}^{r-1}\sum_{k=\lfloor\frac{a+r-1-j}{r}\rfloor}^{\lfloor\frac{n-j}{r}\rfloor} (-1)^{rk+j} f(rk+j). \tag{1.18}$$

**Identity 9:** (basic telescoping formula) Let $r$ and $n$ be fixed positive integers. Then

$$\sum_{k=1}^{n}\left(f(k) - f(k+r)\right) = \sum_{k=1}^{r}\left(f(k) - f(k+n)\right). \tag{1.19}$$

Both sums of Equation (1.19) add and subtract elements from $S = \{f(1), ... f(n+r)\}$. The focus of Identity 9 is combining the $n+r$ elements of $S$ in two different ways. On the left side we select $f(k)$, where $1 \leq k \leq n$, and for each selection subtract $f(k+r)$. On the right side we select $f(k)$, where $1 \leq k \leq r$, and then subtract $f(k+r)$. Identity 9 claims that these two different ways of telescopically combining various elements from $S$ provides the same sum. Clearly if $n = r$, Identity 9 is trivially true. So without loss of generality assume that $r < n$. Then

$$
\begin{aligned}
\sum_{k=1}^{n} (f(k) - f(k+r)) &= \sum_{k=1}^{n} f(k) - \sum_{k=1}^{n} f(k+r) \\
&= \sum_{k=1-r}^{k=n-r} f(k+r) - \sum_{k=1}^{n} f(k+r) \\
&= \sum_{k=1-r}^{0} f(k+r) - \sum_{k=n-r+1}^{n} f(k+r) \\
&= \sum_{k=1}^{r} f(k) - \sum_{k=1}^{r} f(k+n) \\
&= \sum_{k=1}^{r} (f(k) - f(k+n)).
\end{aligned}
$$

A useful interpretation of Identity 9 may be derived as follows. Rewrite Equation (1.19) as

$$
r \sum_{k=1}^{n} \frac{f(k+r) - f(k)}{r} = n \sum_{k=1}^{r} \frac{f(k+n) - f(k)}{n}. \tag{1.20}
$$

The summands of Equation (1.20) are commonly written in terms of the difference operator $\Delta_{k,p} := \frac{f(k+p) - f(k)}{p}$. Hence Equation (1.20) becomes

$$
\frac{1}{n} \sum_{k=1}^{n} \Delta_{k,r} f(k) = \frac{1}{r} \sum_{k=1}^{r} \Delta_{k,n} f(k). \tag{1.21}
$$

Equation (1.21) implies that for fixed positive integers $n$ and $r$, the average of $n$ differences in increments of $r$ equals the average of $r$ differences in increments of $n$.

**Identity 10:** (doubling formula) Let $n$ be a fixed positive integer. Then

$$
2 \sum_{k=1}^{n} f(k) = \sum_{k=1}^{2n} f\left(\left\lfloor \frac{k+1}{2} \right\rfloor\right). \tag{1.22}
$$

An equivalent form of Identity 10 is obtained by setting $k \to k+1$ in the right hand sum.

**Identity 10a:** (doubling formula) Let $n$ be a fixed positive integer. Then

$$2\sum_{k=1}^{n} f(k) = \sum_{k=0}^{2n-1} f\left(\left\lfloor \frac{k}{2}\right\rfloor + 1\right). \qquad (1.23)$$

We let $f(k) \to (-1)^{k-1} f(k)$ in Equation (1.23) to obtain the following useful variation.

**Identity 11:** (alternating series doubling formula) Let $n$ be a positive integer. Then

$$2\sum_{k=1}^{n} (-1)^{k-1} f(k) = \sum_{k=0}^{2n-1} (-1)^{\lfloor \frac{k}{2}\rfloor} f\left(\left\lfloor \frac{k}{2}\right\rfloor + 1\right). \qquad (1.24)$$

**Identity 12:** (index shift property) Let $k$ and $n$ be nonnegative integers such that $n \geq a$. Then

$$\sum_{k=a}^{n} f(k) = \sum_{k=0}^{n-a} f(n-k). \qquad (1.25)$$

When computing $\sum_{k=a}^{n} f(k)$, we run the index from $a$ to $n$. If we shift the index by $k \to n - k$, we still have the same sum, but have inverted the order of summation.

A useful variation of Equation (1.25) is

$$\sum_{k=a}^{n} f(k) = \sum_{k=a}^{n} f(n+a-k), \qquad (1.26)$$

where the right side of Equation (1.26) is the right side of Equation (1.25) with $k \to k - a$.

## 1.2   Pascal's Identity in Evaluation of Series

In the final section of this first chapter we demonstrate how the basic identities are used to algebraically manipulate finite series whose summands involve binomial coefficients. In particular we focus on Pascal's identity which says

$$\binom{x}{j} = \binom{x-1}{j} + \binom{x-1}{j-1}, \qquad (1.27)$$

whenever $x$ is a complex number and $j$ is a nonnegative integer. Take Equation (1.27), let $x \to x + 1$, and solve for $\binom{x}{j}$ to obtain

$$\binom{x}{j} = \binom{x+1}{j} - \binom{x}{j-1}. \tag{1.28}$$

We will work with repeated iterations of Equation (1.28), substituting an equivalence for the second term on the right side given by Equation (1.28), and continuing recursively. The following calculations demonstrate this technique:

$$\binom{x}{j} = \binom{x+1}{j} - \binom{x}{j-1} = \binom{x+1}{j} - \left[\binom{x+1}{j-1} - \binom{x}{j-2}\right]$$

$$= \binom{x+1}{j} - \binom{x+1}{j-1} + \left[\binom{x+1}{j-2} - \binom{x}{j-3}\right]$$

$$= \binom{x+1}{j} - \binom{x+1}{j-1} + \binom{x+1}{j-2} - \left[\binom{x+1}{j-3} - \binom{x}{j-4}\right]$$

$$\cdots$$

By continuing this iterative process $r$ times we deduce that

$$\sum_{k=0}^{r-1} (-1)^k \binom{x+1}{j-k} = \binom{x}{j} - (-1)^r \binom{x}{j-r}, \qquad r \geq 1. \tag{1.29}$$

Equation (1.29) is verified through induction on $r$. To establish the base case of $r = 1$ notice that Equation (1.28) implies

$$\binom{x}{j} = \binom{x+1}{j} - \binom{x}{j-1} = \sum_{k=0}^{0} (-1)^k \binom{x+1}{j-k} + (-1)^1 \binom{x}{j-1}.$$

Now assume Equation (1.29) is true for all positive integers less than $r$. Then

$$\sum_{k=0}^{r} (-1)^k \binom{x+1}{j-k} = (-1)^r \binom{x+1}{j-r} + \sum_{k=0}^{r-1} (-1)^k \binom{x+1}{j-k}$$

$$= (-1)^r \binom{x+1}{j-r} + \binom{x}{j} - (-1)^r \binom{x}{j-r}$$

$$= \binom{x}{j} + (-1)^r \left[\binom{x+1}{j-r} - \binom{x}{j-r}\right]$$

$$= \binom{x}{j} + (-1)^r \binom{x}{j-r-1}$$

$$= \binom{x}{j} - (-1)^{r+1} \binom{x}{j-r-1}.$$

If we replace $x$ with $x - 1$, $j \to n$, $r \to n + 1$, and apply the index shift property, we discover that Equation (1.29) is equivalent to

$$\sum_{k=0}^{n}(-1)^k \binom{x}{k} = (-1)^n \binom{x-1}{n} = \binom{n-x}{n}. \qquad (1.30)$$

Equation (1.30) has a colorful history. Professor Gould first discovered this identity during his undergraduate studies at the University of Virginia. One night while working as a DJ at the student radio station, which broadcasts at frequency 640 AM, he used this exact Pascal identity technique to derive Equation (1.30). In private conversations he often refers to this particular binomial identity as Identity 640 . To obtain Professor Gould's original formulation of Equation (1.30) we need to derive a product expression equivalent to $(-1)^n \binom{x-1}{n}$. Let $\{a_i\}_{i=1}^{n}$ be a set of $n$ not necessarily distinct complex numbers. Define $\prod_{k=1}^{n} a_k := a_1 a_2 \ldots a_n$ to be the product of these $n$ elements. For example, if $a_i = i$, $\prod_{i=1}^{n} i = (1)(2)\ldots(n) = n!$. By definition

$$\prod_{k=1}^{n}\left(1 - \frac{x}{k}\right) = \left(1 - \frac{x}{1}\right)\left(1 - \frac{x}{2}\right)\ldots\left(1 - \frac{x}{n}\right)$$

$$= \left(\frac{1-x}{1}\right)\left(\frac{2-x}{2}\right)\ldots\left(\frac{n-x}{n}\right)$$

$$= (-1)^n\frac{(x-1)(x-2)\ldots(x-n)}{n!} = (-1)^n\binom{x-1}{n}.$$

These calculations imply that

$$\sum_{k=0}^{n}(-1)^k\binom{x}{k} = \prod_{k=1}^{n}\left(1 - \frac{x}{k}\right) = (-1)^n\binom{x-1}{n} = \binom{n-x}{n}. \qquad (1.31)$$

Equation (1.31) is Professor Gould's Identity 640 in its entirety. Equation (1.31) is valid for any complex number $x$, the sum and $\binom{n-x}{n}$ are valid for any nonnegative integer $n$, while the product requires that $n \geq 1$.

Equation (1.31) is a versatile identity and is key to deriving many of the identities presented in Table 1/0 of [Gould, 1972]. For example, an application of the Committee/Chair identity, when combined with Equation (1.31) shows that

$$\sum_{k=1}^{n}(-1)^k\binom{x}{k}k = x\sum_{k=1}^{n}(-1)^k\binom{x-1}{k-1} = -x\sum_{k=0}^{n-1}(-1)^k\binom{x-1}{k}$$

$$= (-1)^n x\binom{x-2}{n-1}. \qquad (1.32)$$

Equation (1.31) may also be combined with the $-\frac{1}{2}$-Transformation. Set $x = \frac{1}{2}$ to obtain

$$\sum_{k=0}^{n}(-1)^{k}\binom{\frac{1}{2}}{k} = (-1)^{n}\binom{-\frac{1}{2}}{n}. \tag{1.33}$$

We simplify the right side of Equation (1.33) through the $-\frac{1}{2}$-Transformation. Techniques similar to those used to prove the $-\frac{1}{2}$-Transformation show that

$$\binom{\frac{1}{2}}{n} = -\frac{1}{2n-1}\binom{-\frac{1}{2}}{n} = (-1)^{n+1}\binom{2n}{n}\frac{1}{2^{2n}(2n-1)}. \tag{1.34}$$

If we take these two transformations, place them into Equation (1.33), and simplify our results, we obtain

$$\sum_{k=0}^{n}\binom{2k}{k}\frac{1}{2^{2k}(2k-1)} = -\binom{2n}{n}\frac{1}{2^{2n}}. \tag{1.35}$$

In Equation (1.31) let $x = -\frac{1}{2}$ to obtain

$$\sum_{k=0}^{n}(-1)^{k}\binom{-\frac{1}{2}}{k} = (-1)^{n}\binom{-\frac{3}{2}}{n}. \tag{1.36}$$

We may use the $-\frac{1}{2}$-Transformation to simplify the left side summand of Equation (1.36). For the right side we claim that

$$\binom{-\frac{3}{2}}{n} = (-1)^{n}\binom{2n}{n}\frac{2n+1}{2^{2n}} = (-1)^{n}\binom{2n+1}{n}\frac{n+1}{2^{2n}}. \tag{1.37}$$

Take the $-\frac{1}{2}$-Transformation and Equation (1.37), substitute them into Equation (1.36), and simplify to discover that

$$\sum_{k=0}^{n}\binom{2k}{k}\frac{1}{2^{2k}} = \binom{2n}{n}\frac{2n+1}{2^{2n}}, \tag{1.38}$$

the complement to Equation (1.35).

Equations (1.32), (1.35), and (1.38) are just a small sampling of what can be done through Equation (1.31). The importance of Equation (1.31) will be seen in our work with Stirling numbers of the first kind.

# Chapter 2

# The Binomial Theorem

In this chapter we will discuss what is perhaps the most important combinatorial identity, the binomial theorem. It is the first identity Professor Gould lists in his seminal work *Combinatorial Identities: A Standardized Set of Tables Listing 500 Binomial Coefficient Summations* [Gould, 1972]. On Page 1 of that work the reader finds

$$\sum_{k=0}^{\infty} \binom{\alpha}{k} z^k = (1+z)^{\alpha}, \qquad \text{valid for complex } \alpha \text{ and complex } |z| < 1,$$

(2.1)

where the principal value of $(1+z)^{\alpha}$ is taken. Convergence is irrelevant when we think of this as a generating function.

Equation (2.1) is the famous **binomial theorem**, sometimes referred to as **Newton's binomial theorem** or the **binomial series**. For the time being we restrict ourselves to the special case of Equation (2.1) in which $\alpha$ is a positive integer $n$. Since $\binom{n}{k} = 0$ whenever $k > n$, we may rewrite Equation (2.1) as

$$\sum_{k=0}^{n} \binom{n}{k} z^k = (1+z)^n, \qquad z \text{ an arbitrary complex number.} \quad (2.2)$$

Here is a combinatorial proof of Equation (2.2). By definition the right side of Equation (2.2) is $(1+z)(1+z)...(1+z) = \prod_{k=1}^{n}(1+z)$. How can we obtain $z^k$ from the product of these $n$ factors? For each such factor the distributive law implies that we must choose either a $z$ or a 1. In order to obtain a $z^k$, we must select $z$ from exactly $k$ of these $n$ factors. Thus, the total number of $z^k$ is the number of ways we can choose a $k$-subset from these $n$ factors, namely $\binom{n}{k}$. By varying $0 \leq k \leq n$, we obtain Equation (2.2).

Equation (2.2) can also be proven via induction on $n$. To establish the base case let $n = 1$ and note that

$$(1+z)^1 = 1 + z = \binom{1}{0} z^0 + \binom{1}{1} z^1 = \sum_{k=0}^{1} \binom{1}{k} z^k.$$

Now assume Equation (2.2) is true for all positive integers less than or equal to $n$. We then obtain

$$(1+z)^{n+1} = (1+z)(1+z)^n = (1+z) \sum_{k=0}^{n} \binom{n}{k} z^k$$

$$= \sum_{k=0}^{n} \binom{n}{k} z^k + \sum_{k=0}^{n} \binom{n}{k} z^{k+1} = \sum_{k=0}^{n} \binom{n}{k} z^k + \sum_{k=1}^{n+1} \binom{n}{k-1} z^k$$

$$= \sum_{k=0}^{n+1} \left[ \binom{n}{k} + \binom{n}{k-1} \right] z^k = \sum_{k=0}^{n+1} \binom{n+1}{k} z^k,$$

where the last equality follows from Pascal's identity. The final sum is identical to the left side of Equation (2.2) when $n \to n+1$. Hence the induction proof is complete and Equation (2.2) is true for all positive $n$.

By adapting the proofs of Equation (2.2), we can replace the one with an arbitrary complex number $y$ and obtain

$$\sum_{k=0}^{n} \binom{n}{k} z^k y^{n-k} = (y+z)^n, \qquad n \text{ a nonnegative integer.} \qquad (2.3)$$

Two important special cases of Equation (2.3) occur when $z = y = 1$ and $z = -1$, $y = 1$. The first case implies that

$$\sum_{k=0}^{n} \binom{n}{k} = (1+1)^n = 2^n, \qquad (2.4)$$

while the second implies that

$$\sum_{k=0}^{n} (-1)^k \binom{n}{k} = \begin{cases} 0, & n \geq 1 \\ 1, & n = 0. \end{cases} \qquad (2.5)$$

These identities are best remembered in terms of Pascal's triangle. Equation (2.4) is the row sum while Equation (2.5) is the alternating row sum. Another identity associated with Pascal's is the equality between the number of even size subsets and the number of odd size subsets. To obtain this identity apply Identities 5 and 7 with $f(k) = \binom{n}{k}$. Identity 5 implies that

$$2^n = \sum_{k=0}^{n} \binom{n}{k} = \sum_{k=0}^{\lfloor \frac{n}{2} \rfloor} \binom{n}{2k} + \sum_{k=0}^{\lfloor \frac{n-1}{2} \rfloor} \binom{n}{2k+1}, \qquad (2.6)$$

while Identity 7 implies that

$$0 = \sum_{k=0}^{n} (-1)^k \binom{n}{k} = \sum_{k=0}^{\lfloor \frac{n}{2} \rfloor} \binom{n}{2k} - \sum_{k=0}^{\lfloor \frac{n-1}{2} \rfloor} \binom{n}{2k+1}, \qquad n \geq 1. \qquad (2.7)$$

From Equation (2.7) we deduce that $\sum_{k=0}^{\lfloor \frac{n}{2} \rfloor} \binom{n}{2k} = \sum_{k=0}^{\lfloor \frac{n-1}{2} \rfloor} \binom{n}{2k+1}$ whenever $n \geq 1$. Substituting this equality into Equation (2.6) implies that

$$\sum_{k=0}^{\lfloor \frac{n}{2} \rfloor} \binom{n}{2k} = \sum_{k=0}^{\lfloor \frac{n-1}{2} \rfloor} \binom{n}{2k+1} = \sum_{k=0}^{n} \binom{2n}{2k-1} = 2^{n-1}, \qquad n \geq 1. \qquad (2.8)$$

The left sum in Equation (2.8) is the total number of even subsets while the right is total number of odd subsets. If $n \to 2n$, Equation (2.8) becomes

$$\sum_{k=0}^{n} \binom{2n}{2k} = \sum_{k=0}^{n} \binom{2n}{2k+1} = 2^{2n-1}, \qquad n \geq 1. \qquad (2.9)$$

Pascal's identity implies

$$\binom{2n+1}{2k+1} = \binom{2n}{2k+1} + \binom{2n}{2k}. \qquad (2.10)$$

Take Equation (2.10) and sum over $k$, where $0 \leq k \leq n$, and combine with Equation (2.9) to obtain

$$\sum_{k=0}^{n} \binom{2n+1}{2k+1} = \sum_{k=0}^{n} \binom{2n}{2k+1} + \sum_{k=0}^{n} \binom{2n}{2k} = 2 \sum_{k=0}^{n} \binom{2n}{2k} = 2^{2n}. \qquad (2.11)$$

Although Equation (2.9) requires $n \geq 1$, Equation (2.11) holds for $n \geq 0$. A similar calculation involving Pascal's Identity and Equation (2.8) implies that

$$2^{2n} = \sum_{k=0}^{n} \binom{2n}{2k} + \sum_{k=0}^{n} \binom{2n}{2k-1} = \sum_{k=0}^{n} \binom{2n+1}{2k}. \qquad (2.12)$$

Identities (2.4) through (2.12) are basic applications of the binomial theorem. Although there are many other applications of Equation (2.2), we demonstrate just one more which is related to Melzak's theorem. Let $n$ be a nonnegative integer. Define $S_n := \sum_{k=0}^{n} \binom{n}{k} \frac{x^k}{k+1}$, where $x$ is a nonzero complex number. Equation (2.2) implies that

$$\begin{aligned}
S_n &= \sum_{k=0}^{n} \binom{n}{k} \frac{x^k}{k+1} = \frac{1}{n+1} \sum_{k=0}^{n} \binom{n+1}{k+1} x^k \\
&= \frac{1}{n+1} \sum_{k=1}^{n+1} \binom{n+1}{k} x^{k-1} = \frac{1}{x(n+1)} \sum_{k=1}^{n+1} \binom{n+1}{k} x^k \\
&= \frac{1}{x(n+1)} \left[ \sum_{k=0}^{n+1} \binom{n+1}{k} x^k - 1 \right] = \frac{(x+1)^{n+1} - 1}{x(n+1)}. \qquad (2.13)
\end{aligned}$$

Two special cases of Equation (2.13) are

$$\sum_{k=0}^{n} \binom{n}{k} \frac{1}{k+1} = \frac{2^{n+1}-1}{n+1}, \qquad x = 1 \qquad (2.14)$$

$$\sum_{k=0}^{n} (-1)^k \binom{n}{k} \frac{1}{k+1} = \frac{1}{n+1}, \qquad x = -1. \qquad (2.15)$$

Equation (2.15) is Melzak's theorem with $f(x) = 1$ and $y = 1$.

## 2.1  Newton's Binomial Theorem and the Geometric Series

We now proceed to the formulation of Equation (2.1) discovered by Newton in 1676. Newton showed that

$$\sum_{k=0}^{\infty} \binom{\alpha}{k} z^k = (1 + z)^{\alpha}, \qquad \alpha \text{ real, } z \text{ complex with } |z| < 1. \qquad (2.16)$$

There are many proofs of Equation (2.16). Most utilize Taylor series expansions and various Taylor remainder theorems. A particularly elegant such proof is found in [Ross, 1980, p.180]. However, we would like to prove Equation (2.16) using techniques which are more in keeping with the algebraic series manipulations espoused by Professor Gould. This approach is based on an exercise found in [Stewart, 2007, Chap.8]. To begin we are given the power series $g(z) = \sum_{k=0}^{\infty} \binom{\alpha}{k} z^k$. Our goal is to eventually show $g(z) = (1 + z)^{\alpha}$. Using the Ratio test we can establish that $g(z)$ is absolutely convergent for $|z| < 1$. This means we can differentiate $g(z)$ whenever $|z| < 1$ and obtain $g'(z) = \sum_{k=1}^{\infty} k \binom{\alpha}{k} z^{k-1}$ [Conway, 1978]. Whenever $|z| < 1$ we have

$$(1+z)g'(z) = (1+z)\sum_{k=1}^{\infty} k\binom{\alpha}{k}z^{k-1} = \sum_{k=1}^{\infty} k\binom{\alpha}{k}z^{k-1} + \sum_{k=1}^{\infty} k\binom{\alpha}{k}z^k$$

$$= \sum_{k=0}^{\infty} (k+1)\binom{\alpha}{k+1}z^k + \sum_{k=0}^{\infty} k\binom{\alpha}{k}z^k$$

$$= \alpha \sum_{k=0}^{\infty} \binom{\alpha-1}{k}z^k + \alpha \sum_{k=0}^{\infty} \binom{\alpha-1}{k-1}z^k$$

$$= \alpha \sum_{k=0}^{\infty} \left[ \binom{\alpha-1}{k} + \binom{\alpha-1}{k-1} \right] z^k = \alpha \sum_{k=0}^{\infty} \binom{\alpha}{k}z^k = \alpha g(z).$$

From these calculations we are able to conclude that

$$g'(z) = \frac{\alpha g(z)}{1+z}, \qquad \text{whenever } |z| < 1. \qquad (2.17)$$

Next define $h(z) = (1+z)^{-\alpha}g(z)$ and observe that

$$h'(z) = \frac{g'(z)}{(1+z)^\alpha} - \frac{\alpha g(z)}{(1+z)^{\alpha+1}} = \frac{\alpha g(z)}{(1+z)^{\alpha+1}} - \frac{\alpha g(z)}{(1+z)^{\alpha+1}} = 0. \quad (2.18)$$

Equation (2.18) implies $h(z) = (1+z)^{-\alpha}g(z) = C$, where $C$ is a constant. To determine $C$ let $z = 0$ and observe that $h(0) = g(0) = 1$. Thus we have $1 = (1+z)^{-\alpha}g(z)$, or equivalently $(1+z)^\alpha = g(z) = \sum_{k=0}^{\infty} \binom{\alpha}{k} z^k$.

Take a careful look at these calculations. The proof is still valid for $\alpha$ complex, the principal branch of $(1+z)^\alpha$, and proves Equation (2.1).

Equation (2.1) often occurs as

$$\sum_{k=0}^{\infty} \binom{\alpha}{k} x^k y^{\alpha-k} = (x+y)^\alpha, \qquad \text{complex } \alpha, x, \text{ and } y \text{ with } |x| < |y|. \tag{2.19}$$

Equation (2.19) is equivalent to Equation (2.1) since $z = \frac{x}{y}$ implies that

$$(1+z)^\alpha = \left(1 + \frac{x}{y}\right)^\alpha = y^{-\alpha}(x+y)^\alpha = \sum_{k=0}^{\infty} \binom{\alpha}{k} x^k y^{-k}.$$

Equation (2.1) provides a series expansion for $\frac{1}{(1-x)^{r+1}}$ where $r$ is an arbitrary complex number since

$$\frac{1}{(1-x)^{r+1}} = (1 + (-x))^{-r-1} = \sum_{k=0}^{\infty} \binom{-r-1}{k}(-x)^k$$

$$= \sum_{k=0}^{\infty} \binom{r+k}{k} x^k. \tag{2.20}$$

If $r$ is a nonnegative integer $n$, we may rewrite Equation (2.20) as

$$\frac{1}{(1-x)^{n+1}} = \sum_{k=0}^{\infty} \binom{n+k}{k} x^k = \sum_{k=0}^{\infty} \binom{n+k}{n} x^k = \sum_{k=n}^{\infty} \binom{k}{n} x^{k-n}. \tag{2.21}$$

Equation (2.21) is called the companion binomial theorem.

If $n = 0$, Equation (2.21) becomes

$$\frac{1}{1-x} = \sum_{k=0}^{\infty} x^k, \qquad |x| < 1, \tag{2.22}$$

the famous geometric series.

We end this chapter by discussing another special instance of Equation (2.21). Let $x = \frac{1}{2}$ to obtain

$$\sum_{k=0}^{\infty} \binom{n+k}{n} \frac{1}{2^k} = \sum_{k=0}^{\infty} \binom{n+k}{k} \frac{1}{2^k} = 2^{n+1}. \tag{2.23}$$

How much do the first $n+1$ terms contribute to the infinite sum of Equation (2.23)? To answer this question define $S_n := \sum_{k=0}^{n} \binom{n+k}{k} \frac{1}{2^k}$ and observe that

$$S_n = \sum_{k=0}^{n} \binom{n+k}{k} \frac{1}{2^k} = \sum_{k=0}^{n} \left[ \binom{n+1+k}{k} - \binom{n+k}{k-1} \right] \frac{1}{2^k}$$

$$= \sum_{k=0}^{n+1} \binom{n+1+k}{k} \frac{1}{2^k} - \binom{2n+2}{n+1} \frac{1}{2^{n+1}} - \sum_{k=1}^{n} \binom{n+k}{k-1} \frac{1}{2^k}$$

$$= \sum_{k=0}^{n+1} \binom{n+1+k}{k} \frac{1}{2^k} - \binom{2n+2}{n+1} \frac{1}{2^{n+1}} - \sum_{k=0}^{n-1} \binom{n+1+k}{k} \frac{1}{2^{k+1}}$$

$$= S_{n+1} - \binom{2n+2}{n+1} \frac{1}{2^{n+1}} - \sum_{k=0}^{n+1} \binom{n+1+k}{k} \frac{1}{2^{k+1}}$$

$$\quad + \binom{2n+1}{n} \frac{1}{2^{n+1}} + \binom{2n+2}{n+1} \frac{1}{2^{n+2}}$$

$$= S_{n+1} - \binom{2n+2}{n+1} \frac{1}{2^{n+1}} - \frac{1}{2} S_{n+1} + \binom{2n+1}{n} \frac{1}{2^{n+1}} + \binom{2n+2}{n+1} \frac{1}{2^{n+2}}$$

$$= \frac{1}{2} S_{n+1} + \binom{2n+2}{n+1} \frac{1}{2^{n+1}} \left[ -1 + \frac{1}{2} \right] + \binom{2n+1}{n} \frac{1}{2^{n+1}}$$

$$= \frac{1}{2} S_{n+1} - \binom{2n+2}{n+1} \frac{1}{2^{n+2}} + \binom{2n+1}{n} \frac{1}{2^{n+1}}$$

$$= \frac{1}{2} S_{n+1} - \frac{2n+2}{n+1} \binom{2n+1}{n} \frac{1}{2^{n+2}} + \binom{2n+1}{n} \frac{1}{2^{n+1}}$$

$$= \frac{1}{2} S_{n+1} - \binom{2n+1}{n} \frac{1}{2^{n+1}} + \binom{2n+1}{n} \frac{1}{2^{n+1}} = \frac{1}{2} S_{n+1}.$$

These calculations imply $S_n = \frac{1}{2} S_{n+1}$. By iterating this basic relationship we see that

$$S_{n+r} = 2 S_{n+r-1} = 2^r S_n, \qquad \text{whenever } r \text{ is a positive integer.} \tag{2.24}$$

We now let $n = 0$ in Equation (2.24) to obtain

$$S_r = 2^r S_0 = 2^r \sum_{k=0}^{0} \binom{0+k}{k} \frac{1}{2^k} = 2^r \cdot 1 = 2^r. \tag{2.25}$$

Replacing $r$ by $n$ in Equation (2.25) shows us that

$$S_n = \sum_{k=0}^{n} \binom{n+k}{k} \frac{1}{2^k} = 2^n. \tag{2.26}$$

Compare Equation (2.26) with Equation (2.23) and notice that

$$\sum_{k=0}^{\infty} \binom{n+k}{k} \frac{1}{2^k} = 2^{n+1} = 2 \cdot 2^n = 2 \sum_{k=0}^{n} \binom{n+k}{k} \frac{1}{2^k}.$$

Thus the sum of the first $n+1$ terms, namely $\sum_{k=0}^{n} \binom{n+k}{k} \frac{1}{2^k}$, is precisely one half of the total value of the infinite sum.

# Chapter 3

# Iterative Series

For the past two chapters our primary object of study involved sums $\sum_i a_i$ where $a_i$ is a complex number involving a binomial coefficient. We now discuss combinatorial identities involving series whose summands contain two or more indices of summation. Such sums are called iterative series. The simplest iterative series is $\sum_{i=b}^{n} \sum_{j=c}^{m} a_{i,j}$. To calculate this double series, fix the outer index $i$ and sum over $j$ to find that

$$
\begin{aligned}
\sum_{i=b}^{n} \sum_{j=c}^{m} a_{i,j} &= \sum_{j=c}^{m} a_{b,j} + \sum_{j=c}^{m} a_{b+1,j} + \cdots + \sum_{j=c}^{m} a_{n,j} \\
&= \left( a_{b,c} + a_{1,c+1} + \cdots + a_{b,m} \right) + \left( a_{b+1,c} + a_{b+1,c+1} + \cdots + a_{b+1,m} \right) \\
&\quad + \cdots + \left( a_{n,c} + a_{n,c+1} + \cdots + a_{n,m} \right) \\
&= \left( a_{b,c} + a_{b+1,c} + \cdots + a_{n,c} \right) + \left( a_{b,c+1} + a_{b+1,c+1} + \cdots + a_{n,c+1} \right) \\
&\quad + \cdots + \left( a_{b,m} + a_{b+1,m} + \cdots + a_{n,m} \right) \\
&= \sum_{i=b}^{n} a_{i,c} + \sum_{i=b}^{n} a_{i,c+1} + \cdots + \sum_{i=b}^{n} a_{i,m} = \sum_{j=c}^{m} \sum_{i=b}^{n} a_{i,j}.
\end{aligned}
$$

We record the results of this calculation as

**Independent Double Series Identity:** Let $n$, $m$, $b$, and $c$ be nonnegative integers such that $b \leq n$ and $c \leq m$. Then

$$
\sum_{i=b}^{n} \sum_{j=c}^{m} a_{i,j} = \sum_{j=c}^{m} \sum_{i=b}^{n} a_{i,j}. \tag{3.1}
$$

Equation (3.1) can easily be generalized to iterative series consisting of $p$ sums.

**Independent Iterative Series Identity:** Let $p$ be a positive integer with $p \geq 2$. Assume $n_k$ and $b_k$ are arbitrary nonnegative integers such that $n_k \leq b_k$ for all $1 \leq k \leq p$. Then

$$\sum_{i_1=b_1}^{n_1} \sum_{i_2=b_2}^{n_2} \cdots \sum_{i_p=b_p}^{n_p} a_{i_1,i_2,\ldots,i_p} = \sum_{i_{\sigma(1)}=b_{\sigma(1)}}^{n_{\sigma(1)}} \sum_{i_{\sigma(2)}=b_{\sigma(2)}}^{n_{\sigma(2)}} \cdots \sum_{i_{\sigma(p)}=b_{\sigma(p)}}^{n_{\sigma(p)}} a_{i_1,i_2,\ldots,i_p}, \tag{3.2}$$

where $\sigma(1)\sigma(2)\ldots\sigma(p)$ is any permutation of $123\ldots p$.

Equation (3.1) is quite useful. For example it is used in the proof of the Taylor series expansion of $f(x) = \sum_{i=0}^{n} a_i x^i$. Let $k$ be a nonnegative integer. Define $f^{(k)}(x)$ to be the $k^{th}$ derivative of $f$ with respect to $x$. Term by term differentiation implies that

$$f^{(k)}(x) = k! \sum_{i=0}^{n} a_i \binom{i}{k} x^{i-k}. \tag{3.3}$$

An application of the binomial theorem implies

$$\left(\frac{x}{y}\right)^i = \left(1 + \left(\frac{x}{y} - 1\right)\right)^i = \sum_{k=0}^{i} \binom{i}{k} \left(\frac{x}{y} - 1\right)^k. \tag{3.4}$$

Combining Equation (3.3) with Equation (3.4) implies that

$$\begin{aligned}
f(x) &= \sum_{i=0}^{n} a_i x^i = \sum_{i=0}^{n} a_i y^i \left(\frac{x}{y}\right)^i = \sum_{i=0}^{n} a_i y^i \sum_{k=0}^{i} \binom{i}{k} \left(\frac{x}{y} - 1\right)^k \\
&= \sum_{i=0}^{n} a_i y^i \sum_{k=0}^{i} \binom{i}{k} (x-y)^k y^{-k} = \sum_{k=0}^{n} \frac{(x-y)^k}{k!} k! \sum_{i=0}^{n} a_i \binom{i}{k} y^{i-k} \\
&= \sum_{k=0}^{n} \frac{(x-y)^k}{k!} f^{(k)}(y).
\end{aligned}$$

We record this result as

**Taylor's Polynomial Expansion:** Let $f(x) = \sum_{i=0}^{n} a_i x^i$. Then

$$f(x) = \sum_{k=0}^{n} \frac{(x-y)^k}{k!} f^{(k)}(y), \tag{3.5}$$

where $f^{(k)}(y)$ denotes the $k^{th}$ derivative of $f$ with respect to $x$ evaluated at $y$.

Equation (3.5) appears in many guises. If we let $x \to x + y$, Equation (3.5) becomes

$$f(x + y) = \sum_{k=0}^{n} \frac{x^k}{k!} f^{(k)}(y).$$ (3.6)

In Equation (3.6) set $y = 0$ to obtain

$$f(x) = \sum_{k=0}^{n} \frac{x^k}{k!} f^{(k)}(0),$$ (3.7)

a finite polynomial form of the Maclaurin series.

## 3.1 Two Summation Interchange Formulas

Not all iterative series have independent indices. As a case in point, take $\sum_{i=1}^{4} \sum_{j=i}^{4} ij^2$. Notice that the index of the inner sum does depend on $i$. The purpose of this section is to demonstrate techniques for interchanging the order of summation for a double series where the index of the inner summation does depend on the choice of outer index. Our first example involves

$$\sum_{k=0}^{n} \sum_{i=0}^{\lfloor \frac{k}{2} \rfloor} a_{i,k} = \sum_{i=0}^{0} a_{i,0} + \sum_{i=0}^{0} a_{i,1} + \sum_{i=0}^{1} a_{i,2} + \sum_{i=0}^{1} a_{i,3} + \dots + \sum_{i=0}^{\lfloor \frac{n}{2} \rfloor} a_{i,n}.$$ (3.8)

We expand each individual sum on the right side of Equation (3.8) and record the results in the following two-dimensional array. Each row of the array corresponds to a sum on the right side of (3.8), where the first row is the expansion of $\sum_{i=0}^{0} a_{i,0}$, the second row is the expansion of $\sum_{i=0}^{0} a_{i,1}$, etc. In particular we have

$$\sum_{k=0}^{n} \sum_{i=0}^{\lfloor \frac{k}{2} \rfloor} a_{i,k} = \begin{array}{lllll} a_{0,0} & & & & \\ +a_{0,1} & & & & \\ +a_{0,2} & +a_{1,2} & & & \\ +a_{0,3} & +a_{1,3} & & & \\ +a_{0,4} & +a_{1,4} & +a_{2,4} & & \\ +\dots & \dots & \dots & \dots & \dots \\ +a_{0,n} & +a_{1,n} & +a_{2,n} & +\dots & +a_{\lfloor \frac{n}{2} \rfloor,n} \end{array}$$

Equation (3.8) corresponds to fixing $i$ and reading along the rows. But it would be just as natural to fix $k$, read down each column, and find as follows:

**First Summation Interchange Formula:**

$$\sum_{k=0}^{n}\sum_{i=0}^{\lfloor\frac{k}{2}\rfloor} a_{i,k} = \sum_{k=0}^{n} a_{0,k} + \sum_{k=2}^{n} a_{1,k} + \sum_{k=4}^{n} a_{2,k} + ... + \sum_{k=n}^{n} a_{\lfloor\frac{n}{2}\rfloor,k}$$

$$= \sum_{i=0}^{\lfloor\frac{n}{2}\rfloor}\sum_{k=2i}^{n} a_{i,k}. \tag{3.9}$$

The derivation of Equation (3.9) illustrates how the original order of summation of a double series corresponds to reading along rows of a two-dimensional array, while the reverse order of summation corresponds to reading down the columns. This concept of visually representing a double series as a two-dimensional array can be generalized to iterative series with $p$ indices, $p \geq 2$, by forming a $p$-dimensional array. Visualizing such an array for $p \geq 4$ is somewhat difficult. Thus, when dealing with $p$-fold sums, where $p \geq 4$, we usually obtain interchange identities via a string of inequalities. Here is an example of using inequalities to obtain Equation (3.9). The left side of Equation (3.9) corresponds to the two inequalities $0 \leq k \leq n$ and $0 \leq i \leq \lfloor\frac{k}{2}\rfloor$. Since $k \leq n$, we conclude that $0 \leq i \leq \lfloor\frac{k}{2}\rfloor \leq \lfloor\frac{n}{2}\rfloor$ and that $0 \leq i \leq \lfloor\frac{n}{2}\rfloor$, which is the inequality associated with the outer sum on the right side of Equation (3.9). Next observe that $i \leq \lfloor\frac{k}{2}\rfloor \leq \lfloor\frac{n}{2}\rfloor$ is equivalent to $2i \leq k \leq 2\lfloor\frac{n}{2}\rfloor$. Either $2\lfloor\frac{n}{2}\rfloor = n$ or $2\lfloor\frac{n}{2}\rfloor = n - 1$. However, the left side of Equation (3.9) shows that $k$ does obtain the value of $n$. Hence, we conclude $2i \leq k \leq 2n$, the inequality associated with the inner sum on the right side of Equation (3.9).

Equation (3.9) has the following generalization which is derived in a similar manner.

**Generalized First Summation Interchange Formula:** Let $r$ be a positive integer and $n$ and $b$ be a nonnegative integers such that $0 \leq b \leq n$. Then

$$\sum_{k=b}^{n}\sum_{i=b}^{\lfloor\frac{k}{r}\rfloor} a_{i,k} = \sum_{i=b}^{\lfloor\frac{n}{r}\rfloor}\sum_{k=ri}^{n} a_{i,k}. \tag{3.10}$$

We provide an application of Equation (3.9) with connections to the Fibonacci numbers. Fix $a_{i,k} = \binom{k-i}{i}$. Write the right side of Equation (3.9) as

$$\sum_{i=0}^{\lfloor\frac{n}{2}\rfloor}\sum_{k=2i}^{n}\binom{k-i}{i} = \sum_{i=0}^{\lfloor\frac{n}{2}\rfloor}\left[\sum_{k=0}^{n}\binom{k-i}{i} - \sum_{k=0}^{2i-1}\binom{k-i}{i}\right]. \tag{3.11}$$

We claim that for a fixed nonnegative integer $r$ and a fixed complex number $a$,

$$\sum_{k=0}^{n}\binom{k-a}{r} = \binom{n-a+1}{r+1} + (-1)^r \binom{a+r}{r+1}. \qquad (3.12)$$

To prove Equation (3.12), apply Pascal's identity, and observe that

$$\sum_{k=0}^{n}\binom{k-a}{r} = \sum_{k=0}^{n}\left[\binom{k-a+1}{r+1} - \binom{k-a}{r+1}\right]$$

$$= \sum_{k=0}^{n}\binom{k-a+1}{r+1} - \sum_{k=-1}^{n-1}\binom{k-a+1}{r+1}$$

$$= \binom{n-a+1}{r+1} - \binom{-a}{r+1} = \binom{n-a+1}{r+1} + (-1)^r \binom{a+r}{r+1}.$$

Equation (3.12) will allow us to simplify the right side of Equation (3.11). Let $a = r = i$, $n \to 2i - 1$, and observe that

$$\sum_{i=0}^{\lfloor \frac{n}{2} \rfloor} \sum_{k=2i}^{n} \binom{k-i}{i}$$

$$= \sum_{i=0}^{\lfloor \frac{n}{2} \rfloor}\left[\binom{n-i+1}{i+1} + (-1)^i \binom{2i}{i+1} - \binom{2i-1-i+1}{i+1} - (-1)^i \binom{2i}{i+1}\right]$$

$$= \sum_{i=0}^{\lfloor \frac{n}{2} \rfloor}\binom{n-i+1}{i+1} = \sum_{i=1}^{\lfloor \frac{n}{2} \rfloor+1}\binom{n+2-i}{i}$$

$$= \sum_{i=1}^{\lfloor \frac{n+2}{2} \rfloor}\binom{n+2-i}{i} = \sum_{i=0}^{\lfloor \frac{n+2}{2} \rfloor}\binom{n+2-i}{i} - 1 = \sum_{k=0}^{n} \sum_{i=0}^{\lfloor \frac{k}{2} \rfloor}\binom{k-i}{i},$$

where the final equality is the left side of Equation (3.9). These calculations show that

$$\sum_{i=0}^{\lfloor \frac{n+2}{2} \rfloor}\binom{n+2-i}{i} - 1 = \sum_{k=0}^{n} \sum_{i=0}^{\lfloor \frac{k}{2} \rfloor}\binom{k-i}{i}. \qquad (3.13)$$

To discover how Equation (3.13) is connected to the Fibonacci numbers, define $f_r = \sum_{k=0}^{\lfloor \frac{r}{2} \rfloor}\binom{r-k}{k}$. An induction proof utilizing Pascal's identity shows that $\{f_r\}_{r=0}^{\infty}$ is recursively defined by $f_0 = 1, f_1 = 1$, and $f_r = f_{r-1} + f_{r-2}, r \geq 2$. Leonardo Fibonacci first discovered a variation these numbers in 1202 when studying rabbit populations [Gould, 1987]. Using this notation, we can rewrite Equation (3.13) as

$$f_{n+2} - 1 = \sum_{k=0}^{n} f_k. \tag{3.14}$$

Our second interchange identity involves $\sum_{k=b}^{n} \sum_{i=b}^{rk} a_{i,k}$. By adjusting the derivation of Equation (3.9), we readily show as follows:

**Second Summation Interchange Formula:** Let $n$ and $b$ be nonnegative integers such that $0 \leq b \leq n$. Let $r$ be a positive integer. Then

$$\sum_{k=b}^{n} \sum_{i=b}^{rk} a_{i,k} = \sum_{i=b}^{rn} \sum_{k=\lfloor \frac{i+r-1}{r} \rfloor}^{n} a_{i,k}. \tag{3.15}$$

The special case of Equation (3.15) with $r = 1$ appears so often in Professor Gould's calculations that we record it as

**Standard Interchange Formula:** Let $n$ and $b$ be nonnegative integers such that $0 \leq b \leq n$. Then

$$\sum_{k=b}^{n} \sum_{i=b}^{k} a_{i,k} = \sum_{i=b}^{n} \sum_{k=i}^{n} a_{i,k}, \qquad b \geq 0. \tag{3.16}$$

There are many variations of the standard interchange formula which involve shifting the indices on both sides of Equation (3.16). For example, if we let $k \to n - k$ in the left side of Equation (3.16), we find that

$$\sum_{k=b}^{n} \sum_{i=b}^{k} a_{i,k} = \sum_{k=0}^{n-b} \sum_{i=b}^{n-k} a_{i,n-k} = \sum_{i=b}^{n} \sum_{k=i}^{n} a_{i,k}.$$

We now let $k \to k + i$ in the right side of Equation (3.16) to discover

$$\sum_{k=b}^{n} \sum_{i=b}^{k} a_{i,k} = \sum_{i=b}^{n} \sum_{k=i}^{n} a_{i,k} = \sum_{i=b}^{n} \sum_{k=0}^{n-i} a_{i,k+i}.$$

By combining the previous two results, we obtain our first variation of the standard interchange formula, namely

$$\sum_{k=0}^{n-b} \sum_{i=b}^{n-k} a_{i,n-k} = \sum_{i=b}^{n} \sum_{k=0}^{n-i} a_{i,k+i}, \qquad b \geq 0. \tag{3.17}$$

Another variation occurs if we let $i \to n - i$ on the right side of Equation (3.17). We obtain

$$\sum_{k=0}^{n-b}\sum_{i=b}^{n-k} a_{i,n-k} = \sum_{i=0}^{n-b}\sum_{k=0}^{i} a_{n-i,k+n-i}. \tag{3.18}$$

We then apply the standard interchange formula to the right side of Equation (3.18) to obtain

$$\sum_{i=0}^{n-b}\sum_{k=0}^{i} a_{n-i,k+n-i} = \sum_{k=0}^{n-b}\sum_{i=k}^{n-b} a_{n-i,k+n-i} = \sum_{k=b}^{n}\sum_{i=k-b}^{n-b} a_{n-i,k-b+n-i}$$

$$= \sum_{k=b}^{n}\sum_{i=k}^{n} a_{n-i+b,k+n-i}. \tag{3.19}$$

Equations (3.17), (3.18), and (3.19) are just three of the many variations of the standard interchange formula. Another useful variation is the identity

$$\sum_{k=b}^{n}\sum_{i=0}^{k} a_{i,k} = \sum_{i=0}^{n-b}\sum_{k=0}^{i} a_{n-i,n-k}, \tag{3.20}$$

whose proof is as follows:

$$\sum_{k=b}^{n}\sum_{i=b}^{k} a_{i,k} = \sum_{i=b}^{n}\sum_{k=i}^{n} a_{i,k} = \sum_{i=0}^{n-b}\sum_{k=n-i}^{n} a_{n-i,k}$$

$$= \sum_{i=0}^{n-b}\sum_{k=-i}^{0} a_{n-i,n+k} = \sum_{i=0}^{n-b}\sum_{k=0}^{i} a_{n-i,n-k}.$$

If $b = 0$, Equation (3.20) has the symmetrical form

$$\sum_{k=0}^{n}\sum_{i=0}^{k} a_{i,k} = \sum_{i=0}^{n}\sum_{k=0}^{i} a_{n-i,n-k}.$$

We end this section with an example of how to use the standard interchange formula in the derivation of a combinatorial identity. Let $a_{i,k} = f(i)$. Equation (3.16) becomes

$$\sum_{k=b}^{n}\sum_{i=b}^{k} f(i) = \sum_{i=b}^{n} f(i) \sum_{k=i}^{n} 1 = \sum_{i=b}^{n} f(i)(n - i + 1),$$

which is equivalent to

$$\sum_{i=a}^{n} i f(i) = (n+1)\sum_{i=b}^{n} f(i) - \sum_{k=b}^{n}\sum_{i=b}^{k} f(i). \tag{3.21}$$

Assume $m$ is a nonnegative integer. Set $f(i) = \binom{i}{m}$. Equation (3.21), when combined with Equation (3.12) implies that

$$\sum_{i=0}^{n} i\binom{i}{m} = (n+1)\sum_{i=0}^{n}\binom{i}{m} - \sum_{k=0}^{n}\sum_{i=0}^{k}\binom{i}{m}$$

$$= (n+1)\binom{n+1}{m+1} - \sum_{k=0}^{n}\binom{k+1}{m+1}, \qquad \text{Eq. (3.12) with } a = 0$$

$$= (n+1)\binom{n+1}{m+1} - \binom{n+2}{m+2}, \qquad \text{Eq. (3.12) with } a = -1.$$

In summary we have proven the binomial identity

$$\sum_{i=0}^{n} i\binom{i}{m} = (n+1)\binom{n+1}{m+1} - \binom{n+2}{m+2} = \frac{nm+m+n}{m+2}\binom{n+1}{m+1}.$$

## 3.2 Gould's Convolution Formula

Throughout this chapter we have dealt with summations consisting of two or more series placed iteratively. We now ask a related question of how to multiply two series. Suppose we want to calculate $\sum_{i=1}^{n} a_i \sum_{j=1}^{n} b_j$. By defining $a_{i,j} = a_i b_j$, we convert $\sum_{i=1}^{n} a_i \sum_{j=1}^{n} b_j$ into $\sum_{i=1}^{n}\sum_{j=1}^{n} a_{i,j}$ and apply the techniques used to derive the iterative series identities of the previous section. In particular, given the iterative series $\sum_{i=1}^{n}\sum_{j=1}^{n} a_{i,j}$ we form the two-dimensional array

| | | | | | |
|---|---|---|---|---|---|
| $a_{1,1}$ | $+a_{1,2}$ | $+a_{1,3}$ | $+\ldots$ | $+a_{1,n-1}$ | $\mathbf{+a_{1,n}}$ |
| $+a_{2,1}$ | $+a_{2,2}$ | $+a_{2,3}$ | $+\ldots$ | $\mathbf{+a_{2,n-1}}$ | $+a_{2,n}$ |
| $+a_{3,1}$ | $+a_{3,2}$ | $+\ldots$ | $\mathbf{+a_{3,n-2}}$ | $+a_{3,n-1}$ | $+a_{3,n}$ |
| $+\ldots$ | .... | .... | .... | .... | .... |
| $+a_{n-1,1}$ | $\mathbf{+a_{n-1,2}}$ | $+a_{n-1,3}$ | $+\ldots$ | $+a_{n-1,n-1}$ | $+a_{n-1,n}$ |
| $\mathbf{+a_{n,1}}$ | $+a_{n,2}$ | $+a_{n,3}$ | $+\ldots$ | $+a_{n,n-1}$ | $+a_{n,n}$ |

Instead of summing along rows and columns, we sum along diagonal rows parallel to the off-diagonal. The terms in boldface are the off-diagonal of the array and consists of those $a_{i,j}$ such that $i + j = n + 1$. Let $1 \leq r \leq n - 1$. The $r^{th}$ slanted row above the off-diagonal contains those $a_{i,j}$ such that $i + j = r + 1$. For the $n - 1$ slanted rows below the off-diagonal, we index them via $n+r$ where $1 \leq r \leq n-1$. The $(n+r)$-th such slanted row contains all $a_{i,j}$ such that $i + j = n + r - 1$. Using this three part decomposition we find that

$$\sum_{i=1}^{n}\sum_{j=1}^{n}a_{i,j}$$

$$= \left[\sum_{i=1}^{1}a_{i,2-i} + \sum_{i=1}^{2}a_{i,3-i} + ... + \sum_{i=1}^{n-1}a_{i,n-i}\right] + \sum_{i=1}^{n}a_{i,n+1-i}$$

$$+ \left[\sum_{i=2}^{n}a_{i,n+2-i} + \sum_{i=3}^{n}a_{i,n+3-i} + ... + \sum_{i=n}^{n}a_{i,2n-i}\right]$$

$$= \sum_{k=1}^{n-1}\sum_{i=1}^{k}a_{i,k+1-i} + \sum_{i=1}^{n}a_{i,n+1-i} + \sum_{k=n+1}^{2n-1}\sum_{i=k+1-n}^{n}a_{i,k+1-i}. \quad (3.22)$$

We claim that

$$\sum_{i=1}^{n}\sum_{j=1}^{n}a_{i,j} = \sum_{k=1}^{2n-1}\sum_{i=1+\lfloor\frac{k-1}{n}\rfloor(k-n)}^{k-\lfloor\frac{k}{n}\rfloor(k-n)}a_{i,k-i+1}. \quad (3.23)$$

To understand why the right side of Equation (3.23) is equivalent to the right side of Equation (3.22), we need to analyze some inequalities. First, let $1 \le k \le n-1$. Then $\lfloor\frac{k-1}{n}\rfloor = 0 = \lfloor\frac{k}{n}\rfloor$. Hence, $i = 1+\lfloor\frac{k-1}{n}\rfloor(k-n) = 1$ and $k - \lfloor\frac{k}{n}\rfloor(k-n) = k$, which corresponds to the first sum on the right side of Equation (3.22). Now let $k = n$ and observe that $i = 1+\lfloor\frac{n-1}{n}\rfloor(n-n) = 1$, while $n - \lfloor\frac{n}{n}\rfloor(n-n) = n$, which corresponds to the second sum on the right side of Equation (3.22). To obtain the third sum on the right side of Equation (3.22), we let $n + 1 \le k \le 2n - 1$. Notice that $\lfloor\frac{k-1}{n}\rfloor = 1 = \lfloor\frac{k}{n}\rfloor$. Hence, $i = 1 + \lfloor\frac{k-1}{n}\rfloor(k-n) = 1 + 1(k-n) = k+n-1$ and $k - \lfloor\frac{k}{n}\rfloor(k-n) = k - 1(k-n) = n$.

We now set $a_{i,j} = f(i)\varphi(j)$ in Equation (3.23) to obtain

$$\sum_{i=1}^{n}f(i)\sum_{j=1}^{n}\varphi(j) = \sum_{k=1}^{2n-1}\sum_{i=1+\lfloor\frac{k-1}{n}\rfloor(k-n)}^{k-\lfloor\frac{k}{n}\rfloor(k-n)}f(i)\varphi(k-i+1), \quad n \ge 1. \quad (3.24)$$

Since many combinatorial identities have indices beginning at zero, it would be beneficial to have a convolution formula for $\sum_{i=0}^{n}f(i)\sum_{j=0}^{n}\varphi(j)$. Take Equation (3.24) and let $n \to n + 1$ to obtain

$$\sum_{i=1}^{n+1}f(i)\sum_{j=1}^{n+1}\varphi(j) = \sum_{k=1}^{2n+1}\sum_{i=1+\lfloor\frac{k-1}{n+1}\rfloor(k-n-1)}^{k-\lfloor\frac{k}{n+1}\rfloor(k-n-1)}f(i)\varphi(k-i+1). \quad (3.25)$$

Now set $i \to i+1$, $j \to j+1$, and $k \to k+1$ in Equation (3.25) to obtain

$$\sum_{i=0}^{n} f(i+1) \sum_{j=0}^{n} \varphi(j+1) = \sum_{k=0}^{2n} \sum_{i=\lfloor \frac{k}{n+1}\rfloor(k-n)}^{k-\lfloor \frac{k+1}{n+1}\rfloor(k-n)} f(i+1)\varphi(k-i+1). \quad (3.26)$$

Define $\bar{f}(i) := f(i+1)$ and $\bar{\varphi}(j) := \varphi(j+1)$. Equation (3.26) becomes

$$\sum_{i=0}^{n} f(i) \sum_{j=0}^{n} \varphi(j) = \sum_{k=0}^{2n} \sum_{i=\lfloor \frac{k}{n+1}\rfloor(k-n)}^{k-\lfloor \frac{k+1}{n+1}\rfloor(k-n)} f(i)\varphi(k-i), \quad (3.27)$$

where we have suppressed the bars after substitution. Equation (3.27) is called **Gould's convolution formula** for the product of two finite series.

Gould's convolution formula is useful for calculating the product of two polynomials. Let $f(i) = A_i x^i$ and $\varphi(j) = B_j x^j$. Equation (3.27) implies

$$\sum_{i=0}^{n} A_i x^i \sum_{j=0}^{n} B_j x^j = \sum_{k=0}^{n} x^k \sum_{i=\lfloor \frac{k}{n+1}\rfloor(k-n)}^{k-\lfloor \frac{k+1}{n+1}\rfloor(k-n)} A_i B_{k-i}. \quad (3.28)$$

Another application of the Gould convolution formula calculates the square of a series. Let $\varphi(j) = f(j)$. Equation (3.27) becomes

$$\left[\sum_{k=0}^{n} f(k)\right]^2 = \sum_{k=0}^{2n} \sum_{i=\lfloor \frac{k}{n+1}\rfloor(k-n)}^{k-\lfloor \frac{k+1}{n+1}\rfloor(k-n)} f(i)f(k-i). \quad (3.29)$$

In particular, if $f(k) = \binom{n}{k}x^k$, Equation (3.29) implies that

$$\left[\sum_{k=0}^{n} \binom{n}{k}x^k\right]^2 = \sum_{k=0}^{2n} x^k \sum_{i=\lfloor \frac{k}{n+1}\rfloor(k-n)}^{k-\lfloor \frac{k+1}{n+1}\rfloor(k-n)} \binom{n}{i}\binom{n}{k-i}$$

$$= \sum_{k=0}^{2n} x^k \sum_{i=0}^{k} \binom{n}{i}\binom{n}{k-i}, \quad (3.30)$$

since $\binom{n}{k-i} = 0$ if $i > k$ and $\binom{n}{i} = 0$ if $i > n$. Because the binomial theorem states that $(1+x)^n = \sum_{k=1}^{n} \binom{n}{k}x^k$, we have

$$\left[\sum_{k=0}^{n} \binom{n}{k}x^k\right]^2 = (1+x)^{2n} = \sum_{k=0}^{2n} \binom{2n}{k}x^k. \quad (3.31)$$

Combining Equations (3.30) and (3.31) implies that

$$\sum_{k=0}^{2n} x^k \left[ \binom{2n}{k} - \sum_{i=0}^{k} \binom{n}{i}\binom{n}{k-i} \right] = 0,$$

which is equivalent to

$$\sum_{i=0}^{k} \binom{n}{i}\binom{n}{k-i} = \binom{2n}{k}. \tag{3.32}$$

Another application of Equation (3.29) occurs if we let $f(k) = a_k x^k$, in which case we obtain

$$\left[ \sum_{k=0}^{n} a_k x^k \right]^2 = \sum_{k=0}^{2n} x^k \sum_{i=\lfloor \frac{k}{n+1} \rfloor (k-n)}^{k - \lfloor \frac{k+1}{n+1} \rfloor (k-n)} a_i a_{k-i}. \tag{3.33}$$

Many times Professor Gould applies the infinite versions of Equations (3.27), (3.28), and (3.33) for determining the closed forms of combinatorial identities. For a fixed $k$, $\left\lfloor \frac{k}{n+1} \right\rfloor = 0$ if $n \geq k$. This fact implies that

$$\sum_{i=0}^{\infty} f(i) \sum_{j=0}^{\infty} \varphi(j) = \sum_{k=0}^{\infty} \sum_{i=0}^{k} f(i)\varphi(k-i) \tag{3.34}$$

$$\sum_{i=0}^{\infty} A_i x^i \sum_{j=0}^{\infty} B_j x^j = \sum_{k=0}^{\infty} x^k \sum_{i=0}^{k} A_i B_{k-i} \tag{3.35}$$

$$\left[ \sum_{k=0}^{\infty} a_k x^k \right]^2 = \sum_{k=0}^{\infty} x^k \sum_{i=0}^{k} a_i a_{k-i}. \tag{3.36}$$

The second of these equations is called the **Cauchy convolution formula** and is helpful in calculating the coefficients of formal power series [Wilf, 2014].

We end this chapter by discussing generalizations of Equation (3.24). Fix positive integers $r$ and $q$ and assume that $q \leq r$. We want to calculate $\sum_{i=1}^{r} \sum_{j=1}^{q} a_{i,j}$. For ease of exposition fix $r = 5$ and $q = 3$. Note that $\sum_{i=1}^{5} \sum_{j=1}^{3} a_{i,j}$ is a rectangular array with $(q-1) + 1 + (r-1) = 7$ slanted rows. Let $1 \leq k \leq 7$. The $k^{th}$ slanted row is the sum of all $a_{i,j}$ such that $i + j = k + 1$. In particular, we see that the first slanted row is $a_{1,1}$; the second is $a_{1,2} + a_{2,1}$; the third is $a_{1,3} + a_{2,2} + a_{3,1}$; the fourth is $a_{2,3} + a_{3,2} + a_{4,1}$; the fifth is $a_{3,3} + a_{4,2} + a_{5,1}$; the sixth is $a_{4,3} + a_{4,2}$; the seventh is $a_{5,3}$. In order to combine the expression for the sums of these seven slanted rows into a double sum similar to that on the right side of Equation (3.23), we need to embed the $5 \times 3$ array into the isosceles triangle we display below.

$$
\begin{array}{lll}
a_{1,1} & +a_{1,2} & +a_{1,3} \quad \bigcirc \quad \bigcirc \quad \bigcirc \quad \bigcirc \\
+a_{2,1} & +a_{2,2} & +a_{2,3} \quad \bigcirc \quad \bigcirc \quad \bigcirc \\
+a_{3,1} & +a_{3,2} & +a_{3,3} \quad \bigcirc \quad \bigcirc \\
+a_{4,1} & +a_{4,2} & +a_{4,3} \quad \bigcirc \\
+a_{5,1} & +a_{5,2} & +a_{5,3} \\
\square & \square \\
\square
\end{array}
$$

The rightmost circle corresponds to $a_{1,7} = a_{1,r+q-1}$ while the lowest square corresponds to $a_{7,1} = a_{r+q-1,1}$. If we sum along the seven slanted rows of the isosceles triangle, we have the double sum $\sum_{k=1}^{7} \sum_{i=1}^{k} a_{i,k-i+1}$. Our given array is missing terms, which we denote by either circles or squares. The first three slanted rows are complete. Let $4 \le k \le 7$. Since the $k^{th}$ slanted row differs from the slanted row of the triangle by $(k-3)$ circles, the index of $i$ starts at $1 + (k-3)$. When $6 \le k \le 7$, the $k^{th}$ slanted row differs from the slanted row of the triangle by an additional $(k-5)$ squares, and the index of $i$ obtains a maximum of $k - (k-r) = r$. For real numbers $a$ and $b$ define

$$
\left\lceil \frac{a}{b} \right\rceil^{*} = \begin{cases} 1, & \text{if } a \ge b \\ 0, & \text{if } a < b. \end{cases}
$$

The above discussion implies that

$$
\sum_{i=1}^{5} \sum_{j=1}^{3} a_{i,j} = \sum_{k=1}^{7} \sum_{i=1+[\frac{k-1}{3}]^{*}(k-3)}^{k-[\frac{k}{5}]^{*}(k-5)} a_{i,k-i+1}.
$$

The preceding argument applies to arbitrary $r$ and $q$ with $q \le r$. Replace all instances of 5 with $r$, all instances of 3 with $q$, and all instances of 7 with $r + q - 1$ to conclude that

$$
\sum_{i=1}^{r} \sum_{j=1}^{q} a_{i,j} = \sum_{k=1}^{r+q-1} \sum_{i=1+[\frac{k-1}{q}]^{*}(k-q)}^{k-[\frac{k}{r}]^{*}(k-r)} a_{i,k-i+1}, \qquad q \le r. \tag{3.37}
$$

Note that if $q = r$, Equation (3.37) becomes Equation (3.23).

If we let $a_{i,j} = f(i)\varphi(j)$, Equation (3.37) becomes

$$
\sum_{i=1}^{r} f(i) \sum_{j=1}^{q} \varphi(j) = \sum_{k=1}^{r+q-1} \sum_{i=1+[\frac{k-1}{q}]^{*}(k-q)}^{k-[\frac{k}{r}]^{*}(k-r)} f(i)\varphi(k-i+1), \quad q \le r, \tag{3.38}
$$

the desired generalization of Equation (3.24). A useful variation of Equation (3.38) is

$$\sum_{i=0}^{r} f(i) \sum_{j=0}^{q} \varphi(j) = \sum_{k=0}^{r+q} \sum_{i=[\frac{k}{q+1}]^*(k-q)}^{k-[\frac{k+1}{r+1}]^*(k-r)} f(i)\varphi(k-i), \qquad q \le r. \tag{3.39}$$

The derivation of Equation (3.39) from Equation (3.38) is similar to the derivation of Equation (3.27) and hence omitted.

We use Equation (3.39) to obtain the Vandermonde convolution. Let $f(i) = \binom{r}{i}x^i$ and $\varphi(j) = \binom{q}{j}x^j$. Equation (3.39) becomes

$$\sum_{i=0}^{r} \binom{r}{i}x^i \sum_{j=0}^{q} \binom{q}{j}x^j = \sum_{k=0}^{r+q} x^k \sum_{i=[\frac{k}{q+1}]^*(k-q)}^{k-[\frac{k+1}{r+1}]^*(k-r)} \binom{r}{i}\binom{q}{k-i}. \tag{3.40}$$

The factor $\binom{r}{i}$ implies the sum is zero if $i > r$, while the factor $\binom{q}{k-i}$ shows the sum is zero if $k - i > q$. Hence, we can replace $k - [\frac{k+1}{r+1}]^*(k-r)$ with $k$, $i = [\frac{k}{q+1}]^*(k-q)$ with zero, and rewrite Equation (3.40) as

$$\sum_{i=0}^{r} \binom{r}{i}x^i \sum_{j=0}^{q} \binom{q}{j}x^j = \sum_{k=0}^{r+q} x^k \sum_{i=0}^{k} \binom{r}{i}\binom{q}{k-i}. \tag{3.41}$$

It is now a matter of using the binomial theorem to simplify Equation (3.41). The left hand side becomes $\sum_{i=0}^{r} \binom{r}{i}x^i \sum_{j=0}^{q} \binom{q}{j}x^j = (1+x)^r(1+x)^q = (1+x)^{r+q} = \sum_{k=0}^{r+q} \binom{r+q}{k}x^k$, and Equation (3.41) becomes

$$\sum_{k=0}^{r+q} \binom{r+q}{k}x^k = \sum_{k=0}^{r+q} x^k \sum_{i=0}^{k} \binom{r}{i}\binom{q}{k-i},$$

which is equivalent to

$$\sum_{i=0}^{k} \binom{r}{i}\binom{q}{k-i} = \binom{r+q}{k}, \qquad \text{for integral } k, r, q \ge 0. \tag{3.42}$$

Equation (3.42) is the **Vandermonde convolution**. The reader may notice that Equation (3.39) implicitly implies $q \le r$, while Equation (3.42) has no such restriction. We justify dropping this restriction by interchanging the roles of $r$ and $q$ in Equation (3.39).

Chapter 4

# Two of Professor Gould's Favorite Algebraic Techniques

We introduce two algebraic techniques Professor Gould often applies in his verifications of combinatorial identities. The first is coefficient comparison while the second is an application of the fundamental theorem of algebra.

## 4.1 Coefficient Comparison

In this section we will work with a technique called **coefficient comparison**. Suppose we have two polynomials in $x$ of degree $n$, $\sum_{k=0}^{n} a_k x^k$ and $\sum_{k=0}^{n} b_k x^k$. Since $\{x^k\}_{k=0}^{n}$ is a basis for the vector space of all polynomials in $x$ of degree $n$, $\sum_{k=0}^{n} a_k x^k = \sum_{k=0}^{n} b_k x^k$ implies that $\sum_{k=0}^{n} [a_k - b_k] x^k = 0$, or equivalently $a_k = b_k$ for all $0 \leq k \leq n$. In other words we compared the coefficients of $x^k$. This coefficient comparison technique is also applicable to formal power series. Given two formal power series $\sum_{n=0}^{\infty} a_n x^n$ and $\sum_{n=0}^{\infty} b_n x^n$, we say $\sum_{n=0}^{\infty} a_n x^n = \sum_{n=0}^{\infty} b_n x^n$ if and only if $a_n = b_n$ for all nonnegative integers $n$. If we assume $\sum_{n=0}^{\infty} a_n x^n$ and $\sum_{n=0}^{\infty} b_n x^n$ are two ways of counting the same combinatorial quantity, or two ways of expanding the same function, we can use the equality of the coefficients to obtain new combinatorial identities [Wilf, 2014].

We have already seen an example of coefficient comparison in the verification of Equation (3.32). For another example of this technique, take the function $(1 + x)^n (a + x)^n$, where $n$ is a nonnegative integer, $x$ is a nonzero complex number, and $a$ is any complex number. Since $(1 + x)^n (a + x)^n = (1 + x)^n [(1 + x) + (a - 1)]^n$, we use the binomial theorem to expand the right side and find that

$$(1 + x)^n(a + x)^n = (1 + x)^n [(1 + x) + (a - 1)]^n$$

$$= (1 + x)^n \sum_{k=0}^{n} \binom{n}{k}(a - 1)^k(1 + x)^{n-k}$$

$$= \sum_{k=0}^{n} \binom{n}{k}(a - 1)^k(1 + x)^{2n-k}$$

$$= \sum_{k=0}^{n} \binom{n}{k}(a - 1)^k \sum_{j=0}^{2n-k} \binom{2n - k}{j} x^{2n-k-j}$$

$$= \sum_{k=0}^{n} \binom{n}{k}(a - 1)^k \sum_{j=k}^{2n} \binom{2n - k}{j - k} x^{2n-j}$$

$$= \sum_{j=0}^{2n} \sum_{k=0}^{j} \binom{n}{k}(a - 1)^k \binom{2n - k}{j - k} x^{2n-j}.$$

Given a formal power series $\sum_{n=0}^{\infty} a_n x^n$, define $[x^n] \sum_{n=0}^{\infty} a_n x^n$ to the coefficient of $x^n$ in this series expansion, i.e. $[x_n] \sum_{n=0}^{\infty} a_n x^n = a_n$. The previous calculations imply that

$$[x^n] \sum_{j=0}^{2n} \sum_{k=0}^{j} \binom{n}{k}(a - 1)^k \binom{2n - k}{j - k} x^{2n-j} =$$

$$\sum_{k=0}^{n} \binom{n}{k}\binom{2n - k}{n - k}(a - 1)^k = \sum_{k=0}^{n} \binom{n}{k}\binom{2n - k}{n}(a - 1)^k. \quad (4.1)$$

We now expand $(1 + x)^n(a + x)^n$ via the Gould convolution formula

$$\sum_{i=0}^{n} \sum_{j=0}^{n} a_{i,j} = \sum_{k=0}^{2n} \sum_{i=\lfloor \frac{k}{n+1} \rfloor (k-n)}^{k - \lfloor \frac{k+1}{n+1} \rfloor (k-n)} a_{i,k-i}, \quad (4.2)$$

where $a_{i,j} = \binom{n}{i} x^{n-i} \binom{n}{j} a^j x^{n-j}$ to obtain

$$(1 + x)^n(a + x)^n = \left[ \sum_{i=0}^{n} \binom{n}{i} x^{n-i} \right] \left[ \sum_{j=0}^{n} \binom{n}{j} a^j x^{n-j} \right]$$

$$= \sum_{k=0}^{2n} \sum_{i=\lfloor \frac{k}{n+1} \rfloor (k-n)}^{k - \lfloor \frac{k+1}{n+1} \rfloor (k-n)} \binom{n}{i} x^{n-i} \binom{n}{k - i} a^{k-i} x^{n-k+i}$$

$$= \sum_{k=0}^{2n} x^{2n-k} \sum_{i=\lfloor \frac{k}{n+1} \rfloor (k-n)}^{k - \lfloor \frac{k+1}{n+1} \rfloor (k-n)} \binom{n}{i}\binom{n}{k - i} a^{k-i}.$$

Then

$$[x^n] \sum_{k=0}^{2n} x^{2n-k} \sum_{i=\lfloor \frac{k}{n+1} \rfloor (k-n)}^{k - \lfloor \frac{k+1}{n+1} \rfloor (k-n)} \binom{n}{i} \binom{n}{k-i} a^{k-i} =$$

$$\sum_{i=0}^{n} \binom{n}{i} \binom{n}{n-i} a^{n-i} = \sum_{i=0}^{n} \binom{n}{i}^2 a^{n-i}. \tag{4.3}$$

Equation (4.3) is the second side of the identity. Since Equations (4.1) and (4.3) provide the coefficient of $x^n$ in the unique power series expansion of $(1+x)^n(a+x)^n$ about $x = 0$, we conclude that

$$\sum_{k=0}^{n} \binom{n}{k} \binom{2n-k}{n} (a-1)^k = \sum_{k=0}^{n} \binom{n}{k}^2 a^{n-k}. \tag{4.4}$$

We now discuss the problem which originally piqued Professor Gould's interest in the technique of coefficient comparison. In 1948 Leo Moser posed the following problem in the Elementary Problem Section of the *American Math Monthly* [Moser, 1948]. Show that

$$\sum_{r=0}^{n} \left[ 2^{n-r} \binom{n}{r}^2 - \binom{2n-r}{n} \binom{n}{r} \right] = 0, \qquad \text{for all nonnegative integers } n.$$

$$\tag{4.5}$$

To prove Equation (4.5) Professor Gould developed Equation (4.4) and set $a = 2$.

Other interesting cases of Equation (4.4) occur when $a = 0$, $a = 1$, and $a = -1$. Define $0^0 = 1$. If $a = 0$, Equation (4.4) becomes

$$\sum_{k=0}^{n} (-1)^k \binom{n}{k} \binom{2n-k}{n} = 1. \tag{4.6}$$

By letting $k \to n - k$, we transform Equation (4.6) into

$$\sum_{k=0}^{n} (-1)^k \binom{n}{k} \binom{n+k}{n} = (-1)^n. \tag{4.7}$$

Equation (4.7) is an example of an $n^{th}$ difference identity. If $a = 1$, Equation (4.4) becomes

$$\sum_{k=0}^{n} \binom{n}{k}^2 = \binom{2n}{n}. \tag{4.8}$$

If $a = -1$, Equation (4.4) with $n \to 2n$ becomes

$$\sum_{k=0}^{2n}(-1)^k \binom{2n}{k}\binom{4n-k}{2n}2^k = \sum_{k=0}^{2n}(-1)^k \binom{2n}{k}^2. \tag{4.9}$$

We claim that

$$\sum_{k=0}^{2n}(-1)^k \binom{2n}{k}^2 = (-1)^n \binom{2n}{n}, \tag{4.10}$$

which implies that

$$\sum_{k=0}^{2n}(-1)^k \binom{2n}{k}\binom{4n-k}{2n}2^k = (-1)^n \binom{2n}{n}.$$

In order to prove Equation (4.10), we apply the coefficient comparison technique to $(1+x)^{2n}(1-x)^{2n}$ and expand this function in two different ways. The first expansion utilizes the Gould convolution formula of Equation (4.2) with $n$ replaced by $2n$. Since the binomial theorem implies that

$$(1+x)^{2n} = \sum_{j=0}^{2n}\binom{2n}{j}x^{2n-j}, \qquad (1-x)^{2n} = \sum_{i=0}^{2n}(-1)^{2n-i}\binom{2n}{i}x^{2n-i},$$

we define the $a_{i,j}$ of Equation (4.2) to be

$$a_{i,j} = (-1)^{2n-i}\binom{2n}{i}x^{2n-i}\binom{2n}{j}x^{2n-j},$$

and thus obtain

$$(1+x)^{2n}(1-x)^{2n} = \sum_{k=0}^{4n}x^{4n-k}\sum_{i=\lfloor\frac{k}{2n+1}\rfloor(k-2n)}^{k-\lfloor\frac{k+1}{2n+1}\rfloor(k-2n)}(-1)^i\binom{2n}{i}\binom{2n}{k-i}x^{4n-k}. \tag{4.11}$$

Therefore

$$[x^{2n}]\sum_{k=0}^{4n}x^{4n-k}\sum_{i=\lfloor\frac{k}{2n+1}\rfloor(k-2n)}^{k-\lfloor\frac{k+1}{2n+1}\rfloor(k-2n)}(-1)^i\binom{2n}{i}\binom{2n}{k-i}x^{4n-k} = \sum_{i=0}^{2n}(-1)^i\binom{2n}{i}^2. \tag{4.12}$$

The second way of expanding $(1+x)^{2n}(1-x)^{2n}$ relies on the observation that

$$(1+x)^{2n}(1-x)^{2n} = (1-x^2)^{2n} = \sum_{k=0}^{2n}(-1)^k\binom{2n}{k}x^{4n-2k}. \tag{4.13}$$

From Equation (4.13) we conclude that

$$[x^{2n}] \sum_{k=0}^{2n} (-1)^k \binom{2n}{k} x^{4n-2k} = (-1)^n \binom{2n}{n}. \qquad (4.14)$$

Equation (4.10) follows from the equality of Equations (4.12) and (4.14).

We obtain a generalization of Equation (4.10) by expanding $(1+x)^n(1-x)^n$ in the same two ways we expanded $(1+x)^{2n}(1-x)^{2n}$. First use the binomial theorem and Equation (4.2) to obtain

$$(1-x)^n(1+x)^n = \left[ \sum_{i=0}^{n} (-1)^i \binom{n}{i} x^i \right] \left[ \sum_{j=0}^{n} \binom{n}{j} x^j \right]$$

$$= \sum_{k=0}^{2n} x^k \sum_{i=\lfloor \frac{k}{n+1} \rfloor (k-n)}^{k-\lfloor \frac{k+1}{n+1} \rfloor (k-n)} (-1)^i \binom{n}{i} \binom{n}{k-i}. \qquad (4.15)$$

We also use the observation that $(1-x)^n(1+x)^n = (1-x^2)^n$ to obtain

$$(1-x)^n(1+x)^n = (1-x^2)^n = \sum_{j=0}^{n} (-1)^j \binom{n}{j} x^{2j}. \qquad (4.16)$$

The powers of $x$ in the right sum of Equation (4.16) are all even. Since Equation (4.15) must equal Equation (4.16), we deduce that odd powers of $k$ in Equation (4.15) vanish. In other words,

$$\sum_{i=0}^{k} (-1)^i \binom{n}{i} \binom{n}{k-i} = 0, \qquad \text{for all positive odd integers } k. \qquad (4.17)$$

What can we say about the even powers of $k$ in Equation (4.15)? The monomial $x^k$, with $k$ even, corresponds to the monomial $x^{2j}$ of Equation (4.16) where $j = \frac{k}{2}$. Therefore, Equations (4.15) and (4.16) imply that

$$\sum_{i=0}^{k} (-1)^i \binom{n}{i} \binom{n}{k-i} = (-1)^{\frac{k}{2}} \binom{n}{\frac{k}{2}}, \qquad \text{for nonnegative even integers } k. \qquad (4.18)$$

**Remark 4.1.** We mention the elegant reformulation of Equations (4.17) and (4.18), namely,

$$\sum_{k=0}^{r} (-1)^k \binom{n}{k} \binom{n}{r-k} = (-1)^{\lfloor \frac{r}{2} \rfloor} \binom{n}{\lfloor \frac{r}{2} \rfloor} \frac{1 + (-1)^r}{2}. \qquad (4.19)$$

Equation (4.19) appears in [Gould, 1972] as (3.32), where $n$ is replaced by an arbitrary complex number $x$.

We may use coefficient comparison to further generalize Equation (4.19). Assume $m$ and $n$ are nonnegative integers such that $0 \leq m \leq 2n$. We look at three expansions of $F(x) = (1+x)^{2n-m}(1-x)^n$. For the first expansion, we apply the binomial theorem with the Gould convolution formula to obtain

$$
F(x) = \left[ \sum_{i=0}^{m} (-1)^i \binom{m}{i} x^i \right] \cdot \left[ \sum_{j=0}^{2n-m} \binom{2n-m}{j} x^j \right]
$$

$$
= \sum_{i=0}^{2n} \sum_{j=0}^{2n} (-1)^i \binom{m}{i} x^i \binom{2n-m}{j} x^j
$$

$$
= \sum_{k=0}^{4n} x^k \sum_{i=\lfloor \frac{k}{2n+1} \rfloor (k-2n)}^{k-\lfloor \frac{k+1}{2n+1} \rfloor (k-2n)} (-1)^i \binom{m}{i} \binom{2n-m}{k-i}.
$$

Hence

$$
[x^n] \sum_{k=0}^{4n} x^k \sum_{i=\lfloor \frac{k}{2n+1} \rfloor (k-2n)}^{k-\lfloor \frac{k+1}{2n+1} \rfloor (k-2n)} (-1)^i \binom{m}{i} \binom{2n-m}{k-i} = \sum_{i=0}^{n} (-1)^i \binom{m}{i} \binom{2n-m}{n-i}.
$$

$$(4.20)$$

For the second expansion, write $F(x) = (1-x)^m (1-x+2x)^{2n-m}$ and observe that

$$
F(x) = (1-x)^m \sum_{i=0}^{2n-m} \binom{2n-m}{i} (1-x)^{2n-m-j} 2^i x^i
$$

$$
= \sum_{i=0}^{2n-m} \binom{2n-m}{i} (1-x)^{2n-i} 2^i x^i
$$

$$
= \sum_{i=0}^{2n-m} \binom{2n-m}{i} 2^i x^i \sum_{j=0}^{2n-i} \binom{2n-i}{j} (-1)^j x^j
$$

$$
= \sum_{i=0}^{2n} \sum_{j=0}^{2n} \binom{2n-m}{i} 2^i x^i \binom{2n-i}{j} (-1)^j x^j
$$

$$
= \sum_{k=0}^{4n} x^k \sum_{i=\lfloor \frac{k}{2n+1} \rfloor (k-2n)}^{k-\lfloor \frac{k+1}{2n+1} \rfloor (k-2n)} (-1)^{k-i} 2^i \binom{2n-m}{i} \binom{2n-i}{k-i}.
$$

Therefore

$$[x^n] \sum_{k=0}^{4n} x^k \sum_{i=\lfloor \frac{k}{2n+1} \rfloor (k-2n)}^{k-\lfloor \frac{k+1}{2n+1} \rfloor (k-2n)} (-1)^{k-i} 2^i \binom{2n-m}{i} \binom{2n-i}{k-i}$$

$$= (-1)^n \sum_{i=0}^{n} (-1)^i \binom{2n-m}{i} \binom{2n-i}{n-i} 2^i. \qquad (4.21)$$

For the third expansion, observe that

$$F(x) = (1-x)^m (1+x)^m (1+x)^{2n-2m} = (1-x^2)^m (1+x)^{2n-2m}$$

and apply the binomial theorem with the Gould convolution formula as follows:

$$F(x) = \sum_{i=0}^{m} (-1)^i \binom{m}{i} x^{2i} \sum_{j=0}^{2n} \binom{2n-2m}{j} x^j$$

$$= \sum_{i=0}^{2m} (-1)^i \binom{m}{i} x^{2i} \sum_{j=i}^{2n+i} \binom{2n-2m}{j-i} x^{j-i}$$

$$= \sum_{i=0}^{4n} (-1)^i \binom{m}{i} x^{2i} \sum_{j=0}^{4n} \binom{2n-2m}{j-i} x^{j-i}$$

$$= \sum_{i=0}^{8n} x^k \sum_{i=\lfloor \frac{k}{4n+1} \rfloor (k-4n)}^{k-\lfloor \frac{k+1}{4n+1} \rfloor (k-4n)} (-1)^i \binom{m}{i} \binom{2n-2m}{k-2i}.$$

Hence

$$[x^n] \sum_{i=0}^{8n} x^k \sum_{i=\lfloor \frac{k}{4n+1} \rfloor (k-4n)}^{k-\lfloor \frac{k+1}{4n+1} \rfloor (k-4n)} (-1)^i \binom{m}{i} \binom{2n-2m}{k-2i} = \sum_{i=0}^{\lfloor \frac{n}{2} \rfloor} (-1)^i \binom{m}{i} \binom{2n-2m}{n-2i}. \quad (4.22)$$

The last equality follows from the fact that $\binom{2n-2m}{n-2i} = 0$ if $n > 2i$.

By combining Equations (4.20), (4.21), and (4.22), we conclude that

$$\sum_{i=0}^{n} (-1)^i \binom{m}{i} \binom{2n-m}{n-i} = (-1)^n \sum_{i=0}^{n} (-1)^i \binom{2n-m}{i} \binom{2n-i}{n-i} 2^i$$

$$= \sum_{i=0}^{\lfloor \frac{n}{2} \rfloor} (-1)^i \binom{m}{i} \binom{2n-2m}{n-2i}. \qquad (4.23)$$

Through a collection of hand calculations Professor Gould conjectured that

$$\sum_{j=0}^{n} (-1)^j \binom{m}{j} \binom{2n-m}{n-j} = (-1)^n 2^{2n} \binom{\frac{m-1}{2}}{n}. \qquad (4.24)$$

He then proved his conjecture by applying the fundamental theorem of algebra. We will study his proof in the next section. Assuming the validity of Equation (4.24), we may rewrite Equation (4.23) as

$$\sum_{k=0}^{n}(-1)^k\binom{x}{k}\binom{2n-x}{n-k} = (-1)^n\sum_{k=0}^{n}(-1)^k\binom{2n-k}{n-k}\binom{2n-x}{k}2^k$$

$$= \sum_{k=0}^{n}\binom{-n-1}{n-k}\binom{2n-x}{k}2^k$$

$$= \sum_{k=0}^{\lfloor\frac{n}{2}\rfloor}(-1)^k\binom{x}{k}\binom{2n-2x}{n-2k}$$

$$= (-1)^n 2^{2n}\binom{\frac{x-1}{2}}{n}, \ n \geq 0 \qquad (4.25)$$

$$= \frac{2^n}{n!}\prod_{k=0}^{n-1}(2k+1-x), \ n \geq 1$$

where $n$ is a nonnegative integer and $x$ is a complex number. We justify this substitution of $m$ via the fundamental theorem of algebra.

## 4.2   The Fundamental Theorem of Algebra

In this section we discuss the fundamental theorem of algebra (FTA) and show how a corollary of FTA provides a means of verifying a conjectured binomial identity. In its simplest form, the fundamental theorem of algebra states that every non-constant polynomial has one zero or root in the complex plane. There are many proofs of this fact, most of which rely on some basic theorems of complex analysis. An excellent survey of such proofs is found in [File and Miller, 2003]. By using the fundamental theorem of algebra, we are able to write each non-constant polynomial as a product of its roots. More precisely, let $f(z) = \sum_{i=0}^{n} a_i z^i$ be a polynomial of degree $n$ in $z$. Assume that $\{a_i\}_{i=0}^{n}$ is a set of complex numbers such that $a_n \neq 0$. We claim that $f(z) = a_n \prod_{i=1}^{n}(z - \alpha_i)$, where $f(\alpha_i) = 0$ for each $1 \leq i \leq n$ and $\alpha_i \in \mathbb{C}$ [Spiegel, 1964, p.125]. This is all the preliminary information we need to state and prove the following corollary:

**Corollary 4.1.** *Let* $f(x) = \sum_{i=0}^{n} a_i x^i$ *be a polynomial in* $x$ *of degree* $n$. *If* $f(x)$ *vanishes for more than* $n$ *values of* $x$, *then* $a_i = 0$ *for* $0 \leq i \leq n$.

**Proof:** Let $f(x)$ be the polynomial defined above. Assume $\{r_i\}_{i=0}^{n}$ is a set of $n+1$ distinct complex numbers such that $f(r_i) = 0$. By the fundamental theorem of algebra, $f(x) = a_n \prod_{i=1}^{n} (x - r_i)$, and $f(r_0) = a_n \prod_{i=1}^{n} (r_0 - r_i)$. By assumption, $f(r_0) = 0 = a_n \prod_{i=1}^{n} (r_0 - r_i)$. From our hypothesis, we know that $r_0 \neq r_i$ whenever $1 \leq i \leq n$. Therefore, $a_n = 0$. This implies that $f(x) = \sum_{i=0}^{n-1} a_i x^i = a_{n-1} \prod_{i=1}^{n-1} (x - r_i)$. By repeating this process, we see that $a_{n-1} = 0$ and $f(x) = \sum_{i=0}^{n-2} a_i x^i = a_{n-2} \prod_{i=1}^{n-2} (x - r_i)$. Continue inductively until we obtain $f(x) = a_0$. Then, $f(r_0) = 0 = a_0$, and $\{a_i\}_{i=0}^{n} = \{0\}$. □

Corollary 4.1 can be restated as follows:

**Corollary 4.2.** *Let $f(x) = \sum_{i=0}^{n} a_i x^i$ be a polynomial in $x$ of degree $n$. If $f(x) = 0$ for more than $n$ values of $x$, it follows that $f(x) = 0$ for every value of $x$. In other words, the relation $f(x) = 0$ is an identity.*

Corollary 4.2 is often used to show that two polynomials of degree $n$ are identically the same. Suppose $f(x) = \sum_{i=0}^{n} a_i x^i$ and $g(x) = \sum_{i=0}^{n} b_i x^i$ are two polynomials of degree $n$. We form $h(x) = f(x) - g(x) = \sum_{i=0}^{n} (a_i - b_i) x^i$, another polynomial of degree $n$. By Corollary 4.2 if we can show that $h(x) = 0$ for more than $n$ values of $x$, we can conclude that $h(x) = 0$ for all $x$, and that $a_i = b_i$ for $0 \leq i \leq n$. In other words, $f(x) = g(x)$. We summarize this paragraph in Corollary 4.3.

**Corollary 4.3.** *Let $f(x) = \sum_{i=0}^{n} a_i x^i$ and $g(x) = \sum_{i=0}^{n} b_i x^i$ be two polynomials of degree $n$. Suppose that $f(x) = g(x)$ for at least $n + 1$ distinct values of $x$. Then $f(x) = g(x)$ is an identity which is true for all complex $x$.*

In order to apply Corollary 4.3 to combinatorial identities, recall that $\binom{x}{k} = \frac{x(x-1)\cdots(x-k+1)}{k!}$ is a polynomial in $x$ of degree $k$. This observation allows us to interpret many binomial coefficient series as polynomials, apply Corollary 4.3, and show that two seemingly different binomial coefficient expressions are identical. We demonstrate this technique with an FTA style proof of Equation (4.24) and show that

$$\sum_{j=0}^{n} (-1)^j \binom{x}{j} \binom{2n - x}{n - j} = (-1)^n 2^{2n} \binom{\frac{x-1}{2}}{n}, \qquad x \text{ complex.} \qquad (4.26)$$

For a fixed nonnegative integer $n$, both sides of Equation (4.26) are polynomials in $x$ of degree $n$. Therefore, if we can show that the left hand side of Equation (4.26) equals the right hand side for $n + 1$ distinct values of $x$, Corollary 4.3 implies that Equation (4.26) must be true for *all* possible values of $x$. The trick to any such FTA style proof is to select the correct $n+1$ distinct values. Clearly Equation (4.26) is trivially true if $n = 0$ since it reduces to $1 = 1$. Without loss of generality assume that $n \geq 1$. The right hand side of Equation (4.26) can be rewritten as

$$(-1)^n 2^{2n} \binom{\frac{x-1}{2}}{n} = \frac{2^n}{n!} \prod_{j=0}^{n-1} (2j + 1 - x)$$

$$= \frac{2^n}{n!}(1 - x)(3 - x)\dots(2n - 1 - x). \qquad (4.27)$$

Equation (4.27) is zero for all $x \in \{1, 3, 5, 7, \dots, 2n - 1\} = \{2m + 1\}_{m=0}^{n-1}$. We show that the left side of Equation (4.26) is zero for these same $n$ values. Let $x = 2m + 1$ for $0 \leq m \leq n - 1$. The left hand sum of Equation (4.26) becomes $\sum_{j=0}^{n}(-1)^j \binom{2m+1}{j}\binom{2n-2m-1}{n-j}$. If $2m + 1 > n$, we can extend the range of summation to $2m + 1$ since $\binom{2n-2m-1}{n-j}$ vanishes whenever $j > n$. If $2m + 1 < n$, we can truncate the range of summation at $2m + 1$ since $\binom{2m+1}{j} = 0$ whenever $j > 2m + 1$. If $2m + 1 = n$, the range of summation stays the same. This analysis implies that $\sum_{j=0}^{n}(-1)^j \binom{2m+1}{j}\binom{2n-2m-1}{n-j} = \sum_{j=0}^{2m+1}(-1)^j \binom{2m+1}{j}\binom{2n-2m-1}{n-j}$. We split this sum into two sums, where one sum contains the first $m + 1$ summands while the other contains the remaining $m + 1$ summands. Then it becomes relatively easy to show that the second sum is the negative of the first. In particular

$$\sum_{j=0}^{2m+1}(-1)^j \binom{2m+1}{j}\binom{2n-2m-1}{n-j} =$$

$$\sum_{j=0}^{m}(-1)^j \binom{2m+1}{j}\binom{2n-2m-1}{n-j}$$

$$+ \sum_{j=m+1}^{2m+1}(-1)^j \binom{2m+1}{j}\binom{2n-2m-1}{n-j}$$

$$= \sum_{j=0}^{m}(-1)^j \binom{2m+1}{j}\binom{2n-2m-1}{n-j}$$

$$+ \sum_{j=0}^{m}(-1)^{2m+1+j}\binom{2m+1}{2m+1-j}\binom{2n-2m-1}{n-(2m+1-j)}$$

$$= \sum_{j=0}^{m}(-1)^j \binom{2m+1}{j}\binom{2n-2m-1}{n-j}$$

$$- \sum_{j=0}^{m}(-1)^j \binom{2m+1}{j}\binom{2n-2m-1}{n-j} = 0.$$

So far we have shown that if $x \in \{2m+1\}_{m=0}^{n-1}$, both sides of Equation (4.26) are zero. According to Corollary 4.3, since we are working with polynomials of degree $n$, we need to show agreement for one more value of $x$. If $x = 0$, the right hand side of Equation (4.26) becomes

$$(-1)^n 2^{2n}\binom{-\frac{1}{2}}{n} = \binom{2n}{n}, \qquad \text{by the } -\frac{1}{2}\text{-Transformation,}$$

while the left side of Equation (4.26) is

$$\sum_{j=0}^{n}(-1)^j \binom{0}{j}\binom{2n}{n-j} = \binom{2n}{n}.$$

Because we have shown that Equation (4.26) is valid for $x \in S = \{0, 1, 3, 5, \ldots, 2n-1\}$, where $|S| = n+1$, Corollary 4.3 implies that Equation (4.26) is true for *all* complex $x$.

In his original proof sketch Professor Gould noted that Equation (4.26) is valid for $x = -1$ and $x = n$. These two cases, while not necessary, are not difficult to verify. Suppose $x = -1$. The right hand side of Equation (4.26) becomes

$$(-1)^n 2^{2n}\binom{-1}{n} = 2^{2n}, \qquad \text{by the } -1\text{-Transformation,}$$

while the left hand side of Equation (4.26) becomes

$$\sum_{j=0}^{n}(-1)^j \binom{-1}{j}\binom{2n+1}{n-j} = \sum_{j=0}^{n}\binom{2n+1}{n-j} = \sum_{j=0}^{n}\binom{2n+1}{j}.$$

If we can show $\sum_{j=0}^{n}\binom{2n+1}{j} = 2^{2n}$, we will be done. To do so, we will make use of the following application of the standard interchange formula. Assume $f(x)$ is a complex valued function defined for all nonnegative integers. For $n \geq 2$, we have

$$\sum_{k=1}^{n} f(k) = \sum_{k=1}^{n} f(n-k+1) = \frac{1}{2}\sum_{k=1}^{n}[f(k) + f(n-k+1)]$$

$$= \sum_{k=1}^{\lfloor \frac{n}{2}\rfloor}[f(k) + f(n-k+1)] + \frac{1-(-1)^n}{2}f\left(\left\lfloor\frac{n+1}{2}\right\rfloor\right). \qquad (4.28)$$

In Equation (4.28) let $n \to n - 1$ and obtain

$$\sum_{k=1}^{n-1} f(k) = \sum_{k=1}^{\lfloor \frac{n-1}{2} \rfloor} [f(k) + f(n-k)] + \frac{1+(-1)^n}{2} f\left(\left\lfloor \frac{n}{2} \right\rfloor\right). \qquad (4.29)$$

Add $f(0) + f(n)$ to both sides of Equation (4.29) to find that

$$\sum_{k=0}^{n} f(k) = \sum_{k=0}^{\lfloor \frac{n-1}{2} \rfloor} [f(k) + f(n-k)] + \frac{1+(-1)^n}{2} f\left(\left\lfloor \frac{n}{2} \right\rfloor\right), \qquad n \geq 1.$$

$$(4.30)$$

In Equation (4.30), let $f(k) = \binom{n}{k}$ and recall that $\sum_{k=0}^{n} \binom{n}{k} = 2^n$. After simplification, Equation (4.30) becomes

$$\sum_{k=0}^{\lfloor \frac{n-1}{2} \rfloor} \binom{n}{k} = 2^{n-1} - \frac{1+(-1)^n}{4} \binom{n}{\lfloor \frac{n}{2} \rfloor}, \qquad n \geq 1. \qquad (4.31)$$

Take Equation (4.31) and let $n \to 2n + 1$ to obtain $\sum_{k=0}^{n} \binom{2n+1}{k} = 2^{2n}$, the desired result.

To finish Professor Gould's proof sketch, we need to verify Equation (4.26) for $x = n$. If we let $x = n$, the right side of Equation (4.26) becomes $(-1)^n 2^{2n} \binom{\frac{n-1}{2}}{n}$, while the left side becomes

$$\sum_{j=0}^{n} (-1)^j \binom{n}{j} \binom{n}{n-j} = (-1)^{\frac{n}{2}} \binom{n}{\frac{n}{2}} \frac{1+(-1)^n}{2}, \qquad \text{by Eq. (4.19).}$$

Therefore it is a matter of showing that $(-1)^{\frac{n}{2}} \binom{n}{\frac{n}{2}} \frac{1+(-1)^n}{2} = (-1)^n 2^{2n} \binom{\frac{n-1}{2}}{n}$. We will do this by analyzing two separate cases. First assume $n$ is odd, i.e. $n = 2m + 1$ for some nonnegative integer $m$. Clearly $(-1)^{\frac{n}{2}} \binom{n}{\frac{n}{2}} \frac{1+(-1)^n}{2} = 0$. On the other hand

$$(-1)^n 2^{2n} \binom{\frac{n-1}{2}}{n} = -2^{4m+2} \binom{\frac{2m+1-1}{2}}{2m+1} = -2^{4m+2} \binom{m}{2m+1} = 0.$$

Now assume $n$ is even, i.e. $n = 2m$ for some nonnegative integer $m$. Then $(-1)^{\frac{n}{2}} \binom{n}{\frac{n}{2}} \frac{1+(-1)^n}{2} = (-1)^m \binom{2m}{m}$. On the other hand

$$(-1)^n 2^{2n} \binom{\frac{n-1}{2}}{n} = 2^{4m} \binom{\frac{2m-1}{2}}{2m}$$

$$= \frac{2^{4m} \left(\frac{2m-1}{2}\right)\left(\frac{2m-1}{2} - 1\right) \ldots \left(\frac{2m-1}{2} - m + 1\right)\left(\frac{2m-1}{2} - m\right) \ldots \left(\frac{2m-1}{2} - 2m + 1\right)}{(2m)!}$$

$$= \frac{2^{4m} \left(\frac{2m-1}{2}\right)\left(\frac{2m-3}{2}\right) \ldots \left(\frac{1}{2}\right)\left(-\frac{1}{2}\right)\left(-\frac{3}{2}\right) \ldots \left(\frac{-2m+1}{2}\right)}{(2m)!}$$

$$= \frac{2^{2m}(2m-1)(2m-3)\ldots(1)(-1)(-3)\ldots(-2m+1)}{(2m-1)(2m-3)\ldots(3)(1)(2m)(2m-2)\ldots(4)(2)}$$

$$= \frac{2^{2m}(-1)(-3)\ldots(-2m+1)}{2^m m!} = \frac{(-1)^m 2^m (1)(3)\ldots(2m-1)}{m!} \cdot \frac{m!}{m!}$$

$$= (-1)^m \frac{(2m)!}{m!m!} = (-1)^m \binom{2m}{m}.$$

# Chapter 5

# Vandermonde Convolution

In the study of combinatorial identities there are two fundamental identities, the binomial theorem and the Vandermonde convolution. Chapter 2 discussed the binomial theorem. We use this chapter to discuss the Vandermonde convolution. We begin our analysis by stating the basic integer version Vandermonde convolution.

**Theorem 5.1.** *(basic Vandermonde convolution) Let $r$, $q$, and $n$ be nonnegative integers. Then*

$$\sum_{k=0}^{n} \binom{r}{k}\binom{q}{n-k} = \binom{r+q}{n}. \tag{5.1}$$

Theorem 5.1 is a restatement of Equation (3.42) and was proven via the Gould convolution formula. For the reader who is interested in the history of Equation (5.1) and its various proofs we refer them to [Gould and Glatzer, 1979]. We present another proof which is combinatorial in nature. The right side of Equation (5.1) counts the number of $n$-sets of $[r+q] = \{1, 2, \ldots, r+q\}$, i.e. the number of subsets of $[r+q]$ of cardinality $n$. We claim the left side also counts the number of $n$-sets of $[r+q]$. Write $[r+q] = \{1, 2, \ldots, r\} \cup \{r+1, r+2, \ldots, r+q\}$. A typical $n$-set of $[r+q]$ is the disjoint union of two subsets, one from $\{1, 2, \ldots, r\}$, and the other from $\{r+1, r+2, \ldots, r+q\}$. Fix $k$, where $0 \leq k \leq n$. The number of ways of choosing a $k$-set from $\{1, 2, \ldots, r\}$ is $\binom{r}{k}$. To expand a $k$-set of $\{1, 2, \ldots, r\}$ into an $n$-set of $[r+q]$, we must choose $n-k$ elements from $\{r+1, r+2, \ldots, r+q\}$. The number of ways to make this choice is $\binom{q}{n-k}$. Thus, for a fixed $k$, the rule of products implies there are $\binom{r}{k}\binom{q}{n-k}$ $n$-sets of $[r+q]$ which have $k$ elements from $\{1, 2, \ldots, r\}$. If we vary $k$, for $0 \leq k \leq n$, we will uniquely account for all the $n$-sets of $[r+q]$. According to the rule of rum, this variation of $k$ corresponds to $\sum_{k=0}^{n} \binom{r}{k}\binom{q}{n-k}$.

Equation (5.1) appears in various guises. By expanding each binomial coefficient in terms of factorials and manipulating the results, we find that

$$\sum_{k=0}^{n} \binom{n}{k} \binom{r+q-n}{r-k} = \binom{r+q}{r} = \binom{r+q}{q}. \tag{5.2}$$

We then let $k \to n - k$ in Equation (5.2) to obtain

$$\sum_{k=0}^{n} \binom{n}{k} \binom{r+q-n}{q-k} = \binom{r+q}{r}. \tag{5.3}$$

We now turn to the identity know as the (general) Vandermonde convolution.

**Theorem 5.2.** *(Vandermonde convolution) Let $x$ and $y$ be complex numbers. Let $n$ be a nonnegative integer. Then*

$$\sum_{k=0}^{n} \binom{x}{k} \binom{y}{n-k} = \binom{x+y}{n}. \tag{5.4}$$

If $x$ and $y$ are nonnegative integers, Equation (5.4) becomes Equation (5.1). We prove Equation (5.4) via the fundamental theorem of algebra. Start with Equation (5.1) and fix $r$. Observe that both sides of Equation (5.1) are polynomials in $q$ of degree $n$ which agree for all nonnegative integer values of $q$. By Corollary 4.3 Equation (5.1) is true for all complex values $q$. Now vary $r$, observe that both sides of Equation (5.1) are polynomials of degree $n$ in $r$, and then use Corollary 4.3 to conclude that Equation (5.1) is also valid for all complex $r$.

## 5.1  Five Basic Applications of the Vandermonde Convolution

We now focus on using Equation (5.4) as a tool for obtaining combinatorial identities and present five specific examples. The first example uses $-1$-Transformation to transform Equation (5.4) and provide a closed form for $\sum_{k=0}^{n} \binom{x+k}{k} \binom{y+n-k}{n-k}$. Equation (5.4) implies that

$$\sum_{k=0}^{n} \binom{-x-1}{k} \binom{-y-1}{n-k} = \binom{-x-y-2}{n}, \qquad x \text{ and } y \text{ complex}, n \geq 0. \tag{5.5}$$

To each binomial coefficient in Equation (5.5), we apply the $-1$-Transformation and obtain

$$\sum_{k=0}^{n}\binom{x+k}{k}\binom{y+n-k}{n-k} = \binom{x+y+n+1}{n}, \qquad x \text{ and } y \text{ complex}, n \geq 0.$$
(5.6)

If we assume $x = j$ and $y = m$ where $m$ and $j$ are two arbitrary nonnegative integers, Equation (5.6) becomes

$$\sum_{k=0}^{n}\binom{j+k}{k}\binom{m+n-k}{n-k} = \sum_{k=0}^{n}\binom{j+k}{j}\binom{m+n-k}{m}$$

$$= \sum_{k=j}^{n+j}\binom{k}{j}\binom{n-k+j+m}{m} = \binom{j+m+n+1}{n}.$$

If we let $n \to n-j-m$, the previous calculation implies that

$$\sum_{k=j}^{n-m}\binom{k}{j}\binom{n-k}{m} = \binom{n+1}{n-j-m} = \binom{n+1}{j+m+1}, \qquad n-m \geq j. \quad (5.7)$$

Equation (5.7) is dual to Equation (5.4) since it has the convolution in the *upper* locations of the binomial coefficient summand unlike Equation (5.4) which has the convolution in the lower positions.

The second example provides a closed form for $\sum_{k=0}^{n}\binom{x}{k}\binom{y}{n-k}\binom{k}{\alpha}$, where $x$ and $y$ are complex numbers while $n$ and $\alpha$ are nonnegative integers. The cancellation identity when combined wtih Equation (5.4) implies that

$$\sum_{k=0}^{n}\binom{x}{k}\binom{y}{n-k}\binom{k}{\alpha} = \binom{x}{\alpha}\sum_{k=0}^{n}\binom{y}{n-k}\binom{x-\alpha}{k-\alpha}$$

$$= \binom{x}{\alpha}\sum_{k=0}^{n}\binom{y}{k}\binom{x-\alpha}{n-k-\alpha}$$

$$= \binom{x}{\alpha}\sum_{k=0}^{n-\alpha}\binom{y}{k}\binom{x-\alpha}{n-\alpha-k}$$

$$= \binom{x}{\alpha}\binom{y+x-\alpha}{n-\alpha}.$$

In summary we have shown that

$$\sum_{k=0}^{n}\binom{x}{k}\binom{y}{n-k}\binom{k}{\alpha} = \binom{x}{\alpha}\binom{y+x-\alpha}{n-\alpha}, \qquad 0 \leq \alpha \leq n. \quad (5.8)$$

We remark that Equation (5.8) is true if $\alpha > n$ since both sides equal zero.

Here is a special case of Equation (5.8). Let $\alpha \to j$ and $y \to n - x$ to obtain

$$\sum_{k=0}^{n}\binom{x}{k}\binom{n-x}{n-k}\binom{k}{j} = \binom{x}{j}\binom{x+n-x-j}{n-j} = \binom{x}{j}. \qquad (5.9)$$

Our third application provides a closed form for $\sum_{k=0}^{n}\binom{n}{k}\binom{x}{2n-k}$. For nonnegative integers $n$ and complex $x$ we find that

$$
\begin{aligned}
\sum_{k=0}^{n}\binom{x}{k}\binom{x}{2n-k} &= \sum_{k=0}^{n}\binom{x}{n-k}\binom{x}{n+k} \\
&= \sum_{k=0}^{2n}\binom{x}{k}\binom{x}{2n-k} - \sum_{k=n+1}^{2n}\binom{x}{k}\binom{x}{2n-k} \\
&= \binom{2x}{2n} - \sum_{k=n+1}^{2n}\binom{x}{k}\binom{x}{2n-k} \\
&= \binom{2x}{2n} - \sum_{k=1}^{n}\binom{x}{k+n}\binom{x}{n-k} \\
&= \binom{2x}{2n} - \sum_{k=0}^{n}\binom{x}{k+n}\binom{x}{n-k} + \binom{x}{n}^{2}.
\end{aligned}
$$

We solve for the bold faced sum to obtain

$$
\begin{aligned}
\sum_{k=0}^{n}\binom{x}{k}\binom{x}{2n-k} &= \sum_{k=0}^{n}\binom{x}{n-k}\binom{x}{n+k} \\
&= \frac{1}{2}\left[\binom{2x}{2n} + \binom{x}{n}^{2}\right]. \qquad (5.10)
\end{aligned}
$$

A related result is

$$
\begin{aligned}
\sum_{k=0}^{n-1}\binom{x}{k}\binom{x}{2n-1-k} &= \sum_{k=0}^{n-1}\binom{x}{n-1-k}\binom{x}{n+k} \\
&= \frac{1}{2}\binom{2x}{2n-1}, \qquad n \geq 1. \qquad (5.11)
\end{aligned}
$$

The proof of Equation (5.11) is similar in nature to the proof of Equation (5.10) and is omitted.

Let $y$ and $z$ be complex numbers and $r$ be a positive integer. The fourth application of the Vandermonde convolution proves the transformation

$$\sum_{k=0}^{r-1}\binom{z}{k}y^{r-k-1}=\sum_{j=1}^{r}\binom{z-j}{r-j}(y+1)^{j-1}, \qquad z,\, y \text{ complex, } r \geq 1, \quad (5.12)$$

since

$$\sum_{k=0}^{r-1}\binom{z}{k}y^{r-k-1}=\sum_{k=0}^{r-1}\binom{z}{k}\sum_{j=0}^{r-k-1}\binom{r-k-1}{j}(y+1)^{j}(-1)^{r-k+1-j}$$

$$=\sum_{j=0}^{r-1}(-1)^{r-1-j}(y+1)^{j}\sum_{k=0}^{r-j-1}(-1)^{k}\binom{z}{k}\binom{r-k-1}{j}$$

$$=\sum_{j=1}^{r}(-1)^{r-j}(y+1)^{j-1}\sum_{k=0}^{r-j}(-1)^{k}\binom{z}{k}\binom{r-k-1}{j-1}$$

$$=\sum_{j=1}^{r}(-1)^{r-j}(y+1)^{j-1}\sum_{k=0}^{r-j}(-1)^{k}\binom{z}{k}\binom{r-k-1}{r-k-j}$$

$$=\sum_{j=1}^{r}(y+1)^{j-1}\sum_{k=0}^{r-j}\binom{z}{k}\binom{-j}{r-j-k}$$

$$=\sum_{j=1}^{r}\binom{z-j}{r-j}(y+1)^{j-1}.$$

There are two cases of Equation (5.12) which often appear in the literature. The first is $y = 0$. If $y = 0$, the only nonzero term in the left sum occurs when $k = r - 1$, in which case Equation (5.12) reduces to

$$\binom{z}{r-1}=\sum_{j=1}^{r}\binom{z-j}{r-j}.$$

The second case is $y = -1$. The only nonzero term in the right sum of Equation (5.12) occurs when $j = 1$, and Equation (5.12) becomes

$$\sum_{k=0}^{r-1}(-1)^{r-k-1}\binom{z}{k}=\binom{z-1}{r-1},$$

a result equivalent to Equation (1.30).

Our fifth example evaluates another two variable sum similar in nature to Equation (5.12). Assume $x$ and $y$ are complex numbers while $n$ is a nonnegative integer. Then

$$\sum_{k=0}^{n}\binom{x}{k}y^k = \sum_{k=0}^{n}(-1)^k\binom{-x+k-1}{k}y^k$$

$$= \sum_{k=0}^{n}(-1)^k y^k \sum_{j=0}^{k}\binom{n-x}{j}\binom{k-n-1}{k-j}$$

$$= \sum_{j=0}^{n}\binom{n-x}{j}\sum_{k=j}^{n}(-1)^k\binom{k-n-1}{k-j}y^k$$

$$= \sum_{j=0}^{n}(-1)^j\binom{n-x}{j}\sum_{k=j}^{n}\binom{n-j}{k-j}y^k$$

$$= \sum_{j=0}^{n}(-1)^j\binom{n-x}{j}y^j\sum_{k=0}^{n-j}\binom{n-j}{k}y^k$$

$$= \sum_{j=0}^{n}(-1)^j\binom{n-x}{j}y^j(1+y)^{n-j}.$$

In summary we have derived the transformation

$$\sum_{k=0}^{n}\binom{x}{k}y^k = \sum_{k=0}^{n}(-1)^k\binom{n-x}{k}y^k(1+y)^{n-k}, \qquad x \text{ and } y \text{ complex, } n \geq 0.$$

$$(5.13)$$

We have seen special instances of Equation (5.13). If $y = -1$, Equation (5.13) reduces to

$$\sum_{k=0}^{n}(-1)^k\binom{x}{k} = \binom{n-x}{n} = (-1)^n\binom{x-1}{n},$$

which is exactly Equation (1.30).

Another special case occurs when $x = n$ since Equation (5.13) becomes

$$\sum_{k=0}^{n}\binom{n}{k}y^k = (1+y)^n,$$

which is the binomial theorem.

Equation (5.13) is a special instance of this more general theorem.

**Theorem 5.3.** *Let $x$ be a complex number. Let $n$ be a nonnegative integer. Let $f(k)$ be a complex valued function defined for all nonnegative integers $k$. Then*

$$\sum_{k=0}^{n} \binom{x}{k} f(k) = \sum_{j=0}^{n} (-1)^j \binom{n-x}{j} \sum_{k=0}^{n-j} \binom{n-j}{k} f(k+j). \qquad (5.14)$$

The proof of Theorem 5.3 follows the derivation of Equation (5.13) with $y^k$ replaced by $f(k)$, details of which are left to the reader. Certain $f(k)$ provide closed forms for the inner sum on the right side of Equation (5.14). Once such case is $f(k) = y^k$. Another such is $f(k) = \binom{y}{k}$. Equation (5.14) becomes

$$\sum_{k=0}^{n-j} \binom{x}{k} \binom{y}{k} = \sum_{j=0}^{n} (-1)^j \binom{n-x}{j} \sum_{k=0}^{n-j} \binom{n-j}{k} \binom{y}{k+j}. \qquad (5.15)$$

We claim that

$$\sum_{k=0}^{n} \binom{n-j}{k} \binom{y}{k+j} = \binom{y+n-j}{n}, \qquad 0 \le j \le n, \qquad (5.16)$$

since

$$\begin{aligned}
\sum_{k=0}^{n-j} \binom{n-j}{k} \binom{y}{k+j} &= \sum_{k=0}^{n-j} \binom{n-j}{n-j-k} \binom{y}{n-k} \\
&= \sum_{k=0}^{n-j} \binom{n-j}{k} \binom{y}{n-k} = \sum_{k=0}^{n} \binom{n-j}{k} \binom{y}{n-k} \\
&= \binom{n-j+y}{n}, \qquad \text{by Eq. (5.4).}
\end{aligned}$$

Hence Equation (5.15) is equivalent to

$$\sum_{k=0}^{n} \binom{x}{k} \binom{y}{k} = \sum_{k=0}^{n} (-1)^k \binom{n-x}{k} \binom{y+n-k}{n}. \qquad (5.17)$$

## 5.2 An In-Depth Investigation Involving Equation (5.4)

In this section we use the Vandermonde convolution to derive the identity

$$\sum_{k=0}^{n} \binom{n}{k} \binom{x}{k+\alpha} = \binom{n+x}{n+\alpha}, \qquad x \text{ complex, } n \text{ and } \alpha \text{ nonnegative integers.} \qquad (5.18)$$

We then use Equation (5.18) to develop the transformation

$$\sum_{j=0}^{n} \binom{n}{j}\binom{x}{j} z^j = \sum_{\alpha=0}^{n} \binom{n}{\alpha}\binom{x+n-\alpha}{n}(z-1)^\alpha, \qquad (5.19)$$

and prove the combinatorial identity

$$\sum_{k=0}^{n} \binom{n}{k}^2 \binom{x+k}{2n} = \binom{x}{n}^2. \qquad (5.20)$$

The proof of Equation (5.18) follows from application of the index shift formula to Equation (5.4). Let $\alpha$ be a nonnegative integer. Then

$$\sum_{k=0}^{n} \binom{n}{k}\binom{x}{k+\alpha} = \sum_{k=0}^{n} \binom{n}{n-k}\binom{x}{n+\alpha-k}$$

$$= \sum_{k=0}^{n+\alpha} \binom{n}{k}\binom{x}{n+\alpha-k} = \binom{n+x}{n+\alpha}.$$

Equation (5.16) is a special case of Equation (5.18).

A useful variation of Equation (5.18) occurs if we let $\alpha \to -\alpha$, in which case we have

$$\sum_{k=0}^{n} \binom{n}{k}\binom{x}{k-\alpha} = \binom{n+x}{n-\alpha}. \qquad (5.21)$$

To prove Equation (5.21) directly from Equation (5.4) observe that

$$\sum_{r=0}^{n-\alpha} \binom{n}{r+\alpha}\binom{x}{r} = \sum_{r=\alpha}^{n} \binom{n}{r}\binom{x}{r-\alpha}$$

$$= \sum_{r=\alpha}^{n} \binom{n}{n-r}\binom{x}{r-\alpha}$$

$$= \sum_{r=0}^{n-\alpha} \binom{n}{n-\alpha-r}\binom{x}{r} = \binom{n+x}{n-\alpha}.$$

We are now in a position to use Equation (5.18) and derive the transformation of $\sum_{j=0}^{n} \binom{n}{j}\binom{x}{j} z^j$ provided by Equation (5.19). Assume $x$ and $z$ are complex numbers while $n$ is a nonnegative integer. Then

$$\sum_{j=0}^{n} \binom{n}{j}\binom{x}{j} z^j = \sum_{j=0}^{n} \binom{n}{j}\binom{x}{j}(z-1+1)^j$$

$$= \sum_{j=0}^{n} \binom{n}{j}\binom{x}{j} \sum_{\alpha=0}^{j} \binom{j}{\alpha}(z-1)^\alpha$$

$$= \sum_{\alpha=0}^{n} (z-1)^\alpha \sum_{j=\alpha}^{n} \binom{n}{j}\binom{x}{j}\binom{j}{\alpha}$$

$$= \sum_{\alpha=0}^{n} \binom{n}{\alpha}(z-1)^\alpha \sum_{j=\alpha}^{n} \binom{n-\alpha}{j-\alpha}\binom{x}{j}$$

$$= \sum_{\alpha=0}^{n} \binom{n}{\alpha}(z-1)^\alpha \sum_{j=0}^{n-\alpha} \binom{n-\alpha}{j}\binom{x}{j+\alpha}$$

$$= \sum_{\alpha=0}^{n} \binom{n}{\alpha}\binom{x+n-\alpha}{n}(z-1)^\alpha, \qquad \text{by Eq. (5.18).}$$

Here are two special cases of Equation (5.19). If $z = 0$, the left hand side of Equation (5.19) has only one term, namely that given by $j = 0$, and Equation (5.19) becomes

$$\sum_{\alpha=0}^{n} (-1)^\alpha \binom{n}{\alpha}\binom{x+n-\alpha}{n} = 1.$$

In Chapter 6 we will see how Euler's finite difference theorem provides an independent proof of this identity.

We can also evaluate Equation (5.19) if $z = 1$. The right side which has only one term, that indexed by $\alpha = 0$, and we obtain

$$\sum_{j=0}^{n} \binom{n}{j}\binom{x}{j} = \binom{x+n}{n}, \tag{5.22}$$

a special case of Equation (5.18) with $\alpha = 0$.

We finally turn our attention to an identity which is cited in Volume 17 of the May 1956 issue of the *Mathematical Reviews*. On Pages 459-460 Professor Gould encountered the intriguing identity

$$\sum_{k=0}^{n} \binom{n}{k}^2 \binom{x+k}{2n} = \binom{x}{n}^2, \qquad x \text{ complex number, } n \geq 0. \tag{5.23}$$

Since the *Mathematical Reviews* did not provide a proof of Equation (5.23), Professor Gould worked out his own proof making use of Equations (5.4) and (5.18):

$$\sum_{k=0}^{n}\binom{n}{k}^2\binom{x+k}{2n} = \sum_{k=0}^{n}\binom{n}{k}^2\sum_{j=0}^{2n}\binom{x}{2n-j}\binom{k}{j}$$

$$= \sum_{j=0}^{2n}\binom{x}{2n-j}\sum_{k=0}^{n}\binom{n}{k}^2\binom{k}{j}$$

$$= \sum_{j=0}^{n}\binom{x}{2n-j}\binom{n}{j}\sum_{k=j}^{n}\binom{n}{k}\binom{n-j}{k-j}$$

$$= \sum_{j=0}^{n}\binom{x}{2n-j}\binom{n}{j}\sum_{k=0}^{n-j}\binom{n-j}{k}\binom{n}{k+j}$$

$$= \sum_{j=0}^{n}\binom{n}{j}\binom{x}{2n-j}\binom{2n-j}{n}, \qquad \text{by Eq. (5.18)}$$

$$= \binom{x}{n}\sum_{j=0}^{n}\binom{n}{j}\binom{x-n}{n-j} = \binom{x}{n}^2.$$

By setting $x \to -x - 1$ and applying the $-1$-Transformation, we observe that Equation (5.23) is equivalent to

$$\sum_{k=0}^{n}\binom{n}{k}^2\binom{x+2n-k}{2n} = \binom{x+n}{n}^2. \qquad (5.24)$$

Take Equation (5.24) and let $x = m$, a nonnegative integer. Then

$$\binom{m+n}{n} = \binom{n+m}{m} = \sum_{k=0}^{m}\binom{m}{k}^2\binom{n+2m-k}{2m}$$

$$= \sum_{k=0}^{n}\binom{m}{k}^2\binom{n+2m-k}{2m}$$

$$= \sum_{k=0}^{n}\binom{m}{k}^2\binom{n+2m-k}{n-k}. \qquad (5.25)$$

In the second to last equality, we switched the range of summation to $n$. To explain why this switch is permissible, notice that if $n \geq m$, $\binom{m}{k} = 0$ for $k \geq m$. On the other hand, if $n < m$, we observe that $\binom{n+2m-k}{2m} = 0$ for $k \geq m$. Equation (5.25) is a polynomial identity in $m$ of degree $n$ which is true for all nonnegative integers $m$. The fundamental theorem of algebra tells us that this polynomial identity is true for all complex $x$. Thus, we may let $m \to x$ in Equation (5.25) to obtain

$$\sum_{k=0}^{n} \binom{x}{k}^2 \binom{n+2x-k}{n-k} = \binom{x+n}{n}^2, \qquad x \text{ complex number.} \qquad (5.26)$$

If $x \to x - n$, Equation (5.26) becomes

$$\sum_{k=0}^{n} \binom{x-n}{k}^2 \binom{2x-n-k}{n-k} = \binom{x}{n}^2. \qquad (5.27)$$

The right side of Equation (5.27) is identical to the right side of Equation (5.23). Hence, we have shown that

$$\sum_{k=0}^{n} \binom{n}{k}^2 \binom{x+k}{2n} = \sum_{k=0}^{n} \binom{x-n}{k}^2 \binom{2x-n-k}{n-k}.$$

We are able to generalize Equation (5.23) in two ways. The first generalization is due to T. S. Nanjundiah [1958]. For $x$ and $y$ complex numbers and $m$ and $n$ nonnegative integers, Nanjundiah claims that

$$\sum_{k=0}^{n} \binom{m-x+y}{k} \binom{n+x-y}{n-k} \binom{x+k}{m+n} = \binom{x}{m}\binom{y}{n}. \qquad (5.28)$$

If $m = n$ and $x = y$, Equation (5.28) becomes Equation (5.23).

Equation (5.28) follows from three applications of Equation (5.4) since

$$\sum_{k=0}^{n} \binom{m-x+y}{k} \binom{n+x-y}{n-k} \binom{x+k}{m+n}$$

$$= \sum_{k=0}^{n} \binom{m-x+y}{k} \binom{n+x-y}{n-k} \sum_{j=0}^{k} \binom{k}{j} \binom{x}{m+n-j}$$

$$= \sum_{j=0}^{n} \binom{x}{m+n-j} \sum_{k=j}^{n} \binom{m-x+y}{k} \binom{k}{j} \binom{n+x-y}{n-k}$$

$$= \sum_{j=0}^{n} \binom{x}{m+n-j} \binom{m-x+y}{j} \sum_{k=j}^{n} \binom{m-x+y-j}{k-j} \binom{n+x-y}{n-k}$$

$$= \sum_{j=0}^{n} \binom{x}{m+n-j} \binom{m-x+y}{j} \sum_{k=0}^{n-j} \binom{m-x+y-j}{k} \binom{n+x-y}{n-j-k}$$

$$= \sum_{j=0}^{n} \binom{x}{m+n-j} \binom{m-x+y}{j} \binom{m+n-j}{n-j}$$

$$= \binom{x}{m} \sum_{j=0}^{n} \binom{m-x+y}{j} \binom{x-m}{n-j} = \binom{x}{m}\binom{y}{n}.$$

Another generalization of Equation (5.23) is

$$\sum_{k=0}^{n} \binom{n}{k}\binom{\alpha}{k}\binom{x+n+\alpha-k}{n+\alpha} = \binom{x+\alpha}{\alpha}\binom{x+n}{n}, \qquad (5.29)$$

where $x$ is a complex number and $\alpha$ is a nonnegative integer. If $\alpha = n$, Equation (5.29) becomes Equation (5.24).

To prove Equation (5.29) define $f(x) = \sum_{k=0}^{n} \binom{n}{k}\binom{\alpha}{k}\binom{x+n+\alpha-k}{n+\alpha}$. Two applications of Equation (5.4) show that

$$
\begin{aligned}
f(-x-1) &= \sum_{k=0}^{n} \binom{n}{k}\binom{\alpha}{k}\binom{-x-1+n+\alpha-k}{n+\alpha} \\
&= (-1)^{n+\alpha} \sum_{k=0}^{n} \binom{n}{k}\binom{\alpha}{k}\binom{x+k}{n+\alpha} \\
&= (-1)^{n+\alpha} \sum_{k=0}^{n} \binom{n}{k}\binom{\alpha}{k} \sum_{j=0}^{n+\alpha} \binom{x}{n+\alpha-j}\binom{k}{j} \\
&= (-1)^{n+\alpha} \sum_{j=0}^{n+\alpha} \binom{x}{n+\alpha-j} \sum_{k=0}^{n} \binom{n}{k}\binom{\alpha}{k}\binom{k}{j} \\
&= (-1)^{n+\alpha} \sum_{j=0}^{n+\alpha} \binom{x}{n+\alpha-j}\binom{\alpha}{j} \sum_{k=0}^{n} \binom{n}{k}\binom{\alpha-j}{k-j} \\
&= (-1)^{n+\alpha} \sum_{j=0}^{n+\alpha} \binom{x}{n+\alpha-j}\binom{\alpha}{j}\binom{n+\alpha-j}{n-j}, \quad \text{by Eq. (5.21)} \\
&= (-1)^{n+\alpha} \sum_{j=0}^{n+\alpha} \binom{x}{n+\alpha-j}\binom{n+\alpha-j}{\alpha}\binom{\alpha}{j} \\
&= (-1)^{n+\alpha} \binom{x}{\alpha} \sum_{j=0}^{n+\alpha} \binom{\alpha}{j}\binom{x-\alpha}{n-j} = (-1)^{n+\alpha} \binom{x}{\alpha}\binom{x}{n}.
\end{aligned}
$$

Since $f(-x-1) = (-1)^{n+\alpha}\binom{x}{\alpha}\binom{x}{n}$, we let $x \to -x-1$, apply the $-1$-Transformation, and conclude that

$$
\begin{aligned}
f(x) &= \sum_{k=0}^{n} \binom{n}{k}\binom{\alpha}{k}\binom{x+n+\alpha-k}{n+\alpha} = (-1)^{n+\alpha}\binom{-x-1}{\alpha}\binom{-x-1}{n} \\
&= \binom{x+n}{n}\binom{x+\alpha}{\alpha},
\end{aligned}
$$

which is precisely Equation (5.29).

We remark that Equation (5.29) is a special case of Saalschütz's theorem for hypergeometric series [Carlitz, 1959].

# Chapter 6

# The $n^{th}$ Difference Operator and Euler's Finite Difference Theorem

This chapter is devoted to Euler's finite difference theorem, a theorem which provides a way of evaluating the $n^{th}$ difference of polynomial whose degree is less than or equal to $n$. In order to understand Euler's finite difference theorem, we must first define the notion of a difference operator. Suppose we are given a sequence $\{a_i\}_{i=0}^{\infty}$. One way to transform this sequence is to calculate the new sequence $\{b_i\}_{i=0}^{\infty}$ where $b_i = a_{i+1} - a_i$. In other words we take the difference between two consecutive elements of $\{a_i\}_{i=0}^{\infty}$. We can continue this process *ad infinitum* by calculating the difference of $\{b_i\}_{i=0}^{\infty}$ as $\{c_i\}_{i=0}^{\infty}$ where $c_i = b_{i+1} - b_i = a_{i+2} - 2a_{i+1} + a_i$, and then calculating the difference of $\{c_i\}_{i=0}^{\infty}$ etc. The consecutive differences are best displayed in a difference table (Table 6.1).

| $a_0$ | | $a_1$ | | | $a_2$ | | $a_3$ |
|---|---|---|---|---|---|---|---|
| | $a_1 - a_0$ | | | $a_2 - a_1$ | | $a_3 - a_2$ | |
| | | $a_2 - 2a_1 + a_0$ | | | $a_3 - 2a_2 + a_1$ | | |
| | | | $a_3 - 3a_2 + 3a_1 - a_0$ | | | | |

Table 6.1: A portion of the difference table for $\{a_i\}_{i=0}^{\infty}$. The main diagonal is bold

The main diagonal sequence $\{D_n\}_{n=0}^{\infty}$ is $D_n = \sum_{k=0}^{n}(-1)^{n+k}\binom{n}{k}a_k$, a claim which may be verified via induction on $n$. Furthermore rows parallel to the main diagonal have a similar structure. Working from left to right, the entries along the second diagonal are $\sum_{k=0}^{n}(-1)^{n+k}\binom{n}{k}a_{1+k}$, the entries along the third diagonal are $\sum_{k=0}^{n}(-1)^{n+k}\binom{n}{k}a_{2+k}$, while the entries along the $p^{th}$ diagonal are $\sum_{k=0}^{n}(-1)^{n+k}\binom{n}{k}a_{p-1+k}$. All of these sums are represented as $\sum_{k=0}^{n}(-1)^{n+k}\binom{n}{k}a_{x+k}$, or if we let $a_{x+k} = f(x+k)$, as $\sum_{k=0}^{n}(-1)^{n+k}\binom{n}{k}f(x+k)$. This last sum is known as an $n^{th}$ **difference**

**operator** of $f(x)$ and is represented in the literature as

$$\Delta_1^n f(x) = (-1)^n \sum_{k=0}^{n} (-1)^k \binom{n}{k} f(x+k). \tag{6.1}$$

Equation (6.1) is generalized as

$$\Delta_h^n f(x) = \frac{\Delta_h^{n-1} f(x+h) - \Delta_h^{n-1} f(x)}{h}$$

$$= \frac{(-1)^n}{h^n} \sum_{k=0}^{n} (-1)^k \binom{n}{k} f(x+kh) \tag{6.2}$$

$$\Delta_h^0 f(x) = f(x),$$

where $h$ is an arbitrary nonzero complex number. Equation (6.1) is Equation (6.2) with $h = 1$.

If $f(x)$ is an $n$-times differentiable function, Equation (6.2) implies that

$$\frac{d^n}{dx^n} f(x) = \lim_{h \to 0} \Delta_h^n f(x).$$

The $n^{th}$ difference operator obeys the following five properties [Jordan, 1957]. We provide proofs of Properties 3 through 5 and leave the proof of Properties 1 and 2 to the reader.

**Property 1:**

$$\Delta_h^n (af(x) + bg(x)) = a\Delta_h^n f(x) + b\Delta_h^n g(x), \qquad a, b \text{ constants.} \tag{6.3}$$

Equation (6.3) shows that $\Delta_h^n$ is a linear operator.

**Property 2:**

$$\Delta_h^n \Delta_h^r f(x) = \Delta_h^r \Delta_h^n f(x) = \Delta_h^{n+r} f(x), \qquad n \text{ and } r \text{ nonnegative integers.} \tag{6.4}$$

The third property is an inversion property since it provides a way of writing $f(x)$ in terms of $\Delta_h^n f(x)$.

**Property 3:**

$$f(x + mh) = \sum_{n=0}^{m} \binom{m}{n} h^n \Delta_h^n f(x). \tag{6.5}$$

**Proof of Property 3:**

$$\sum_{n=0}^{m} \binom{m}{n} h^n \Delta_h^n f(x) = \sum_{n=0}^{m} \binom{m}{n} h^n \frac{(-1)^n}{h^n} \sum_{k=0}^{n} (-1)^k \binom{n}{k} f(x+kh)$$

$$= \sum_{k=0}^{m} (-1)^k f(x+kh) \sum_{n=k}^{m} (-1)^n \binom{m}{n} \binom{n}{k}$$

$$= \sum_{k=0}^{m} (-1)^k \binom{m}{k} f(x+kh) \sum_{n=k}^{m} (-1)^n \binom{m-k}{n-k}$$

$$= \sum_{k=0}^{m} \binom{m}{k} f(x+kh) \sum_{n=0}^{m-k} (-1)^n \binom{m-k}{n}$$

$$= \binom{m}{m} f(x+mh) \binom{0}{0} = f(x+mh). \qquad \square$$

**Property 4:**

$$\sum_{j=0}^{n} f(x+jh) = \sum_{k=0}^{n} \binom{n+1}{k+1} h^k \Delta_h^k f(x). \tag{6.6}$$

**Proof of Property 4:** From Equation (6.5) we have

$$\sum_{j=0}^{n} f(x+jh) = \sum_{j=0}^{n} \sum_{k=0}^{j} \binom{j}{k} h^k \Delta_h^k f(x) = \sum_{k=0}^{n} h^k \Delta_h^k f(x) \sum_{j=k}^{n} \binom{j}{k}$$

$$= \sum_{k=0}^{n} \binom{n+1}{k+1} h^k \Delta_h^k f(x), \qquad \text{by Eq. (3.12).} \qquad \square$$

The dual to Equation (6.6) is

**Property 5:**

$$\sum_{j=0}^{n} (-1)^j h^j \Delta_h^j f(x) = \sum_{k=0}^{n} (-1)^k f(x+kh) \binom{n+1}{k+1}. \tag{6.7}$$

**Proof of Property 5:**

$$\sum_{j=0}^{n} (-1)^j h^j \Delta_h^j f(x) = \sum_{j=0}^{n} \sum_{k=0}^{j} (-1)^k \binom{j}{k} f(x+kh)$$

$$= \sum_{k=0}^{n} (-1)^k f(x+kh) \sum_{j=k}^{n} \binom{j}{k}$$

$$= \sum_{k=0}^{n} (-1)^k f(x+kh) \binom{n+1}{k+1}. \qquad \square$$

An operator closely related to $\Delta_h^n$ is the shift operator $E_h^n$ where

$$E_h^n f(x) = f(x + nh), \qquad n \text{ a nonnegative integer.} \qquad (6.8)$$

By definition we have

$$\Delta_h^1 f(x) = \frac{f(x+h) - f(x)}{h} = \frac{1}{h}(E_h^1 - 1)f(x) \qquad (6.9)$$

$$E_h^1 f(x) = f(x) + h\left(\frac{f(x+h) - f(x)}{h}\right) = (1 + h\Delta_h^1)f(x) \qquad (6.10)$$

$$\Delta_h^n f(x) = \frac{(-1)^n}{h^n} \sum_{k=0}^{n} (-1)^k \binom{n}{k} f(x + kh)$$

$$= \frac{(-1)^n}{h^n} \sum_{k=0}^{n} (-1)^k \binom{n}{k} E_h^k f(x)$$

$$= \frac{1}{h^n} \sum_{k=0}^{n} (-1)^k \binom{n}{k} E_h^{n-k} f(x). \qquad (6.11)$$

We may use Equation (6.8) to rewrite Equation (6.5) as

$$E_h^n f(x) = \sum_{k=0}^{n} \binom{n}{k} h^k \Delta_h^k f(x). \qquad (6.12)$$

If $n = 1$, Equation (6.12) becomes Equation (6.10).

The shift operator satisfies

$$E_h^n E_h^m f(x) = E_h^n f(x+mh) = f(x+nh+mh) = f(x+(n+m)h) = E_h^{m+n} f(x),$$

which is the obvious complement to Equation (6.4).

The difference and shift operators commute with each other since

$$E_h^1 \Delta_h^1 f(x) = E_h^1 \left[\frac{f(x+h) - f(x)}{h}\right] = \frac{f(x+2h) - f(x+h)}{h}$$

$$= \Delta_h^1 f(x+h) = \Delta_h^1 E_h^1 f(x).$$

The shift operator allows us to find a formula for the $n^{th}$ difference of a product. Let $H(x) = f(x)g(x)$. Clearly

$$E_h^n[f(x)g(x)] = E_h^n H(x) = H(x+nh)$$

$$= f(x+nh)g(x+nh) = E_h^n f(x)E_h^n g(x). \qquad (6.13)$$

Let $u = f(x)$ and $v = g(x)$. We have

$$\Delta_h^n uv = \frac{1}{h^n} \sum_{k=0}^{n} (-1)^k \binom{n}{k} E_h^{n-k} uv = \frac{1}{h^n} \sum_{k=0}^{n} (-1)^k \binom{n}{k} E_h^{n-k} u E_h^{n-k} v,$$

$$= \frac{1}{h^n} \sum_{j=0}^{n} h^j \Delta_h^j u \sum_{k=0}^{n-j} (-1)^k \binom{n}{k} \binom{n-k}{j} E_h^{n-k} v, \qquad \text{by Eq. (6.12)}$$

$$= \frac{1}{h^n} \sum_{j=0}^{n} \binom{n}{j} h^j \Delta_h^j u \sum_{k=0}^{n-j} (-1)^k \binom{n-j}{k} E_h^{n-k} v$$

$$= \frac{1}{h^n} \sum_{j=0}^{n} \binom{n}{j} h^j \Delta_h^j u \sum_{k=0}^{n-j} (-1)^k \binom{n-j}{k} E_h^{n-k-j} E_h^j v$$

$$= \sum_{j=0}^{n} \binom{n}{j} \Delta_h^j u \Delta_h^{n-j} \left[ E_h^j v \right], \qquad \text{by Eq. (6.11)}.$$

These calculations show that

$$\Delta_h^n [f(x)g(x)] = \sum_{k=0}^{n} \binom{n}{k} \Delta_h^k f(x) \Delta_h^{n-k} \left[ E_h^k g(x) \right]. \qquad (6.14)$$

If $n = 1$ the right side of Equation (6.14) simplifies to

$$\Delta_h^1 [f(x)g(x)] = f(x)\Delta_h^1 g(x) + E_h^1 g(x)\Delta_h^1 f(x)$$
$$= f(x) \left( \frac{g(x+h) - g(x)}{h} \right) + g(x+h) \left( \frac{f(x+h) - f(x)}{h} \right)$$
$$= \frac{f(x+h)g(x+h) - f(x)g(x)}{h},$$

a result which agrees with Equation (6.2). By letting $h \to 0$, Equation (6.14) implies that

$$\lim_{h \to 0} \Delta_h^n [f(x)g(x)] = \frac{d^n}{dx^n} [f(x)g(x)] = \sum_{k=0}^{n} \binom{n}{k} \frac{d^k}{dx^k} f(x) \frac{d^{n-k}}{dx^{n-k}} g(x),$$

which is precisely the Leibniz theorem for derivatives.

We end this section by developing a recurrence formula for $\Delta_h^n f(x+h)$.

By Equation (6.9) we have

$$\Delta_h^n f(x) = \Delta_h^1 \Delta_h^{n-1} f(x) = \frac{1}{h} \left[ E_h^1 - 1 \right] \Delta_h^{n-1} f(x)$$

$$= \frac{1}{h} \left[ E_h^1 \Delta_h^{n-1} f(x) - \Delta_h^{n-1} f(x) \right]$$

$$= \frac{1}{h} \left[ \Delta_h^{n-1} E_h^1 f(x) - \Delta_h^{n-1} f(x) \right]$$

$$= \frac{1}{h} \left[ \Delta_h^{n-1} f(x+h) - \Delta_h^{n-1} f(x) \right].$$

In summary we have shown that

$$\Delta_h^n f(x+h) = h\Delta_h^{n+1} f(x) + \Delta_h^n f(x). \tag{6.15}$$

We have covered the basic properties of the shift and difference operators. This is all the reader will need for the formulation and proof of Euler's finite difference theorem. For those readers who desire to learn more about the difference and shift operators we recommend Charles Jordan's treatise [1957] *Calculus of Finite Differences*.

## 6.1   Euler's Finite Difference Theorem

We are ready to prove Euler's finite difference theorem. Given $f(x) = \sum_{j=0}^r a_j x^j$, Euler's finite difference theorem states that

$$\sum_{k=0}^n (-1)^k \binom{n}{k} f(k) = (-1)^n \Delta_1^n f(x)|_{x=0}$$

$$= \begin{cases} 0, & 0 \le r < n \\ (-1)^n n! a_n, & r = n. \end{cases} \tag{6.16}$$

Our proof of Equation (6.16) begins with the special case $f(x) = x^j$. To verify Equation (6.16) for $f(x) = x^j$, we analyze two expansions of $(e^x - 1)^n$. The binomial theorem implies that

$$(e^x - 1)^n = \sum_{k=0}^n \binom{n}{k} (-1)^{n-k} e^{kx} = \sum_{k=0}^n (-1)^{n-k} \binom{n}{k} \sum_{j=0}^\infty \frac{k^j x^j}{j!}$$

$$= \sum_{j=0}^\infty \frac{x^j}{j!} \sum_{k=0}^n (-1)^{n-k} \binom{n}{k} k^j. \tag{6.17}$$

On the other hand the Taylor series for $e^x$ implies that

$$
\begin{aligned}
(e^x - 1)^n &= \left( \sum_{k=0}^{\infty} \frac{x^k}{k!} - 1 \right)^n = \left( \sum_{k=1}^{\infty} \frac{x^k}{k!} \right)^n \\
&= \left( x \sum_{k=1}^{\infty} \frac{x^{k-1}}{k!} \right)^n = x^n \left( \sum_{k=1}^{\infty} \frac{x^{k-1}}{k!} \right)^n .
\end{aligned}
\tag{6.18}
$$

By comparing the coefficient of $x^j$ in Equations (6.17) and (6.18) we discover that

$$
\sum_{k=0}^{n} (-1)^k \binom{n}{k} k^j = (-1)^n \Delta_1^n x^j \big|_{x=0} =
\begin{cases}
0, & 0 \leq j < n \\
(-1)^n n!, & j = n,
\end{cases}
\tag{6.19}
$$

which is precisely Equation (6.16) with $f(x) = x^j$.

Coefficient comparison of Equations (6.17) and (6.18) does provide a way of obtaining closed forms for $\sum_{k=0}^{n}(-1)^k \binom{n}{k} k^j$ when $j > n$ is fixed, but it becomes quite tedious if $j$ is much larger than $n + 2$. When we study Stirling numbers of the second kind, $S(n, k)$, we will discover that

$$
\sum_{k=0}^{n} (-1)^k \binom{n}{k} k^j = (-1)^n n! S(j, n), \qquad j \geq n,
\tag{6.20}
$$

a formula first attributed to Euler [Gould, 1987, p. 77 ].

To prove Equation (6.16) for $f(x) = \sum_{j=0}^{r} a_j x^j$, observe that

$$
\begin{aligned}
\sum_{k=0}^{n} (-1)^k \binom{n}{k} f(k) &= \sum_{k=0}^{n} (-1)^k \binom{n}{k} \sum_{j=0}^{r} a_j k^j \\
&= \sum_{j=0}^{r} a_j \sum_{k=0}^{n} (-1)^k \binom{n}{k} k^j \\
&=
\begin{cases}
0, & 0 \leq r \leq n \\
(-1)^n n! a_n, & r = n.
\end{cases}
\end{aligned}
$$

where the final equality followed from Equation (6.19).

## 6.2    Applications of Equation (6.16)

We now spend some time exploring how Euler's finite difference theorem leads to various combinatorial identities. A successful application of Euler's finite difference theorem involves the careful specification of the polynomial $f(k)$. For example let $f(k) = (x - bk)^r$, where $r$ is a nonnegative integer. The binomial theorem implies that $(x - bk)^r = \sum_{j=0}^r (-1)^j \binom{r}{j} b^j k^j x^{r-j}$, which in turn implies that the coefficient of $k^r$ is $(-1)^r b^j$. By Equation (6.16) we have

$$\sum_{k=0}^{n} (-1)^k \binom{n}{k} (x - bk)^r = \begin{cases} 0, & r < n \\ b^n n!, & r = n. \end{cases} \qquad (6.21)$$

If $b = 1$, Equation (6.21) becomes

$$\sum_{k=0}^{n} (-1)^k \binom{n}{k} (x - k)^r = \begin{cases} 0, & r < n \\ n!, & r = n. \end{cases} \qquad (6.22)$$

Equation (6.21) allows us to prove the following lemma:

**Lemma 6.1.** *Let* $f(x) = \sum_{i=0}^n a_i x^i$. *Then*

$$\sum_{k=0}^{m} (-1)^k \binom{m}{k} f(x - kz) = \begin{cases} 0, & n < m \\ a_m m! z^m, & n = m. \end{cases} \qquad (6.23)$$

**Proof:** By definition

$$\sum_{k=0}^{m} (-1)^k \binom{m}{k} f(x - kz) = \sum_{k=0}^{m} (-1)^k \binom{m}{k} \sum_{i=0}^n a_i (x - kz)^i$$

$$= \sum_{i=0}^{n} a_i \sum_{i=0}^m (-1)^k \binom{m}{k} (x - kz)^i,$$

and the result immediately follows from Equation (6.21). $\qquad \Box$

For a second example of Euler's finite difference theorem let $f(k) = \binom{ak}{r}$ where $a$ is any nonzero complex number. The definition of the binomial coefficient implies that $f(k)$ is a polynomial of degree $r$ in $k$, i.e. $\binom{ak}{r} = \sum_{j=0}^r a_j k^j$. The coefficient of $k^r$ is $\frac{a^r}{r!}$ since $\binom{ak}{r} = \frac{(ak)(ak-1)...(ak-r+1)}{r!}$. Equation (6.16) implies that

$$\sum_{k=0}^{n} (-1)^k \binom{n}{k} \binom{ak}{r} = \begin{cases} 0, & r < n \\ (-1)^n a^n, & r = n. \end{cases} \qquad (6.24)$$

If $a = 1$, Equation (6.24) becomes

$$\sum_{k=0}^{n} (-1)^k \binom{n}{k} \binom{k}{j} = \begin{cases} 0, & j < n \\ (-1)^n, & j = n. \end{cases} \qquad (6.25)$$

From Equation (6.25) we observe that

$$n^p = \sum_{j=1}^{n} (-1)^{j-1} j^p \sum_{k=1}^{n} (-1)^{k-1} \binom{n}{k} \binom{k}{j}$$

$$= \sum_{k=1}^{n} (-1)^{k-1} \binom{n}{k} \sum_{j=1}^{k} (-1)^{j-1} \binom{k}{j} j^p. \qquad (6.26)$$

Equation (6.26) is valid for any complex number $p$.

If we define $g(n) = n^p$ and $G(k) = \sum_{j=1}^{k} (-1)^{j-1} \binom{k}{j} j^p$, Equation (6.26) becomes $g(n) = \sum_{k=1}^{n} (-1)^{k-1} \binom{n}{k} G(k)$. Then

$$\sum_{i=1}^{n} g(i) = \sum_{i=1}^{n} \sum_{k=1}^{i} (-1)^{k-1} \binom{i}{k} G(k)$$

$$= \sum_{k=1}^{n} (-1)^{k-1} G(k) \sum_{i=k}^{n} \binom{i}{k} = \sum_{k=1}^{n} (-1)^{k-1} \binom{n+1}{k+1} G(k),$$

where the last equality follows from Equation (3.12). This identity implies that

$$\sum_{k=1}^{n} k^p = \sum_{k=1}^{n} (-1)^{k-1} \binom{n+1}{k+1} \sum_{j=1}^{k} (-1)^{j-1} \binom{k}{j} j^p. \qquad (6.27)$$

We now specify $p$ to be a positive integer and use Equation (6.10) to evaluate the inner sum as

$$\sum_{j=1}^{k} (-1)^j \binom{k}{j} j^p = \begin{cases} 0, & p < k \\ (-1)^k k!, & p = k. \end{cases}$$

If $p \leq n$, this evaluation implies that we may truncate the range of the outer sum to $p$. If $p > n$, $\binom{n+1}{k+1} = 0$ whenever $k > n$, and we may extend the range of the outer sum to $p$. Hence Equation (6.27) is equivalent to

$$\sum_{k=1}^{n} k^p = \sum_{k=1}^{p} (-1)^k \binom{n+1}{k+1} \sum_{j=1}^{k} (-1)^j \binom{k}{j} j^p, \qquad p \geq 1. \qquad (6.28)$$

An application of Pascal's identity to $\binom{n+1}{k+1}$ transforms Equation (6.28) as follows:

$$
\begin{aligned}
\sum_{k=1}^{n} k^p &= \sum_{k=1}^{p} (-1)^k \binom{n+1}{k+1} \sum_{j=1}^{k} (-1)^j \binom{k}{j} j^p \\
&= \sum_{k=1}^{p} (-1)^k \binom{n+1}{k+1} \sum_{j=0}^{k} (-1)^j \binom{k}{j} j^p \\
&= \sum_{k=1}^{p} (-1)^k \binom{n}{k} \sum_{j=0}^{k} (-1)^j \binom{k}{j} j^p + \sum_{k=1}^{p} (-1)^k \binom{n}{k+1} \sum_{j=0}^{k} (-1)^j \binom{k}{j} j^p \\
&= \sum_{k=1}^{p} (-1)^k \binom{n}{k} \sum_{j=0}^{k} (-1)^j \binom{k}{j} j^p - \sum_{k=2}^{p+1} (-1)^k \binom{n}{k} \sum_{j=0}^{k-1} (-1)^j \binom{k-1}{j} j^p \\
&= \sum_{k=1}^{p+1} (-1)^k \binom{n}{k} \sum_{j=0}^{k} (-1)^j \left[ \binom{k}{j} - \binom{k-1}{j} \right] j^p \\
&= \sum_{k=1}^{p+1} (-1)^k \binom{n}{k} \sum_{j=1}^{k} (-1)^j \binom{k-1}{j-1} j^p \\
&= \sum_{k=0}^{p} (-1)^k \binom{n}{k+1} \sum_{j=0}^{k} (-1)^j \binom{k}{j} (j+1)^p.
\end{aligned}
$$

In conclusion we have shown that

$$
\sum_{k=1}^{n} k^p = \sum_{k=0}^{p} (-1)^k \binom{n}{k+1} \sum_{j=0}^{k} (-1)^j \binom{k}{j} (j+1)^p, \qquad p \geq 1. \qquad (6.29)
$$

Both Equations (6.28) and (6.29) will appear in the chapter on Bernoulli numbers.

For a third example of Euler's finite difference theorem let $x$ be a complex number, $y$ be a nonzero complex number, and $f(k) = \binom{x+ky}{j}$. Clearly $f(k)$ is a polynomial of degree $j$ in $k$ with leading coefficient $\frac{y^j}{j!}$. By Equation (6.16) we deduce that

$$
\sum_{k=0}^{n} (-1)^k \binom{n}{k} \binom{x+ky}{j} = \begin{cases} 0, & 0 \leq j < n \\ (-1)^n y^n, & j = n. \end{cases} \qquad (6.30)
$$

If $x = 0$, Equation (6.30) reduces to Equation (6.24).

If $y = 1$, Equation (6.30) becomes

$$\sum_{k=0}^{n}(-1)^k\binom{n}{k}\binom{x+k}{j} = \begin{cases} 0, & 0 \le j < n \\ (-1)^n, & j = n. \end{cases} \tag{6.31}$$

There is another way to derive a closed form for the left sum of Equation (6.31). This alternative derivation utilizes the expansion

$$\sum_{k=0}^{n}\binom{n}{k}\binom{x+k}{j}y^k = \sum_{i=0}^{j}\binom{x}{j-i}\binom{n}{i}y^i(1+y)^{n-i}, \tag{6.32}$$

which we prove through an application of the Vandermonde convolution:

$$\sum_{k=0}^{n}\binom{n}{k}\binom{x+k}{j}y^k = \sum_{k=0}^{n}\binom{n}{k}y^k\sum_{i=0}^{j}\binom{k}{i}\binom{x}{j-i}$$

$$= \sum_{i=0}^{j}\binom{x}{j-i}\sum_{k=0}^{n}\binom{n}{k}\binom{k}{i}y^k$$

$$= \sum_{i=0}^{j}\binom{x}{j-i}\binom{n}{i}\sum_{k=i}^{n}\binom{n-i}{k-i}y^k$$

$$= \sum_{i=0}^{j}\binom{x}{j-i}\binom{n}{i}y^i\sum_{k=0}^{n-i}\binom{n-i}{k}y^k$$

$$= \sum_{i=0}^{j}\binom{x}{j-i}\binom{n}{i}y^i(1+y)^{n-i}.$$

Take Equation (6.32) and let $y = -1$. The left side becomes $\sum_{k=0}^{n}(-1)^k\binom{n}{k}\binom{x+k}{j}$. To evaluate the right side assume that $j > n$. We truncate the right sum at $n$ since $\binom{n}{i} = 0$ whenever $i > n$. Therefore

$$\sum_{i=0}^{j}\binom{x}{j-i}\binom{n}{i}y^i(1+y)^{n-i} = \sum_{i=0}^{n}\binom{x}{j-i}\binom{n}{i}y^i(1+y)^{n-i}, \qquad j > n, \tag{6.33}$$

and

$$\sum_{i=0}^{n}\binom{x}{j-i}\binom{n}{i}(-1)^i(1+(-1))^{n-i} = (-1)^n\binom{x}{j-n}, \qquad j > n. \tag{6.34}$$

We combine Equations (6.34) and (6.31) to write

$$\sum_{k=0}^{n}(-1)^k\binom{n}{k}\binom{x+k}{j} = (-1)^n\binom{x}{j-n}, \tag{6.35}$$

which is how this identity appears in [Gould, 1972].

By using the definition of a difference operator, we may rewrite Equation (6.35) as

$$(-1)^n \Delta_1^n \binom{x}{j} = \sum_{k=0}^{n} (-1)^k \binom{n}{k} \binom{x+k}{j} = (-1)^n \binom{x}{j-n},$$

which is equivalent to

$$\Delta_1^n \binom{x}{j} = \binom{x}{j-n}. \tag{6.36}$$

If we take Equation (6.35) and apply the $-1$-Transformation to $\binom{x+k}{j}$, we obtain

$$\sum_{k=0}^{n} (-1)^k \binom{n}{k} \binom{-x+j-1-k}{j} = (-1)^{n+j} \binom{x}{j-n}. \tag{6.37}$$

In Equation (6.37) let $x = j - 1 - y$ to obtain

$$\sum_{k=0}^{n} (-1)^k \binom{n}{k} \binom{y-k}{j} = (-1)^{n+j} \binom{j-1-y}{j-n} = \binom{y-n}{j-n}. \tag{6.38}$$

Equation (6.38) provides a closed form for $\sum_{k=0}^{n} (-1)^k \binom{n}{k} \binom{x+yk}{j}$ when $y = -1$. Here is an alternative one-line proof of Equation (6.38) which does not rely on the $-1$-Transformation. It does however use Equation (6.35). In particular,

$$\sum_{k=0}^{n} (-1)^k \binom{n}{k} \binom{x-k}{j} = \sum_{k=0}^{n} (-1)^{n-k} \binom{n}{n-k} \binom{x-n+k}{j}$$

$$= (-1)^n \cdot (-1)^n \binom{x-n}{j-n} = \binom{x-n}{j-n}.$$

Our final application of Euler's finite difference theorem uses Equation (6.16) to derive two expansions for $x^n$. The first, attributed to Niels Abel, is

$$\sum_{k=1}^{n} \binom{n}{k} \frac{(yk)^{n-k}(x-yk)^k}{k} = \frac{x^n}{n}, \qquad n \geq 2, \qquad y \text{ complex}. \tag{6.39}$$

To prove Equation (6.39) observe that

$$\sum_{k=1}^{n}\binom{n}{k}\frac{(yk)^{n-k}(x-yk)^k}{k} = \sum_{k=1}^{n}\binom{n}{k}\frac{(yk)^{n-k}}{k}\sum_{j=0}^{k}(-1)^{k-j}\binom{k}{j}x^j(yk)^{k-j}$$

$$= \sum_{k=1}^{n}\binom{n}{k}\frac{(yk)^{n-k}}{k}\sum_{j=1}^{k}(-1)^{k-j}\binom{k}{j}x^j(yk)^{k-j} + \sum_{k=1}^{n}(-1)^k\binom{n}{k}\frac{(yk)^{n-k}(yk)^k}{k}$$

$$= \sum_{j=1}^{n}x^j\sum_{k=j}^{n}(-1)^{k-j}\binom{n}{k}\binom{k}{j}\frac{(yk)^{n-j}}{k} + y^n\sum_{k=1}^{n}(-1)^k\binom{n}{k}k^{n-1}$$

$$= \sum_{j=1}^{n}x^j\sum_{k=j}^{n}(-1)^{k-j}\binom{n}{k}\binom{k}{j}\frac{(yk)^{n-j}}{k}, \qquad \text{by Eq. (6.16)}$$

$$= \sum_{j=1}^{n}(-1)^j\binom{n}{j}x^jy^{n-j}\sum_{k=j}^{n}(-1)^k\binom{n-j}{k-j}k^{n-j-1}$$

$$= \sum_{j=1}^{n}\binom{n}{j}x^jy^{n-j}\sum_{k=0}^{n-j}(-1)^k\binom{n-j}{k}(k+j)^{n-j-1}$$

$$= \sum_{j=1}^{n}\binom{n}{j}x^jy^{n-j}\sum_{k=0}^{n-j}(-1)^k\binom{n-j}{k}\sum_{r=0}^{n-j}\binom{n-j-1}{r}k^r j^{n-j-1-r}$$

$$= \sum_{j=1}^{n}\binom{n}{j}x^jy^{n-j}\sum_{r=0}^{n-j}\binom{n-j-1}{r}j^{n-j-1-r}\sum_{k=0}^{n-j}(-1)^k\binom{n-j}{k}k^r.$$

Since $0 \le r \le n-j$, Equation (6.19) tells us that $\sum_{k=0}^{n-j}(-1)^k\binom{n-j}{k}k^r$ vanishes unless $r = n - j$. Therefore the previous line becomes

$$\sum_{k=1}^{n}\binom{n}{k}\frac{(yk)^{n-k}(x-yk)^k}{k} = \sum_{j=1}^{n}\binom{n}{j}x^jy^{n-j}\binom{n-j-1}{n-j}j^{-1}(-1)^{n-j}(n-j)!.$$

Notice that $\binom{n-j-1}{n-j} = 0$ unless $n - j = 0$. Hence

$$\sum_{k=1}^{n}\binom{n}{k}\frac{(yk)^{n-k}(x-yk)^k}{k} = x^n\binom{n}{n}y^{n-n}\binom{-1}{0}n^{-1}(-1)^0 0! = \frac{x^n}{n}.$$

Equation (6.39) is an amazing source of binomial identities. If we differentiate both sides with respect to $x$, we obtain

$$\sum_{k=1}^{n}\binom{n}{k}(yk)^{n-k}(x-yk)^{k-1} = x^{n-1}, \qquad n \ge 1. \qquad (6.40)$$

We set $y = 1$ in Equation (6.40) to obtain

$$\sum_{k=1}^{n} \binom{n}{k} k^{n-k} (x-k)^{k-1} = x^{n-1}. \tag{6.41}$$

We can also let $y \to -y$ in Equation (6.40) to obtain

$$\sum_{k=0}^{n} \binom{n}{k} (-yk)^{n-k} (x+yk)^{k-1} = x^{n-1}, \qquad n \geq 0. \tag{6.42}$$

When verifying Equation (6.42) for $n = 0$, use the convention that $0^0 = 1$.

We extend Equation (6.42) to polynomials. Let $f(x) = \sum_{j=0}^{n} a_j x^j$ be a polynomial of degree $n$ in $x$. By Equation (6.42) we have

$$
\begin{aligned}
f(x) &= \sum_{j=0}^{n} a_j x^j = \sum_{j=0}^{n} a_j \sum_{k=0}^{j+1} \binom{j+1}{k} (-yk)^{j+1-k} (x+yk)^{k-1} \\
&= \sum_{j=0}^{n} a_j \sum_{k=1}^{j+1} \binom{j+1}{k} (-yk)^{j+1-k} (x+yk)^{k-1} \\
&= \sum_{j=0}^{n} a_j \sum_{k=0}^{j} \binom{j+1}{k+1} (-yk-y)^{j-k} (x+yk+y)^{k} \\
&= \sum_{k=0}^{n} (x+yk+y)^{k} \sum_{j=k}^{n} \binom{j+1}{k+1} (-yk-y)^{j-k} a_j.
\end{aligned}
$$

Thus we have the polynomial transformation

$$f(x) = \sum_{j=0}^{n} a_j x^j = \sum_{k=0}^{n} (x+yk+y)^k \sum_{j=k}^{n} \binom{j+1}{k+1} (-yk-y)^{j-k} a_j. \tag{6.43}$$

We next use Equation (6.40) to obtain an expansion for $(x + y + nz)^n$. Assume $x$, $y$, and $z$ are complex numbers while $n$ is any nonnegative integer.

Then

$$(x + y + nz)^n = \sum_{j=0}^{n} \binom{n}{j}(y + nz)^{n-j}x^j$$

$$= x \sum_{j=0}^{n} \binom{n}{j}(y + nz)^{n-j} \sum_{k=0}^{j} \binom{j}{k}(-zk)^{j-k}(x + zk)^{k-1}, \quad \text{by Eq. (6.40)}$$

$$= x \sum_{k=0}^{n} \binom{n}{k}(x + zk)^{k-1} \sum_{j=k}^{n} \binom{n-k}{j-k}(y + nz)^{n-j}(-zk)^{j-k}$$

$$= x \sum_{k=0}^{n} \binom{n}{k}(x + zk)^{k-1} \sum_{j=0}^{n-k} \binom{n-k}{j}(y + nz)^{n-k-j}(-zk)^{j}$$

$$= x \sum_{k=0}^{n} \binom{n}{k}(y + nz - zk)^{n-k}(x + zk)^{k-1}.$$

These calculations derive the Abel formula

$$(x + y + nz)^n = x \sum_{j=0}^{n} \binom{n}{j}(y + (n-j)z)^{n-j}(x + yz)^{j-1}, \qquad x \neq 0. \quad (6.44)$$

The second expansion for $x^n$, which complements Equation (6.41), is

$$\sum_{k=0}^{n} \binom{n}{k}(k + 1)^{k-1}(x - k - 1)^{n-k} = x^n. \quad (6.45)$$

In order to prove Equation (6.45) take the right side, expand via the binomial theorem, and observe that

$$\sum_{k=0}^{n} \binom{n}{k}(x - k - 1)^{n-k}(k + 1)^{k-1}$$

$$= \sum_{k=0}^{n} \binom{n}{k}(k + 1)^{k-1} \sum_{j=0}^{n-k}(-1)^{n-k-j} \binom{n-k}{j}(k + 1)^{n-k-j}x^j$$

$$= \sum_{j=0}^{n}(-1)^{n-j}x^j \sum_{k=0}^{n-j}(-1)^k \binom{n}{n-k}\binom{n-k}{j}(k + 1)^{n-j-1}$$

$$= \sum_{j=0}^{n}(-1)^{n-j}\binom{n}{j}x^j \sum_{k=0}^{n-j}(-1)^k \binom{n-j}{k}(k + 1)^{n-j-1}$$

$$= \sum_{j=0}^{n}(-1)^{n-j}\binom{n}{j}x^j \sum_{k=0}^{n-j}(-1)^k \binom{n-j}{k} \sum_{r=0}^{n-j-1} \binom{n-j-1}{r}k^r$$

$$= \sum_{j=0}^{n}(-1)^{n-j}\binom{n}{j}x^j \sum_{r=0}^{n-j-1} \binom{n-j-1}{r} \sum_{k=0}^{n-j}(-1)^k \binom{n-j}{k}k^r. \quad (6.46)$$

Since $r \leq n - j - 1 < n - j$, Euler's finite difference theorem implies this inner sum vanishes as long as $n - j - 1 \geq 0$, i.e. $n - j \geq 1$. The only term that remains is when $n - j = 0$. When $j = n$ Equation (6.46) becomes

$$s \sum_{k=0}^{n} \binom{n}{k}(x - k - 1)^{n-k}(k + 1)^{k-1} =$$

$$x^n \binom{n}{n}(-1)^{n-n} \sum_{k=0}^{0}(-1)^k \binom{0}{k}(k + 1)^{-1} = x^n.$$

Equation (6.45) often appears as

$$\sum_{k=0}^{n} \binom{n}{k}(x - k)^{n-k}(k + 1)^{k-1} = (x + 1)^n, \qquad (6.47)$$

since, without loss of generality, $x \to x + 1$.

# Chapter 7

# Melzak's Formula

In this chapter we use partial fraction decomposition to prove a binomial identity attributed to Z. A. Melzak and then use this identity to derive other combinatorial identities. Let $n$ be a fixed nonnegative integer and $f(x)$ be a polynomial in $x$ of degree $n$, i.e. $f(x) = \sum_{k=0}^{n} a_k x^k$. Let $y$ be an arbitrary complex number. Melzak's formula in [Melzak, 1953] states

$$f(x+y) = y\binom{y+n}{n} \sum_{k=0}^{n} (-1)^k \binom{n}{k} \frac{f(x-k)}{y+k}, \qquad y \neq 0, -1, -2, \ldots, -n.$$

(7.1)

Our proof of Melzak's formula makes use of Euler's finite difference theorem. Since Equation (7.1) is trivially true for $n = 0$, we assume that $n \geq 1$. Start with the right side of Equation (7.1) to obtain

$$y\binom{y+n}{n} \sum_{k=0}^{n} (-1)^k \binom{n}{k} \frac{f(x-k)}{y+k}$$

$$= y\binom{y+n}{n} \sum_{k=0}^{n} (-1)^k \binom{n}{k} \frac{1}{y+k} \sum_{i=0}^{n} a_i (x-k)^i$$

$$= y\binom{y+n}{n} \sum_{k=0}^{n} (-1)^k \binom{n}{k} \frac{1}{y+k} \sum_{i=0}^{n} a_i \sum_{j=0}^{i} (-1)^j \binom{i}{j} k^j x^{i-j}$$

$$= y\binom{y+n}{n} \sum_{i=0}^{n} a_i \sum_{j=0}^{i} (-1)^j \binom{i}{j} x^{i-j} \sum_{k=0}^{n} (-1)^k \binom{n}{k} \frac{k^j}{y+k}. \quad (7.2)$$

In order to simplify Equation (7.2) we must investigate $\sum_{k=0}^{n}(-1)^k \binom{n}{k} \frac{k^j}{y+k}$. We claim that

$$\frac{k^j}{k+y} = (-1)^j \frac{y^j}{k+y} - (-1)^j \sum_{r=0}^{j-1} (-1)^r y^{j-1-r} k^r, \qquad (7.3)$$

since

$$\frac{k^j}{k+y} = \frac{(-1)^j y^j}{k+y} - \frac{(-1)^j y^j - k^j}{k+y}$$

$$= \frac{(-1)^j y^j}{k+y} - (-1)^j y^{j-1}\left[\frac{1-\left(-\frac{k}{y}\right)^j}{1-\left(-\frac{k}{y}\right)}\right]$$

$$= \frac{(-1)^j y^j}{k+y} - (-1)^j y^{j-1}\sum_{r=0}^{j-1}\left(-\frac{k}{y}\right)^r$$

$$= (-1)^j\frac{y^j}{k+y} - (-1)^j\sum_{r=0}^{j-1}(-1)^r y^{j-1-r}k^r.$$

Equation (7.3) implies that

$$\sum_{k=0}^{n}(-1)^k\binom{n}{k}\frac{k^j}{k+y} = (-y)^j\sum_{k=0}^{n}(-1)^k\binom{n}{k}\frac{1}{k+y}$$

$$-(-1)^j\sum_{r=0}^{j-1}(-1)^r y^{j-r-1}\sum_{k=0}^{n}(-1)^k\binom{n}{k}k^r. \quad (7.4)$$

By applying partial fraction decompositions of $\prod_{j=0}^{n}\frac{1}{k+j}$, we will show that

$$\sum_{k=0}^{n}(-1)^k\binom{n}{k}\frac{1}{k+y} = \frac{n!}{y(y+1)\ldots(y+n)} = \frac{1}{y\binom{y+n}{n}}. \quad (7.5)$$

The particular partial fraction decomposition we use is

$$\prod_{j=0}^{n}\frac{1}{y+j} = \frac{1}{y(y+1)\ldots(y+n)}$$

$$= \frac{1}{ny(y+1)\ldots(y+n-1)} - \frac{1}{n(y+1)(y+2)\ldots(y+n)},$$

where $n$ is a positive integer and $y$ is an arbitrary nonzero complex number which is not a negative integer. By recursively applying this decomposition we deduce that

$$\prod_{j=0}^{n}\frac{1}{y+j} = \sum_{k=0}^{n}(-1)^k\frac{1}{k!(n-k)!(k+y)}, \qquad n\geq 0. \quad (7.6)$$

An independent proof of Equation (7.6) follows from induction on $n$. Assume $y$ is an arbitrary yet fixed nonzero complex number which is not a negative integer. Equation (7.6) is trivially true if $n=0$. If $n=1$, Equation (7.6) implies that

$$\prod_{j=0}^{1} \frac{1}{y+j} = \frac{1}{y(y+1)} = \frac{1}{y} - \frac{1}{y+1} = \sum_{k=0}^{1}(-1)^k \frac{1}{k!(1-k)!(k+y)}.$$

Assume Equation (7.6) is true for all positive integers less than or equal to $n$ and any fixed choice of $y$ for which the product is defined. Then

$$\prod_{j=0}^{n+1} \frac{1}{y+j} = \frac{1}{(n+1)y(y+1)...(y+n)} - \frac{1}{(n+1)(y+1)(y+2)...(y+n+1)}$$

$$= \frac{1}{n+1}\prod_{j=0}^{n} \frac{1}{y+j} - \frac{1}{n+1}\prod_{j=0}^{n} \frac{1}{y+1+j}$$

$$= \frac{1}{n+1}\sum_{k=0}^{n} \frac{(-1)^k}{k!(n-k)!(k+y)} - \frac{1}{n+1}\prod_{j=0}^{n} \frac{1}{y+1+j}, \text{ induction hypothesis}$$

$$= \frac{1}{n+1}\sum_{k=0}^{n} \frac{(-1)^k}{k!(n-k)!(k+y)}$$

$$- \frac{1}{n+1}\sum_{k=0}^{n} \frac{(-1)^k}{k!(n-k)!(k+y+1)}, \text{ induction hypothesis } y \to y+1$$

$$= \frac{1}{n+1}\sum_{k=0}^{n} \frac{(-1)^k}{k!(n-k)!(k+y)} - \frac{1}{n+1}\sum_{k=1}^{n+1} \frac{(-1)^{k-1}}{(k-1)!(n-k+1)!(k+y)}$$

$$= \frac{1}{(n+1)!y} + \frac{(-1)^{n+1}}{(n+1)!(y+n+1)}$$

$$+ \frac{1}{n+1}\sum_{k=1}^{n} \frac{(-1)^k}{(k-1)!(n-k)!(k+y)}\left[\frac{1}{k} + \frac{1}{n-k+1}\right]$$

$$= \frac{1}{(n+1)!y} + \frac{(-1)^{n+1}}{(n+1)!(y+n+1)}$$

$$+ \frac{1}{n+1}\sum_{k=1}^{n} \frac{(-1)^k}{(k-1)!(n-k)!(k+y)}\left[\frac{n+1}{k(n-k+1)}\right]$$

$$= \frac{1}{(n+1)!y} + \frac{(-1)^{n+1}}{(n+1)!(y+n+1)} + \sum_{k=1}^{n} \frac{(-1)^k}{k!(n-k+1)!(k+y)}$$

$$= \sum_{k=0}^{n+1} \frac{(-1)^k}{k!(n-k+1)!(k+y)}.$$

Since we have verified Equation (7.6), we multiply both sides by $n!$ and obtain Equation (7.5).

We remark that since $\frac{n!(y-1)!}{(n+y)!} = \frac{\Gamma(n+1)\Gamma(y)}{\Gamma(n+y-1)} = \beta(n+1,y)$ [Rainville, 1960, p. 19], Equation (7.6) is equivalent to

$$\beta(n+1,y) = \frac{\Gamma(n+1)\Gamma(y)}{\Gamma(n+y-1)} = \sum_{k=0}^{n}(-1)^k\binom{n}{k}\frac{1}{k+y}. \qquad (7.7)$$

Substitute Equation (7.5) into Equation (7.4) and obtain

$$\sum_{k=0}^{n}(-1)^k\binom{n}{k}\frac{k^j}{k+y}$$

$$= (-1)^j\frac{y^{j-1}}{\binom{y+n}{n}} - (-1)^j\sum_{r=0}^{j-1}(-1)^r y^{j-r-1}\sum_{k=0}^{n}(-1)^k\binom{n}{k}k^r. \qquad (7.8)$$

Equation (7.2) becomes

$$y\binom{y+n}{n}\sum_{k=0}^{n}(-1)^k\binom{n}{k}\frac{f(x-k)}{y+k} = \sum_{i=0}^{n}a_i\sum_{j=0}^{i}\binom{i}{j}x^{i-j}y^j$$

$$-y\binom{y+n}{n}\sum_{i=1}^{n}a_i\sum_{j=1}^{i}\binom{i}{j}x^{i-j}\sum_{r=0}^{j-1}(-1)^r y^{r-j-1}\sum_{k=0}^{n}(-1)^k\binom{n}{k}k^r$$

$$= \sum_{i=0}^{n}a_i(x+y)^i$$

$$-y\binom{y+n}{n}\sum_{i=1}^{n}a_i\sum_{j=1}^{i}\binom{i}{j}x^{i-j}\sum_{r=0}^{j-1}(-1)^r y^{r-j-1}\sum_{k=0}^{n}(-1)^k\binom{n}{k}k^r$$

$$= f(x+y)$$

$$-y\binom{y+n}{n}\sum_{i=1}^{n}a_i\sum_{j=1}^{i}\binom{i}{j}x^{i-j}\sum_{r=0}^{j-1}(-1)^r y^{r-j-1}\sum_{k=0}^{n}(-1)^k\binom{n}{k}k^r. \qquad (7.9)$$

We are almost done. The innermost sum on the right side of Equation (7.9) is evaluated by Euler's finite difference theorem. Since $r \leq j-1 \leq i-1 \leq n-1$, Equation (6.19) implies that $\sum_{k=0}^{n}(-1)^k\binom{n}{k}k^r = 0$. The second sum in Equation (7.9) *always* vanishes and we conclude that

$$y\binom{y+n}{n}\sum_{k=0}^{n}(-1)^k\binom{n}{k}\frac{f(x-k)}{y+k} = f(x+y),$$

which is precisely Equation (7.1).

## 7.1 Basic Applications of Melzak's Formula

As the title of this section implies, we shall use Melzak's formula to obtain combinatorial identities. The trick will be to specify the $f(x)$ in Equation (7.1). We will begin with $f(x) = 1$. Equation (7.1) becomes

$$\sum_{k=0}^{n}(-1)^k \binom{n}{k}\frac{1}{y+k} = \frac{1}{y\binom{y+n}{n}}. \tag{7.10}$$

In Equation (7.10) let $y = \frac{z}{\alpha}$. After simplification we obtain

$$\sum_{k=0}^{n}(-1)^k \binom{n}{k}\frac{1}{z+\alpha k} = \frac{1}{z\binom{n+\frac{z}{\alpha}}{n}}, \tag{7.11}$$

whenever $z \neq 0, -\alpha, -2\alpha, \ldots, -\alpha n$. Given Equation (7.11) we may invert this identity via the following well-known inversion theorem,

$$F(n) = \sum_{k=0}^{n}(-1)^k \binom{n}{k}f(k) \quad \text{if and only if} \quad f(n) = \sum_{k=0}^{n}(-1)^k \binom{n}{k}F(k).$$
$$\tag{7.12}$$

The proof of Equation (7.12) is an exercise in substitution. We will prove one implication and leave the other for the reader. Suppose $f(n) = \sum_{k=0}^{n}(-1)^k \binom{n}{k}F(k)$. Then

$$\sum_{k=0}^{n}(-1)^k \binom{n}{k}f(k) = \sum_{k=0}^{n}(-1)^k \binom{n}{k}\sum_{j=0}^{k}(-1)^k \binom{k}{j}F(j)$$

$$= \sum_{j=0}^{n}F(j)\sum_{k=j}^{n}(-1)^{k-j}\binom{n}{k}\binom{k}{j}$$

$$= \sum_{j=0}^{n}\binom{n}{j}F(j)\sum_{k=j}^{n}(-1)^{k-j}\binom{n-j}{k-j}$$

$$= \sum_{j=0}^{n}\binom{n}{j}F(j)\sum_{k=0}^{n-j}(-1)^{k}\binom{n-j}{k}.$$

An application of the binomial theorem implies that $\sum_{k=0}^{n-j}(-1)^k \binom{n-j}{k} = 0$ unless $j = n$, in which case we have

$$\sum_{k=0}^{n}(-1)^k \binom{n}{k}f(k) = \sum_{j=0}^{n}\binom{n}{j}F(j)\sum_{k=0}^{n-j}(-1)^{k}\binom{n-j}{k} = F(n).$$

Equation (7.12) often appears in the literature with the substitution $F(k) \to (-1)^k F(k)$, and implies that

$$F(n) = \sum_{k=0}^{n} (-1)^{n-k} \binom{n}{k} f(k) \quad \text{if and only if} \quad f(n) = \sum_{k=0}^{n} \binom{n}{k} F(k).$$
$$(7.13)$$

Equation (7.12) with $f(k) = \frac{1}{z+\alpha k}$ becomes

$$\sum_{k=0}^{n} (-1)^k \binom{n}{k} \frac{1}{\binom{\frac{z}{\alpha}+k}{k}} = \frac{z}{z+\alpha n}, \qquad (7.14)$$

which is valid for most nonzero complex $\alpha$ and requires that $z \neq 0, -\alpha, -2\alpha, \ldots, -\alpha n$. If $\alpha = 1$ and $z = n$, Equation (7.14) becomes

$$\sum_{k=0}^{n} (-1)^k \binom{n}{k} \frac{1}{\binom{n+k}{k}} = \frac{1}{2}, \qquad n \geq 1. \qquad (7.15)$$

If we take the previous equation, expand out the binomial coefficients, and multiply by $\frac{(2n)!}{n!^2}$, we ultimately obtain

$$\sum_{k=0}^{n} (-1)^k \binom{2n}{n-k} = \frac{1}{2} \binom{2n}{n} = \binom{2n-1}{n}, \qquad n \geq 1.$$

For our second application of Equation (7.1) let $f(x) = x^n$ and observe that

$$\sum_{k=0}^{n} (-1)^k \binom{n}{k} \frac{(x-k)^n}{y+k} = \frac{(x+y)^n}{y\binom{y+n}{n}}, \qquad y \neq 0, -1, -2, \ldots, -n. \quad (7.16)$$

Equation (7.16) is quite a versatile identity. For example, if $x = n = y$ where $n > 0$, it becomes

$$\sum_{k=0}^{n} (-1)^k \binom{n}{k} \frac{(n-k)^n}{n+k} = \frac{2^n \cdot n^{n-1}}{\binom{2n}{n}}. \qquad (7.17)$$

On the other hand, if $x = n$ and $y = 1$, Equation (7.16) evolves into

$$\sum_{k=0}^{n} (-1)^k \binom{n}{k} \frac{(n-k)^n}{1+k} = (n+1)^{n-1}. \qquad (7.18)$$

Equation (7.18) will hold for $n = 0$ if we use the convention that $0^0 = 1$.

So far we have worked with $f(x) = 1$ and $f(x) = x^n$, two simple polynomials in $x$ of degree at most $n$. Another well-known polynomial in $x$ of degree $n$ is $\binom{x}{n}$. Take Equation (7.1) and let $f(x) = \binom{zx}{n}$, where $z$ is an arbitrary complex number, to obtain

$$\sum_{k=0}^{n} (-1)^k \binom{n}{k} \binom{zx - zk}{n} \frac{1}{y+k} = \frac{\binom{zx+zy}{n}}{y\binom{y+n}{n}}, \qquad y \neq 0, -1, -2, \ldots, -n.$$

(7.19)

Here are some useful special cases of Equation (7.19). First take $x = 0$ to obtain

$$\sum_{k=0}^{n} (-1)^k \binom{n}{k} \binom{-zk}{n} \frac{1}{y+k} = \frac{\binom{zy}{n}}{y\binom{y+n}{n}},$$

(7.20)

$$\sum_{k=0}^{n} (-1)^k \binom{n}{k} \binom{zk}{n} \frac{1}{y+k} = (-1)^n \frac{\binom{zy+n-1}{n}}{y\binom{y+n}{n}},$$

(7.21)

where we assume $y \neq 0, -1, -2, \ldots, -n$. Next assume $\alpha$ is a nonzero complex number. Let $z \to \alpha$ and then set $z = \alpha x$ in Equation (7.19) to obtain

$$\sum_{k=0}^{n} (-1)^k \binom{n}{k} \binom{z - \alpha k}{n} \frac{1}{y+k} = \frac{\binom{z+\alpha y}{n}}{y\binom{y+n}{n}}, \qquad y \neq 0, -1, -2, \ldots, -n.$$

(7.22)

Finally let $y \to -\frac{z}{\alpha}$ in Equation (7.22) to obtain

$$\sum_{k=0}^{n} (-1)^k \binom{n}{k} \binom{z - \alpha k}{n} \frac{z}{z - \alpha k} = \begin{cases} 1, & n = 0 \\ 0, & n \geq 1. \end{cases}$$

(7.23)

For another application of Equation (7.1) let $f(x) = \binom{x}{r}^p$ and obtain

$$\sum_{k=0}^{n} (-1)^k \binom{n}{k} \binom{x - k}{r}^p \frac{1}{y+k} = \frac{\binom{x+y}{r}^p}{y\binom{y+n}{n}}, \qquad y \neq 0, -1, -2, \ldots, -n,$$

(7.24)

where $r$, $p$, and $n$ are nonnegative integers such that $rp = n$. If $x = n$ and $y = r$, Equation (7.24) becomes

$$\sum_{k=0}^{n} (-1)^k \binom{n}{k} \binom{n - k}{r}^p \frac{1}{k+r} = \frac{1}{r} \binom{n + r}{r}^{p-1}, \qquad r \neq 0, \ rp = n.$$

(7.25)

If $f(x) = \binom{x+n}{n}$, Equation (7.1) becomes

$$\sum_{k=0}^{n} (-1)^k \binom{n}{k} \binom{x - k + n}{n} \frac{1}{y+k} = \frac{\binom{x+y+n}{n}}{y\binom{y+n}{n}} \qquad y \neq 0, -1, -2, \ldots, -n.$$

(7.26)

If $x = n$, Equation (7.26) implies that

$$\sum_{k=0}^{n}(-1)^k\binom{n}{k}\binom{2n-k}{n}\frac{1}{y+k} = \frac{\binom{2n+y}{n}}{y\binom{y+n}{n}} \tag{7.27}$$

$$\sum_{k=0}^{n}(-1)^k\binom{2n}{k}\binom{2n-k}{n}^2\frac{1}{y+k} = \binom{2n}{n}\sum_{k=0}^{n}(-1)^k\binom{n}{k}\binom{2n-k}{n}\frac{1}{y+k}$$

$$= \frac{\binom{2n}{n}\binom{2n+y}{n}}{y\binom{y+n}{n}}. \tag{7.28}$$

Equation (7.26) is equivalent to

$$\sum_{k=0}^{n}(-1)^k\binom{n}{k}\binom{-x-1+k}{n}\frac{1}{y+k} = (-1)^n\frac{\binom{x+y+n}{n}}{y\binom{y+n}{n}} \tag{7.29}$$

since the $-1$-Transformation implies that $\binom{x-k+n}{n} = (-1)^n\binom{-x+k-1}{n}$. Take Equation (7.29) and let $x = -n-1$ to obtain

$$\sum_{k=0}^{n}(-1)^k\binom{n}{k}\binom{n+k}{n}\frac{1}{y+k} = \sum_{k=0}^{n}(-1)^k\binom{n+k}{2k}\binom{2k}{k}\frac{1}{y+k}$$

$$= (-1)^n\frac{\binom{y-1}{n}}{y\binom{y+n}{n}}. \tag{7.30}$$

If $y = 1$, Equation (7.30) becomes

$$\sum_{k=0}^{n}(-1)^k\binom{n}{k}\binom{n+k}{n}\frac{1}{k+1} = \begin{cases} 1, & n = 0 \\ 0, & n \geq 1. \end{cases} \tag{7.31}$$

## 7.2 Two Advanced Applications of Melzak's Formula

So far our applications of Equation (7.1) have focused on specifying a particular choice of $f(x)$. Now we will focus on using Equation (7.1) as a tool for proving two sophisticated binomial identities. The identity, a transformation attributed to Niels Abel, is

$$\sum_{k=0}^{n}\binom{n}{k}(x-k)^k(1-x+k)^{n-k}\frac{1}{y+k} = \sum_{k=0}^{n}(-1)^k\binom{n}{k}\frac{(x+y)^k}{y\binom{y+k}{k}}, \tag{7.32}$$

where $x$ and $y$ are complex numbers such that $y \neq 0, -1, -2, \ldots, -n$. If $x = 0$ we use the convention that $0^0 = 1$.

To prove Equation (7.32) we start with Equation (7.1), interchange the roles of $x$ and $y$, and let $f(x) = x^n$. This gives us

$$\sum_{k=0}^{n}(-1)^k \binom{n}{k}\frac{(y-k)^n}{x+k} = \frac{(y+x)^n}{x\binom{x+n}{n}}, \qquad x \neq 0, -1, -2, \ldots, -n. \quad (7.33)$$

Equation (7.33) implies that

$$\sum_{n=0}^{p}(-1)^n \binom{p}{n}\frac{(y+x)^n}{x\binom{x+n}{n}} = \sum_{n=0}^{p}(-1)^n \binom{p}{n}\sum_{k=0}^{n}(-1)^k \binom{n}{k}\frac{(y-k)^n}{x+k}$$

$$= \sum_{k=0}^{p}(-1)^k \frac{1}{x+k}\sum_{n=k}^{p}(-1)^n \binom{p}{n}\binom{n}{k}(y-k)^n$$

$$= \sum_{k=0}^{p}(-1)^k \binom{p}{k}\frac{1}{x+k}\sum_{n=k}^{p}(-1)^n \binom{p-k}{n-k}(y-k)^n$$

$$= \sum_{k=0}^{p}\binom{p}{k}\frac{(y-k)^k}{x+k}\sum_{n=0}^{p-k}(-1)^n \binom{p-k}{n}(y-k)^n$$

$$= \sum_{k=0}^{p}\binom{p}{k}\frac{(y-k)^k(1-y+k)^{p-k}}{x+k},$$

a result clearly equivalent to Equation (7.32).

By adapting the proof of Equation (7.33) we can readily show that

$$\sum_{k=0}^{n}\binom{n}{k}\frac{(by-bk)^k(a-by+bk)^{n-k}}{x+k}$$

$$= \sum_{k=0}^{n}(-1)^k \binom{n}{k}a^{n-k}\frac{(by+bx)^k}{x\binom{v\mid k}{k}}, \qquad x \neq 0, -1, -2, \quad , -n, \quad (7.34)$$

whenever $a$ and $b$ are nonzero complex numbers. If $a = 1 = b$, Equation (7.34) becomes Equation (7.32).

The second identity we prove in this section is due to R. Frisch, a famous Scandinavian economist. Assume that $b$ and $c$ are *positive* integers with $b \geq c$. Also assume that $\binom{b+k}{c} \neq 0$ for $k = 0, 1, 2, \ldots, n$. Then

$$\sum_{k=0}^{n}(-1)^k \binom{n}{k}\frac{1}{\binom{b+k}{c}} = \frac{c}{n+c}\frac{1}{\binom{n+b}{b-c}}, \qquad c \neq n. \quad (7.35)$$

Our proof of Equation (7.35) begins by transforming the series on the left side into a equivalent series which does not have $\binom{b+k}{c}$ in the denominator.

Let $S = \sum_{k=0}^{n}(-1)^k\binom{a}{k}\frac{1}{\binom{b+k}{c}}$ where $a$ is a nonnegative integer such that $n \geq a \geq 0$. Then

$$S = \sum_{k=b-c}^{n+b-c}(-1)^{k-b+c}\binom{a}{k-b+c}\frac{1}{\binom{k+c}{c}}. \tag{7.36}$$

We now recall Equation (6.35) which states

$$(-1)^n\binom{x}{j-n} = \sum_{k=0}^{n}(-1)^k\binom{n}{k}\binom{x+k}{j}.$$

By applying this to $\binom{a}{k-(b-c)}$, we convert Equation (7.36) into

$$S = \sum_{k=b-c}^{n+b-c}(-1)^{k-b+c}(-1)^{b-c}\frac{1}{\binom{k+c}{c}}\sum_{j=0}^{b-c}(-1)^j\binom{b-c}{j}\binom{a+j}{k}$$

$$= \sum_{j=0}^{b-c}(-1)^j\binom{b-c}{j}\sum_{k=b-c}^{n+b-c}(-1)^k\frac{\binom{a+j}{k}}{\binom{k+c}{k}}$$

$$= \sum_{j=0}^{b-c}(-1)^j\binom{b-c}{j}\sum_{k=b-c}^{n+b-c}\frac{\binom{a+j}{k}}{\binom{-c-1}{k}}. \tag{7.37}$$

In order to simplify the inner summation of Equation (7.37) we refer to

$$\sum_{k=j}^{n}\frac{\binom{z}{k}}{\binom{x}{k}} = \frac{x+1}{x-z+1}\left[\frac{\binom{z}{j}}{\binom{x+1}{j}} - \frac{\binom{z}{n+1}}{\binom{x+1}{n+1}}\right]. \tag{7.38}$$

Equation (7.38) is not hard to prove. Observe that

$$\frac{\binom{z}{n}}{\binom{x}{n}} = \frac{x+1}{x-z+1}\left[\frac{\binom{z}{n}}{\binom{x+1}{n}} - \frac{\binom{z}{n+1}}{\binom{x+1}{n+1}}\right], \tag{7.39}$$

and use this identity to compute $\sum_{k=j}^{n}\frac{\binom{z}{k}}{\binom{x}{k}}$ as a telescopic series. By applying Equation (7.38) to the inner sum of Equation (7.37) we see that

$$S = \sum_{j=0}^{b-c}(-1)^j\frac{c}{c+a+j}\left[\frac{\binom{a+j}{b-c}}{\binom{-c}{b-c}} - \frac{\binom{a+j}{n+b-c+1}}{\binom{-c}{n+b-c+1}}\right]\binom{b-c}{j}$$

$$= \frac{c}{\binom{-c}{b-c}}\sum_{j=0}^{b-c}(-1)^j\binom{b-c}{j}\binom{a+j}{b-c}\frac{1}{c+a+j}$$

$$- \frac{c}{\binom{-c}{n+b-c+1}}\sum_{j=0}^{b-c}(-1)^j\binom{b-c}{j}\binom{a+j}{n+b-c+1}\frac{1}{c+a+j}. \tag{7.40}$$

Take a look at the second sum in Equation (7.40). There are two important observations. First since $c$ is a positive integer with $c \le b$, $\binom{-c}{b-c+n+1} = (-1)^{b-c+n+1}\binom{b+n}{b+n-c+1} \ne 0$. Secondly, since $n \ge a$, observe that $\binom{a+j}{b-c+n+1} = 0$ whenever $0 \le j \le b - c$. We conclude that the second sum in Equation (7.40) is identically zero and that

$$S = \frac{c}{\binom{-c}{b-c}} \sum_{j=0}^{b-c} (-1)^j \binom{b-c}{j} \binom{a+j}{b-c} \frac{1}{c+a+j}. \tag{7.41}$$

We will evaluate Equation (7.41) with Equation (7.1). In Equation (7.1) take $f(x) = \binom{-x}{n}$ and set $x = -z$ to obtain

$$\sum_{k=0}^{n} (-1)^k \binom{n}{k} \binom{z+k}{n} \frac{1}{y+k} = \frac{\binom{z-y}{n}}{y\binom{y+n}{n}}. \tag{7.42}$$

Take Equation (7.42), let $n \to b - c$, $z \to a$ and $y \to c + a$, and use this result on Equation (7.41) to obtain

$$\begin{aligned}
S &= \frac{c}{\binom{-c}{b-c}} \frac{\binom{a-(c+a)}{b-c}}{(c+a)\binom{c+a+b-c}{b-c}} = \frac{c}{\binom{-c}{b-c}} \frac{\binom{-c}{b-c}}{(c+a)\binom{a+b}{b-c}} \\
&= \frac{c}{c+a} \frac{1}{\binom{a+b}{b-c}}.
\end{aligned}$$

In summary we have shown that

$$\sum_{k=0}^{n} (-1)^k \binom{a}{k} \frac{1}{\binom{b+k}{c}} = \frac{c}{c+a} \frac{1}{\binom{a+b}{b-c}}, \qquad n \ge a. \tag{7.43}$$

If $a = n$, Equation (7.43) becomes Equation (7.35).

## 7.3 Partial Fraction Generalizations of Equation (7.1)

In this section we discuss the series $\sum_{k=0}^{n}(-1)^k \binom{n}{k} \frac{f(x-k)}{\prod_{i=1}^{j}(x_i+k)}$ where we assume $f(x)$ is a polynomial of degree $n - j + 1$ in $x$. Melzak's formula provides a closed form when $j = 1$. We can parlay this information into finding a closed form for *any* positive integer $j$ by writing $\frac{1}{\prod_{i=1}^{j}(x_i+k)}$ in terms of partial fractions.

We begin our analysis with the case of $j = 2$. Let $f(x) = \sum_{k=0}^{n} a_k x^k$. According to Equation (7.1) we have

$$\sum_{k=0}^{n} (-1)^k \binom{n}{k} \frac{xf(y-k)}{(x+k)(x-z)} = \frac{f(x+y)}{(x-z)\binom{x+n}{n}}, \qquad x \ne z. \tag{7.44}$$

Since

$$\frac{x}{(x+k)(x-z)} = \frac{k}{(k+z)(k+x)} + \frac{z}{(k+z)(x-z)}, \qquad (7.45)$$

Equation (7.44) is equivalent to

$$\frac{f(x+y)}{(x-z)\binom{x+n}{n}} = \sum_{k=0}^{n}(-1)^k \binom{n}{k}\frac{kf(y-k)}{(k+z)(x+k)}$$

$$+ \frac{z}{x-z}\sum_{k=0}^{n}(-1)^k \binom{n}{k}\frac{f(y-k)}{k+z}$$

$$= \sum_{k=0}^{n}(-1)^k \binom{n}{k}\frac{kf(y-k)}{(k+z)(x+k)} + \frac{f(z+y)}{(x-z)\binom{z+n}{n}}$$

$$= n\sum_{k=0}^{n}(-1)^k \binom{n-1}{k-1}\frac{f(y-k)}{(k+z)(x+k)} + \frac{f(z+y)}{(x-z)\binom{z+n}{n}}.$$

If we replace $k$ with $k+1$ and transpose terms we obtain

$$\frac{f(x+y)}{(x-z)\binom{x+n}{n}} - \frac{f(z+y)}{(x-z)\binom{z+n}{n}}$$

$$= n\sum_{k=0}^{n-1}(-1)^{k-1}\binom{n-1}{k}\frac{f(y-k-1)}{(k+1+z)(k+1+x)}. \qquad (7.46)$$

Take Equation (7.46) and let $n \to n+1$, $y \to y+1$, $x \to x_1 - 1$, $z \to x_2 - 1$. After simplification we obtain

$$\sum_{k=0}^{n}(-1)^k \binom{n}{k}\frac{f(y-k)}{(x_1+k)(x_2+k)}$$

$$= \frac{1}{(n+1)(x_1-x_2)}\left[\frac{f(y+x_2)}{\binom{x_2+n}{n+1}} - \frac{f(y+x_1)}{\binom{x_1+n}{n+1}}\right], \qquad (7.47)$$

whenever $f(x) = \sum_{k=0}^{n+1} a_k x^k$.

Equation (7.47) is the desired closed form for $j = 2$. We can use it to determine a closed form for $j = 3$ by rewriting the partial fraction decomposition at (7.45) as

$$\frac{k}{(k+u)(k+v)} = \frac{u}{(u-v)(k+u)} - \frac{v}{(u-v)(k+v)}. \qquad (7.48)$$

For $f(x) = \sum_{k=0}^{n+1} a_k x^k$ we have

$$\sum_{k=0}^{n} (-1)^k \binom{n}{k} \frac{k f(y-k)}{(k+u)(k+v)(k+w)} =$$

$$\frac{u}{u-v} \sum_{k=0}^{n} (-1)^k \binom{n}{k} \frac{f(y-k)}{(k+u)(k+w)} - \frac{v}{u-v} \sum_{k=0}^{n} (-1)^k \binom{n}{k} \frac{f(y-k)}{(k+v)(k+w)}$$

$$= \frac{u}{(u-v)(n+1)(u-w)} \left[ \frac{f(y+w)}{\binom{w+n}{n+1}} - \frac{f(y+u)}{\binom{u+n}{n+1}} \right]$$

$$- \frac{v}{(u-v)(n+1)(v-w)} \left[ \frac{f(y+w)}{\binom{w+n}{n+1}} - \frac{f(y+v)}{\binom{v+n}{n+1}} \right],$$

where the last equality follows from two applications of Equation (7.47).

Since

$$\sum_{k=0}^{n} (-1)^k \binom{n}{k} \frac{k f(y-k)}{(k+u)(k+v)(k+w)}$$

$$= n \sum_{k=0}^{n} (-1)^k \binom{n-1}{k-1} \frac{f(y-k)}{(k+u)(k+v)(k+w)}$$

$$= -n \sum_{k=0}^{n-1} (-1)^k \binom{n-1}{k} \frac{f(y-k-1)}{(k+u+1)(k+v+1)(k+w+1)},$$

we conclude that

$$\sum_{k=0}^{n-1} (-1)^{k-1} \binom{n-1}{k} \frac{f(y-k-1)}{(k+u+1)(k+v+1)(k+w+1)}$$

$$= \frac{u}{n(u-v)(n+1)(u-w)} \left[ \frac{f(y+w)}{\binom{w+n}{n+1}} - \frac{f(y+u)}{\binom{u+n}{n+1}} \right]$$

$$- \frac{v}{n(u-v)(n+1)(v-w)} \left[ \frac{f(y+w)}{\binom{w+n}{n+1}} - \frac{f(y+v)}{\binom{v+n}{n+1}} \right]. \qquad (7.49)$$

If we let $n \to n + 1$, $y \to y + 1$, $u \to x_1 - 1$, $v \to x_2 - 1$, and $w \to x_3 - 1$, Equation (7.49) becomes

$$\sum_{k=0}^{n} (-1)^k \binom{n}{k} \frac{f(y-k)}{(k+x_1)(k+x_2)(k+x_3)}$$

$$= \frac{x_1 - 1}{(n+1)(n+2)(x_1 - x_2)(x_1 - x_3)} \left[ \frac{f(y+x_1)}{\binom{x_1+n}{n+2}} - \frac{f(y+x_3)}{\binom{x_3+n}{n+2}} \right]$$

$$- \frac{x_2 - 1}{(n+1)(n+2)(x_1 - x_2)(x_2 - x_3)} \left[ \frac{f(y+x_2)}{\binom{x_2+n}{n+2}} - \frac{f(y+x_3)}{\binom{x_3+n}{n+3}} \right], \quad (7.50)$$

whenever $f(x) = \sum_{k=0}^{n+2} a_k x^k$.

We may continue this process, use Equation (7.48) to decompose $\frac{k}{(k+x_1)(k+x_2)(k+x_3)(k+x_4)}$ into two fractions, and apply Equation (7.50) to the new fractions. However it is more efficient to decompose $\frac{1}{\prod_{i=1}^{j}(k+u_i)}$ as $k$ partial fractions, each of which has only one factor in the denominator which is linear in $k$. In particular we can verify that

$$\frac{1}{\prod_{i=1}^{j}(k+u_i)} = \frac{1}{(k+u_1)(k+u_2)\ldots(k+u_j)}$$

$$= \sum_{i=1}^{j} \frac{1}{(k+u_i) \prod_{\substack{r=1 \\ r \neq i}}^{j}(u_r - u_i)}. \quad (7.51)$$

For example, if $j = 2$, Equation (7.51) implies that

$$\frac{1}{(k+u_1)(k+u_2)} = \frac{1}{(k+u_1)(u_2 - u_1)} + \frac{1}{(k+u_2)(u_1 - u_2)},$$

while if $j = 3$, Equation (7.51) becomes

$$\frac{1}{(k+u_1)(k+u_2)(k+u_3)}$$

$$= \frac{1}{(k+u_1)(u_2 - u_1)(u_3 - u_1)} + \frac{1}{(k+u_2)(u_1 - u_2)(u_3 - u_2)}$$

$$+ \frac{1}{(k+u_3)(u_1 - u_3)(u_2 - u_3)}.$$

If we take Equation (7.51), multiply by $(-1)^k \binom{n}{k} f(y-k)$, sum over $k$, interchange the order of summation on the right side, and apply Equation

(7.1) to each of the $j$ sums, we ultimately obtain

$$\sum_{k=0}^{n}(-1)^k\binom{n}{k}\frac{f(y-k)}{\prod_{i=1}^{j}(k+u_i)} = \sum_{i=1}^{j}\frac{f(y+u_i)}{u_i\prod_{\substack{r=1\\r\neq i}}^{j}(u_r-u_i)\binom{u_i+n}{n}}, \quad (7.52)$$

$$\text{for } f(x) = \sum_{k=0}^{n+j-1} a_k x^k.$$

If $j = 2$, Equation (7.52) becomes

$$\sum_{k=0}^{n}(-1)^k\binom{n}{k}\frac{f(y-k)}{(u_1+k)(u_2+k)} = \frac{f(y+u_1)}{u_1(u_2-u_1)\binom{u_1+n}{n}} + \frac{f(y+u_2)}{u_2(u_1-u_2)\binom{u_2+n}{n}},$$

a result equivalent to Equation (7.47). If $j = 3$ Equation (7.52) becomes

$$\sum_{k=0}^{n}(-1)^k\binom{n}{k}\frac{f(y-k)}{(u_1+k)(u_2+k)(u_3+k)}$$

$$= \frac{f(y+u_1)}{u_1(u_2-u_1)(u_3-u_1)\binom{u_1+n}{n}} + \frac{f(y+u_2)}{u_2(u_1-u_2)(u_3-u_2)\binom{u_2+n}{n}}$$

$$+ \frac{f(y+u_3)}{u_3(u_1-u_3)(u_2-u_3)\binom{u_3+n}{n}},$$

which is equivalent to Equation (7.50).

A special case of Equation (7.52) occurs when $u_i = i$, in which case we obtain, after multiplication by $j!$,

$$\sum_{k=0}^{n}(-1)^k\binom{n}{k}\frac{f(y-k)}{\binom{k+j}{k}} = -\sum_{k=1}^{j}(-1)^k\binom{j}{k}\frac{f(y+k)}{\binom{k+n}{k}}, \quad (7.53)$$

$$\text{for } f(x) = \sum_{k=0}^{n+j-1} a_k x^k.$$

## 7.4 Lagrange Interpolation Theorem

We end this chapter by asking a question on polynomial interpolation which at first glance seems to have no relationship to combinatorial identities. Given $n+1$ distinct points, find a polynomial $P(x)$ of degree $n$ determined by these points. Assume the points lie in the real plane and are denoted by $\{[x_i, y_i]\}_{i=0}^{n}$. By assumption $x_i = x_j$ if and only if $i = j$. In order to specify the coefficients $\{a_i\}_{i=0}^{n}$ associated with $P(x) = \sum_{k=0}^{n} a_k x^k$ where

$P(x_i) = y_i$ for $0 \leq i \leq n$, we use linear algebra. Set $n = 1$. Then $P(x) = a_0 + a_1 x$ with

$$P(x_0) = a_0 + a_1 x_0 = y_0, \qquad P(x_1) = a_0 + a_1 x_1 = y_1, \qquad (7.54)$$

where $[x_0, y_0]$ and $[x_1, y_1]$ are two distinct points. The equations of (7.54) are compactly written as

$$\begin{bmatrix} 1 & x_0 \\ 1 & x_1 \end{bmatrix} \begin{bmatrix} a_0 \\ a_1 \end{bmatrix} = \begin{bmatrix} y_0 \\ y_1 \end{bmatrix}. \qquad (7.55)$$

Since $x_0 \neq x_1$, the $2 \times 2$ matrix is invertible, and we find that

$$\begin{bmatrix} a_0 \\ a_1 \end{bmatrix} = \begin{bmatrix} 1 & x_0 \\ 1 & x_1 \end{bmatrix}^{-1} \begin{bmatrix} y_0 \\ y_1 \end{bmatrix} = \frac{1}{x_1 - x_0} \begin{bmatrix} x_1 & -x_0 \\ -1 & 1 \end{bmatrix} \begin{bmatrix} y_0 \\ y_1 \end{bmatrix}$$

$$= \frac{1}{x_1 - x_0} \begin{bmatrix} x_1 y_0 - x_0 y_1 \\ -y_0 + y_1 \end{bmatrix}.$$

Therefore

$$P(x) = \frac{-y_0 + y_1}{x_1 - x_0} x + \frac{x_1 y_0 - x_0 y_1}{x_1 - x_0}. \qquad (7.56)$$

Let us do a little more algebra and rewrite Equation (7.56) as

$$P(x) = \left( \frac{x - x_1}{x_0 - x_1} \right) y_0 + \left( \frac{x - x_0}{x_1 - x_0} \right) y_1. \qquad (7.57)$$

By using Equation (7.57) it is easy to show that $P(x_0) = y_0$ and $P(x_1) = y_1$.

We answered the question for polynomials of degree one. What about degree 2? The same set up will work, but this time $P(x) = a_0 + a_1 x + a_2 x^2$, the three distinct points are $\{[x_0, y_0], [x_1, y_1], [x_2, y_2]\}$, and Equation (7.55) becomes

$$\begin{bmatrix} 1 & x_0 & x_0^2 \\ 1 & x_1 & x_1^2 \\ 1 & x_2 & x_2^2 \end{bmatrix} \begin{bmatrix} a_0 \\ a_1 \\ a_2 \end{bmatrix} = \begin{bmatrix} y_0 \\ y_1 \\ y_2 \end{bmatrix}. \qquad (7.58)$$

We mention that $\begin{bmatrix} 1 & x_0 & x_0^2 \\ 1 & x_1 & x_1^2 \\ 1 & x_2 & x_2^2 \end{bmatrix}$ is a special case of a Vandermonde matrix $V_n$, where $V_n$ is the $(n+1) \times (n+1)$ matrix $[v_{i,j}]$ with $v_{i,j} = x_{i-1}^{j-1}$. As long as $x_i \neq x_j$ whenever $i \neq j$, it is well known that the determinant of $V_n$ is nonzero, which implies that $V_n$ is invertible.[1] In general if

[1] Wikipedia, *Vandermonde determinant*, http://www.proofwiki.org/wiki/Vandermonde_Determinant

$P(x) = \sum_{k=0}^{n} a_k x^k$, the condition that $P(x_i) = y_i$ for $0 \le i \le n$ leads to the matrix system $V_n A = Y$ where $V_n$ is the $(n+1) \times (n+1)$ Vandermonde matrix, $A$ is the $(n+1) \times 1$ matrix $[a_{i,1}]$ with $a_{i,1} = a_{i-1}$, and $Y$ is the $(n+1) \times 1$ matrix $[y_{i,1}]$ with $y_{i,1} = y_{i-1}$. Then $A = V_n^{-1} Y$, and after collecting with respect to $\{y_i\}_{i=0}^{n}$, we discover that

$$\sum_{k=0}^{2} a_k x^k = \frac{(x - x_1)(x - x_2)}{(x_0 - x_1)(x_0 - x_2)} y_0 + \frac{(x - x_0)(x - x_2)}{(x_1 - x_0)(x_1 - x_2)} y_1$$
$$+ \frac{(x - x_0)(x - x_1)}{(x_2 - x_0)(x_2 - x_1)} y_2 \tag{7.59}$$

$$\sum_{k=0}^{3} a_k x^k = \frac{(x - x_1)(x - x_2)(x - x_3)}{(x_0 - x_1)(x_0 - x_2)(x_0 - x_3)} y_0 + \frac{(x - x_0)(x - x_2)(x - x_3)}{(x_1 - x_0)(x_1 - x_2)(x_1 - x_3)} y_1$$
$$+ \frac{(x - x_0)(x - x_1)(x - x_3)}{(x_2 - x_0)(x_2 - x_1)(x_2 - x_3)} y_2$$
$$+ \frac{(x - x_0)(x - x_1)(x - x_2)}{(x_3 - x_0)(x_3 - x_1)(x_3 - x_2)} y_3 \tag{7.60}$$

$$\sum_{k=0}^{4} a_k x^k = \frac{(x - x_1)(x - x_2)(x - x_3)(x - x_4)}{(x_0 - x_1)(x_0 - x_2)(x_0 - x_3)(x_0 - x_4)} y_0$$
$$+ \frac{(x - x_0)(x - x_2)(x - x_3)(x - x_4)}{(x_1 - x_0)(x_1 - x_2)(x_1 - x_3)(x_1 - x_4)} y_1$$
$$+ \frac{(x - x_0)(x - x_1)(x - x_3)(x - x_4)}{(x_2 - x_0)(x_2 - x_1)(x_2 - x_3)(x_2 - x_4)} y_2$$
$$+ \frac{(x - x_0)(x - x_1)(x - x_2)(x - x_4)}{(x_3 - x_0)(x_3 - x_1)(x_3 - x_2)(x_3 - x_4)} y_3$$
$$+ \frac{(x - x_0)(x - x_1)(x - x_2)(x - x_3)}{(x_4 - x_0)(x_4 - x_1)(x_4 - x_2)(x_4 - x_3)} y_4. \tag{7.61}$$

Equations (7.57), (7.59), (7.60), and (7.61) are succinctly written as

$$P_n(x) = \sum_{k=0}^{n} l_{n,k}(x) y_k, \qquad l_{n,k}(x) = \prod_{\substack{j=0 \\ j \ne k}}^{n} \frac{(x - x_j)}{(x_k - x_j)}, \qquad 1 \le n \le 4.$$
$$\tag{7.62}$$

By definition $l_{n,k}(x_k) = 1$ and $l_{n,k}(x_j) = 0$ whenever $j \ne k$. Therefore Equation (7.62) implies that $P_n(x)$ is a polynomial of degree $n$ in $x$ such that $P_n(x_k) = y_k$ for $0 \le k \le n$, and this is the content of the Lagrange interpolation theorem. Although we have only verified Equation (7.62) for $1 \le n \le 4$, it is true for *any* positive integer. One way to justify Equation (7.62) for all positive integers $n$ is to observe that $\{l_{n,k}\}_{k=0}^{n}$ is a basis for $\mathcal{P}_n(x)$, where $\mathcal{P}_n(x)$ is the set of polynomials in $x$ of degree at most $n$, i.e. $\mathcal{P}_n(x) = \{p(x)|p(x) = \sum_{k=0}^{n} a_k x^k\}$. Here is another way. Let $P_n(x) = \sum_{k=0}^{n} a_k x^k$. Assume that $P_n(x) = \sum_{k=0}^{n} A_k \prod_{\substack{i=0 \\ i \neq k}}^{n} (x - x_i)$, where the $x_i$'s are distinct and $\{A_k\}_{k=0}^{n}$ is to be determined. By definition

$$P_n(x_j) = \sum_{k=0}^{n} A_k \prod_{\substack{i=0 \\ i \neq k}}^{n} (x_j - x_i)$$

$$= A_j \prod_{\substack{i=0 \\ i \neq j}}^{n} (x_j - x_i), \qquad \text{since } \prod_{\substack{i=0 \\ i \neq k}}^{n} (x_j - x_i) = 0 \text{ unless } j = k.$$

Therefore $A_j = \frac{P_n(x_j)}{\prod_{\substack{i=0 \\ i \neq j}}^{n} (x_j - x_i)}$ whenever $0 \leq j \leq n$. This implies that

$$P_n(x) = \sum_{k=0}^{n} P_n(x_k) \prod_{\substack{i=0 \\ i \neq k}}^{n} \frac{x - x_i}{x_k - x_i}. \tag{7.63}$$

Equation (7.63) becomes Equation (7.62) if we set $y_k = P_n(x_k)$.

In summary we have proven the following theorem:

**Theorem 7.1.** *Lagrange Interpolation Formula: Let* $\phi(x) = \sum_{i=0}^{n} a_i x^i$ *with* $n \geq 1$. *Then*

$$\phi(x) = \sum_{k=0}^{n} \phi(x_k) \prod_{\substack{i=0 \\ i \neq k}}^{n} \frac{x - x_i}{x_k - x_i}, \tag{7.64}$$

*whenever* $\{x_i\}_{i=0}^{n}$ *is a set of cardinality* $n + 1$.

The Lagrange interpolation formula has another formulation. Let $f(x) = \prod_{i=0}^{n} (x - x_i)$. An application of the product rule implies that

$$D_x f(x)|_{x=x_k} = \prod_{\substack{i=0 \\ i \neq k}}^{n} (x_k - x_i), \qquad 0 \leq k \leq n, \tag{7.65}$$

where $D_x$ denotes the derivative with respect to $x$. Equation (7.65) may be used to rewrite Equation (7.64) as

$$\phi(x) = \sum_{k=0}^{n} \phi(x_k) \prod_{\substack{i=0 \\ i \neq k}}^{n} \frac{x - x_i}{x_k - x_i}$$

$$= \prod_{i=0}^{n} (x - x_i) \sum_{k=0}^{n} \frac{\phi(x_k)}{(x - x_k) \prod_{\substack{i=0 \\ i \neq k}}^{n} (x_k - x_i)}$$

$$= f(x) \sum_{k=0}^{n} \frac{\phi(x_k)}{(x - x_k) f'(x_k)}.$$

This calculation is summarized by the following theorem:

**Theorem 7.2.** *Lagrange Interpolation Formula, Derivative Form: Let $n$ be a positive integer. Let $\{x_i\}_{i=0}^n$ be a set of $n + 1$ distinct complex numbers. Let $\phi(x) = \sum_{i=0}^n a_i x^i$ and $f(x) = \prod_{i=0}^n (x - x_i)$. Then*

$$\phi(x) = f(x) \sum_{k=0}^n \frac{\phi(x_k)}{(x - x_k) f'(x_k)}. \tag{7.66}$$

If we take Equation (7.66), divide by $f(x)$ and integrate, we obtain

$$\int \frac{\phi(x)}{f(x)} \, dx = \sum_{k=0}^n \frac{\phi(x_k)}{f'(x_k)} \ln(x - x_k) + C. \tag{7.67}$$

Equation (7.66) provides an interpolation for $\phi(x + y)$. By definition

$$
\begin{aligned}
f(x) \sum_{k=0}^n \frac{\phi(y + x_k)}{(x - x_k) f'(x_k)} &= f(x) \sum_{k=0}^n \frac{1}{(x - x_k) f'(x_k)} \sum_{\alpha=0}^n a_\alpha (y + x_k)^\alpha \\
&= \sum_{\alpha=0}^n a_\alpha f(x) \sum_{k=0}^n \frac{(y + x_k)^\alpha}{(x - x_k) f'(x_k)} \\
&= \sum_{\alpha=0}^n a_\alpha f(x) \sum_{k=0}^n \frac{1}{(x - x_k) f'(x_k)} \sum_{j=0}^\alpha \binom{\alpha}{j} x_k^j y^{\alpha - j} \\
&= \sum_{\alpha=0}^n a_\alpha \sum_{j=0}^\alpha \binom{\alpha}{j} y^{\alpha - j} f(x) \sum_{k=0}^n \frac{x_k^j}{(x - x_k) f'(x_k)} \\
&= \sum_{\alpha=0}^n a_\alpha \sum_{j=0}^\alpha \binom{\alpha}{j} y^{\alpha - j} x^j, \quad \text{Eq. (7.66) with } \phi(x) = x^j \\
&= \sum_{\alpha=0}^n a_\alpha (y + x)^\alpha = \phi(x + y).
\end{aligned}
$$

These calculations verify

**Corollary 7.1.** *Let $n$ be a positive integer. Let $\phi(x) = \sum_{i=0}^n a_i x^i$. Define $f(x) = \prod_{i=0}^n (x - x_i)$ where the $x_i$'s are distinct. Then*

$$\phi(x + y) = f(x) \sum_{k=0}^n \frac{\phi(y + x_k)}{(x - x_k) f'(x_k)} = \sum_{k=0}^n \phi(y + x_k) \prod_{\substack{j=0 \\ j \neq k}}^n \frac{x - x_j}{x_j - x_j}. \tag{7.68}$$

We now show how Equations (7.66) and (7.68) are useful tools for proving certain types of combinatorial identities, namely those involving partial fraction decompositions of rational functions. We saw such an example in the previous section, namely Equation (7.51). Equation (7.51) is Equation (7.66) with $\phi(x) = 1$, $x_i = -u_i$ and $x = k$. Melzak's formula is also a corollary of Theorem 7.2. Set $x_i = -i$. By definition

$$f(x) = \prod_{i=0}^{n}(x - x_i) = \prod_{i=0}^{n}(x + i)$$

$$= x(x+1)(x+2)\dots(x+n) = n!x\binom{x+n}{n}. \qquad (7.69)$$

The next step is to compute $f'(x_k) = f'(-k)$.

$$D_x f(x)|_{x=-k} = \prod_{\substack{i=0 \\ i \neq k}}^{n}(-k + i), \qquad \text{by Eq. (7.65)}$$

$$= \prod_{i=0}^{k-1}(-k + i) \cdot \prod_{i=k+1}^{n}(-k + i), \qquad *$$

$$= (-1)^k \prod_{i=0}^{k-1}(k - i) \cdot \prod_{i=k+1}^{n}(i - k)$$

$$= (-1)^k k!(n - k)!. \qquad (7.70)$$

At Line $*$ we assume the first product is 1 if $k = 0$, while the second product is 1 if $k = n$.

If we substitute Equations (7.69) and (7.70) into Equation (7.68) we see that

$$\phi(x + y) = n!x\binom{x+n}{n}\sum_{k=0}^{n}(-1)^k \frac{\phi(y - k)}{(x + k)k!(n - k)!}$$

$$= x\binom{x+n}{n}\sum_{k=0}^{n}(-1)^k \binom{n}{k}\frac{\phi(y - k)}{x + k},$$

a result equivalent to Equation (7.1). This proof of Melzak's identity was suggested by R. V. Parker [1953].

We continue to use Equation (7.68), but this time set $x_i = i$. The second equality of Equation (7.68) with $\phi(x) \to f(x)$ becomes

$$f(x + y) = \sum_{k=0}^{n} f(k + y) \prod_{\substack{j=0 \\ j \neq k}}^{n}\frac{x - j}{k - j} = \sum_{k=0}^{n}\binom{x}{k}\binom{n - x}{n - k}f(k + y). \quad (7.71)$$

The last equality made use of the fact that $\prod_{\substack{j=0 \\ j \neq k}}^{n} \frac{x-j}{k-j} = \binom{x}{k}\binom{n-x}{n-k}$. A noteworthy case of Equation (7.71) occurs when $f(x) = \binom{x}{j}$ with $j \leq n$. Equation (7.71) then becomes

$$\sum_{k=0}^{n} \binom{x}{k}\binom{n-x}{n-k}\binom{k+y}{j} = \binom{x+y}{j}. \tag{7.72}$$

Set $y = 0$ in Equation (7.72) to obtain

$$\sum_{k=0}^{n} \binom{x}{k}\binom{n-x}{n-k}\binom{k}{j} = \binom{x}{j}. \tag{7.73}$$

Using Equation (7.72) it is possible to show that

$$\sum_{k=0}^{n} \binom{x}{k}\binom{n-x}{n-k}\binom{k+z}{j} = \binom{x+z}{j} = \sum_{\alpha=0}^{j} \binom{z}{j-\alpha}\binom{x}{\alpha}, \tag{7.74}$$

where the last equality follows from the Vandermonde convolution. We mention that Equation (7.73) is obtainable from Equation (5.8), namely

$$\sum_{k=0}^{n} \binom{x}{k}\binom{y}{n-k}\binom{k}{j} = \binom{x}{j}\binom{x+y-j}{n-j}. \tag{7.75}$$

If we set $y = n - x$, Equation (7.75) becomes Equation (7.73). Equation (7.75) has the advantage of providing a generalization of Equation (7.74) since an application of the Vandermonde convolution implies

$$\sum_{k=0}^{n} \binom{x}{k}\binom{y}{n-k}\binom{k+z}{j} = \sum_{k=0}^{n} \binom{x}{k}\binom{y}{n-k}\sum_{\alpha=0}^{j}\binom{k}{\alpha}\binom{z}{j-\alpha}$$

$$= \sum_{\alpha=0}^{j}\binom{z}{j-\alpha}\sum_{k=0}^{n}\binom{x}{k}\binom{y}{n-k}\binom{k}{\alpha}$$

$$= \sum_{\alpha=0}^{j}\binom{z}{j-\alpha}\binom{x}{\alpha}\sum_{k=\alpha}^{n}\binom{x-\alpha}{k-\alpha}\binom{y}{n-k}$$

$$= \sum_{\alpha=0}^{j}\binom{z}{j-\alpha}\binom{x}{\alpha}\sum_{k=\alpha}^{n}\binom{x-\alpha}{k}\binom{y}{n-\alpha-k}$$

$$= \sum_{\alpha=0}^{j}\binom{z}{j-\alpha}\binom{x}{\alpha}\binom{y+x-\alpha}{n-\alpha}.$$

Thus we have shown that

$$\sum_{k=0}^{n} \binom{x}{k}\binom{y}{n-k}\binom{k+z}{j} = \sum_{\alpha=0}^{j}\binom{z}{j-\alpha}\binom{x}{\alpha}\binom{y+x-\alpha}{n-\alpha}. \tag{7.76}$$

Equation (7.74) is Equation (7.76) with $y = n - x$.

# Chapter 8

# Generalized Derivative Formulas

In this chapter we focus on two derivative formulas which prove useful for deriving Professor Gould's explicit formulas for Stirling numbers of the first kind. The first is an extension of the product rule while the second is an extension of the chain rule.

## 8.1 Leibniz Rule

The Leibniz rule is a generalization of the product rule which computes higher order derivatives for product functions.

**Leibniz Rule:** Let $r$ be a nonnegative integer and assume $u$ and $v$ are $r$-times differentiable functions of $x$. Then

$$\frac{d^r}{dx^r}(uv) = \sum_{k=0}^{r} \binom{r}{k} \frac{d^{r-k}u}{dx^{r-k}} \frac{d^k v}{dx^k}. \tag{8.1}$$

**Proof:** The proof follows from induction on $r$. Equation (8.1) is trivially true if $r = 0$. If $r = 1$, Equation (8.1) becomes

$$\frac{d}{dx}(uv) = \sum_{k=0}^{1} \binom{1}{k} \frac{d^{1-k}u}{dx^{1-k}} \frac{d^k v}{dx^k} = u'v + v'u,$$

which is simply a restatement of the product rule.

Now assume Equation (8.1) is true for all nonnegative integers less than or equal $r$. We need to prove it holds for $r+1$. By the induction hypothesis we have

$$\frac{d^{r+1}}{dx^{r+1}}(uv) = \frac{d}{dx}\frac{d^r}{dx^r}(uv) = \frac{d}{dx}\left[\sum_{k=0}^{r}\binom{r}{k}\frac{d^{r-k}u}{dx^{r-k}}\frac{d^k v}{dx^k}\right]$$

$$= \sum_{k=0}^{r}\binom{r}{k}\frac{d^{r-k+1}u}{dx^{r-k+1}}\cdot\frac{d^k v}{dx^k} + \sum_{k=0}^{r}\binom{r}{k}\frac{d^{r-k}u}{dx^{r-k}}\frac{d^{k+1}v}{dx^{k+1}}$$

$$= \sum_{k=0}^{r+1}\left[\binom{r}{k}+\binom{r}{k-1}\right]\frac{d^{r-k+1}u}{dx^{r-k+1}}\frac{d^k v}{dx^k}$$

$$= \sum_{k=0}^{r+1}\binom{r+1}{k}\frac{d^{r-k+1}u}{dx^{r-k+1}}\frac{d^k v}{dx^k}. \qquad \square$$

## 8.2    Generalized Chain Rule

In the previous section we generalized the product rule. We now turn our attention to the chain rule. The generalized chain rule will compute higher derivatives of a composite function. Let $y$ be a function of $x$ and $x$ be a function of $z$. We would like to compute $\frac{d^r y}{dz^r}$ for a nonnegative integer $r$. Our discussion will focus on an explicit double summation due to Reinhold Hoppe. For more information about the historical background of Hoppe's formula, we refer the reader to H. W. Gould's paper on the generalized chain rule [Gould, 2002].

To derive Hoppe's formula for the generalized chain rule we start with the chain rule and observe that

$$\frac{dy}{dz} = \frac{dy}{dx}\frac{dx}{dz}. \qquad (8.2)$$

Now use Equation (8.2), apply the product and chain rules, and compute $\frac{d^2 y}{dz^2}$:

$$\frac{d^2 y}{dz^2} = \frac{d}{dz}\left[\frac{dy}{dx}\frac{dx}{dz}\right] = \frac{d}{dz}\left[\frac{dy}{dx}\right]\frac{dx}{dz} + \frac{dy}{dx}\frac{d^2 x}{dz^2}$$

$$= \frac{d^2 y}{dx^2}\left(\frac{dx}{dz}\right)^2 + \frac{dy}{dx}\frac{d^2 x}{dz^2}. \qquad (8.3)$$

Continue this process and use Equation (8.3) to compute $\frac{d^3y}{dz^3}$:

$$\frac{d^3y}{dz^3} = \frac{d}{dz}\frac{d^2y}{dz^2} = \frac{d}{dz}\left[\frac{dy}{dx}\frac{d^2x}{dz^2} + \frac{d^2y}{dx^2}\left(\frac{dx}{dz}\right)^2\right]$$

$$= \frac{d}{dz}\left[\frac{dy}{dx}\frac{d^2x}{dz^2}\right] + \frac{d}{dz}\left[\frac{d^2y}{dx^2}\left(\frac{dx}{dz}\right)^2\right]$$

$$= \frac{d}{dz}\left[\frac{dy}{dx}\right]\frac{d^2x}{dz^2} + \frac{dy}{dx}\frac{d^3x}{dz^3} + \frac{d}{dz}\left[\frac{d^2y}{dx^2}\right]\left(\frac{dx}{dz}\right)^2 + \frac{d}{dz}\left(\frac{dx}{dz}\right)^2\frac{d^2y}{dx^2}$$

$$= \frac{d^2y}{dx^2}\frac{dx}{dz}\frac{d^2x}{dz^2} + \frac{dy}{dx}\frac{d^3x}{dz^3} + \frac{d^3y}{dx^3}\frac{dx}{dz}\left(\frac{dx}{dz}\right)^2 + 2\frac{d^2y}{dx^2}\left(\frac{dx}{dz}\right)\frac{d^2x}{dz^2}$$

$$= \frac{d^3y}{dx^3}\left(\frac{dx}{dz}\right)^3 + 3\frac{d^2y}{dx^2}\frac{dx}{dz}\frac{d^2x}{dz^2} + \frac{dy}{dx}\frac{d^3x}{dz^3}. \tag{8.4}$$

A similar calculation applied to Equation (8.4) shows that

$$\frac{d^4y}{dz^4} = \frac{d^4y}{dx^4}\left(\frac{dx}{dz}\right)^4 + 6\frac{d^3y}{dx^3}\left(\frac{dx}{dz}\right)^2\frac{d^2x}{dz^2}$$

$$+ \frac{d^2y}{dx^2}\left(4\frac{dx}{dz}\frac{d^3x}{dz^3} + 3\left(\frac{d^2x}{dz^2}\right)^2\right) + \frac{dy}{dx}\frac{d^4x}{dz^4}. \tag{8.5}$$

Take a look at the structure of Equations (8.2) through (8.5). Each equation is a sum of a finite number of terms, where each term is a certain derivative of $y$ with respect to $x$ times a coefficient which consists only of derivatives of $x$ with respect to $z$. To make the rest of our calculations easier to read, send $y \to f(x)$ and use the operator notation $\frac{d^r}{dz^r} = D_z^r$. In this notation Equation (8.5) becomes

$$D_z^4 f(x) = D_x^4 f(x)(D_z x)^4 + 6D_x^3 f(x)(D_z x)^2 D_z^2 x$$
$$+ D_x^2 f(x)(4D_z x D_z^3 x + 3(D_z^2 x)^2) + D_x f(x)D_z^4 x. \tag{8.6}$$

We may rewrite Equation (8.6) as

$$D_z^4 f(x) = A_4^4 D_x^4 f(x) + A_3^4 D_x^3 f(x) + A_2^4 D_x^2 f(x) + A_1^4 D_x f(x)$$

$$= \sum_{j=0}^{4} A_j^4 D_x^j f(x), \tag{8.7}$$

where $A_4^4 = (D_z x)^4$, $A_3^4 = 6(D_z x)^2 D_z^2 x$, $A_2^4 = 4D_z x D_z^3 x + 3(D_z^2 x)^2$, $A_1^4 = D_z^4 x$, and $A_0^4 = 0$.

The following lemma will prove that $D_z^n f(x)$ can be described in terms of this relatively simple summation notation whenever $n$ is a nonnegative integer.

**Lemma 8.1.** *Let $f(x)$ be a real or complex valued function that is $n$-times differentiable. Let $x = \phi(z)$ where $\phi(z)$ is also an $n$-times differentiable function. Then*

$$D_z^n f(x) = \sum_{j=0}^n A_j^n D_x^j f(x), \qquad (8.8)$$

*where the $A_j^n$ are independent of $f$ and contains only derivatives of $x$ with respect to $z$. We define $A_j^n = 0$ if $j < 0$ or $j > n$.*

**Proof:** We use induction on $n$. Observe that Equation (8.8) is trivially true if $n = 0$. Our previous calculations have verified Equation (8.8) for $n = 1, 2, 3, 4$. Assume Equation (8.8) is valid for all nonnegative integers less than or equal to $n$, and let us see what happens for $n + 1$.

$$D_z^{n+1} f(x) = D_z D_z^n f(x) = D_z \sum_{j=0}^n A_j^n D_x^j f(x)$$

$$= \sum_{j=0}^n \left[ A_j^n D_z D_x^j f(x) + D_z A_j^n D_x^j f(x) \right]$$

$$= \sum_{j=0}^n A_j^n D_x^{j+1} f(x) D_z x + \sum_{j=0}^n D_x^j f(x) D_z A_j^n$$

$$= \sum_{j=0}^{n+1} \left[ A_{j-1}^n D_z x + D_z A_j^n \right] D_x^j f(x).$$

Since $D_z^{n+1} f(x) = \sum_j^{n+1} A_j^{n+1} D_x^j f(x)$, we conclude that

$$A_j^{n+1} = A_{j-1}^n D_z x + D_z A_j^n. \qquad (8.9)$$

Because we assumed for $0 \le i \le n$ that $A_i^n$ is independent of $f$ and contains only derivatives of $x$ with respect to $z$, we conclude the right side of Equation (8.9) is also independent of $f$ and contains only derivatives of $x$ with respect to $z$. Hence $A_j^{n+1}$ is independent of $f$ and contains only derivatives with respect to $z$. $\square$

We find an explicit formula for $A_j^n$ by evaluating Equation (8.8) for $f(x) = (x - a)^k$, where $k$ is a nonnegative integer, $k > n$, and $a$ is independent of $x$. By Equation (8.8) we have

$$D_z^n (x - a)^k = \sum_{j=0}^n A_j^n D_x^j (x - a)^k.$$

Thus

$$
D_z^n (x-a)^k \big|_{a=x} = \sum_{j=0}^{n} A_j^n D_x^j (x-a)^k \bigg|_{a=x}
$$

$$
= \sum_{j=0}^{n} j! \binom{k}{j} A_j^n (x-a)^{k-j} \bigg|_{a=x} = k! A_k^n.
$$

Hence

$$
A_k^n = \frac{1}{k!} D_z^n (x-a)^k \bigg|_{a=x} = \frac{1}{k!} D_z^n \sum_{j=0}^{k} \binom{k}{j} x^j (-a)^{k-j} \bigg|_{a=x}
$$

$$
= \frac{(-1)^k}{k!} \sum_{j=0}^{k} (-1)^j \binom{k}{j} x^{k-j} D_z^n x^j. \tag{8.10}
$$

In summary we have proven the following theorem:

**Theorem 8.1.** *(Hoppe's Generalized Chain Rule) Let $f(x)$ be a real or complex valued function that is $n$-times differentiable. Let $x = \phi(z)$, where $\phi(z)$ is also an $n$-times differentiable function. Then*

$$
D_z^n f(x) = \sum_{k=0}^{n} D_x^k f(x) \frac{(-1)^k}{k!} \sum_{j=0}^{k} (-1)^j \binom{k}{j} x^{k-j} D_z^n x^j. \tag{8.11}
$$

## 8.3 Five Applications of Hoppe's Formula

We now take the time to demonstrate the usefulness of Equation (8.11) and derive a collection of identities that Professor Gould utilizes throughout Chapters 12 and 13. The first identity is a change of variable formula. Let $z = a + bx$ where $b$ is a nonzero real or complex number. Equation (8.11) yields

$$
D_z^n f(x) = \sum_{k=0}^{n} D_x^k f(x) \frac{(-1)^k}{k!} \sum_{j=0}^{k} (-1)^j \binom{k}{j} \left(\frac{z-a}{b}\right)^{k-j} D_z^n \left(\frac{z-a}{b}\right)^j
$$

$$
= \frac{n!}{b^n} \sum_{k=0}^{n} D_x^k f(x) \frac{(-1)^k}{k!} \left(\frac{z-a}{b}\right)^{k-n} \sum_{j=0}^{k} (-1)^j \binom{k}{j} \binom{j}{n}
$$

$$
= \frac{n!}{b^n} \sum_{k=0}^{n} D_x^k f(x) \frac{(-1)^k}{k!} \left(\frac{z-a}{b}\right)^{k-n} \binom{k}{n} \sum_{j=0}^{k-n} (-1)^{j+n} \binom{k-n}{j}
$$

$$
= \frac{n!}{b^n} D_x^n f(x) \frac{(-1)^n}{n!} \left(\frac{z-a}{b}\right)^{n-n} \binom{n}{n} (-1)^n.
$$

In summary we obtain the change of variable formula

$$D_z^n f(x)\big|_{z=a+bx} = \frac{1}{b^n} D_x^n f(x), \qquad b \neq 0. \tag{8.12}$$

The second identity is another change of variable formula, but this time $x = e^z$. Equation (8.11) becomes

$$
\begin{aligned}
D_z^n f(x) &= \sum_{k=0}^{n} D_x^k f(x) \frac{(-1)^k}{k!} \sum_{j=0}^{k} (-1)^j \binom{k}{j} (e^z)^{k-j} D_z^n e^{jz} \\
&= \sum_{k=0}^{n} D_x^k f(x) \frac{(-1)^k}{k!} \sum_{j=0}^{k} (-1)^j \binom{k}{j} e^{zk-zj} j^n e^{jz} \\
&= \sum_{k=0}^{n} D_x^k f(x) e^{zk} \frac{(-1)^k}{k!} \sum_{j=0}^{k} (-1)^j \binom{k}{j} j^n \\
&= \sum_{k=0}^{n} D_x^k f(x) x^k S(n,k), \tag{8.13}
\end{aligned}
$$

where we let

$$S(n,k) = \frac{(-1)^k}{k!} \sum_{j=0}^{k} (-1)^j \binom{k}{j} j^n.$$

The quantity $S(n,k)$ is called a Stirling number of the second kind. Combinatorially $S(n,k)$ counts the number of ways to form set partitions of $\{1, 2, ..., n\}$ consisting of exactly $k$ subsets. We discuss this important number sequence in the next chapter.

Before progressing to our third application of Equation (8.11), we derive the dual to Equation (8.13). Since $x = e^z$, $z = \ln x$, $\frac{dx}{dz} = e^z$, and $\frac{dz}{dx} = \frac{1}{x} = e^{-z}$. Assume $y$ is a function of $z$ and ultimately of $x$. By the chain rule we have

$$D_x y = D_z y D_x z = e^{-z} D_z y. \tag{8.14}$$

If we reapply the operator $D_x$ to Equation (8.14) we find that

$$
\begin{aligned}
D_x^2 y &= D_x(e^{-z} D_z y) = e^{-z} D_x D_z y + D_z y D_x e^{-z} \\
&= e^{-z} D_z^2 y D_x z - e^{-z} D_z y D_x z \\
&= e^{-2z} D_z^2 y - e^{-2z} D_z y \\
&= e^{-2z} \left( D_z^2 - D_z \right) y = e^{-2z} D_z (D_z - 1) y. \tag{8.15}
\end{aligned}
$$

Applying $D_x$ to Equation (8.15) gives us

$$D_x^3 y = e^{-3z} \left( D_z^3 y - 3 D_z^2 y + 2 D_z y \right) = e^{-3z} D_z (D_z - 1)(D_z - 2) y. \tag{8.16}$$

The pattern in Equations (8.14) to (8.16) implies that

$$D_x^n y = e^{-nz} D_z (D_z - 1) \ldots (D_z - n + 1) y, \qquad x = e^z, \qquad n \geq 1. \quad (8.17)$$

Equation (8.17) is the change of variable formula dual to that of Equation (8.13). The proof of Equation (8.17) follows from induction on $n$. Equation (8.14) establishes the base case of $n = 1$. Now assume Equation (8.17) is true for all positive integers less than or equal to $n$. Then

$$
\begin{aligned}
D_x^{n+1} y &= D_x D_x^n y = D_x \left[ e^{-nz} D_z (D_z - 1) \ldots (D_z - n + 1) \right] y \\
&= D_x e^{-nz} D_z (D_z - 1) \ldots (D_z - n + 1) y \\
&\quad + e^{-nz} D_x \left[ D_z (D_z - 1) \ldots (D_z - n + 1) \right] y \\
&= -n e^{-nz} D_x z D_z (D_z - 1) \ldots (D_z - n + 1) y \\
&\quad + e^{-nz} D_z \left[ D_z (D_z - 1) \ldots (D_z - n + 1) \right] y \cdot D_x z \\
&= -n e^{-(n+1)z} D_z (D_z - 1) \ldots (D_z - n + 1) y \\
&\quad + e^{-(n+1)z} D_z D_z (D_z - 1) \ldots (D_z - n + 1) y \\
&= e^{-(n+1)z} D_z (D_z - 1) \ldots (D_z - n + 1)(D_z - n) y.
\end{aligned}
$$

Since $x = e^z$ Equation (8.17) is often rewritten symbolically as

$$x^n D_x^n y = n! \binom{D_z}{n} y, \qquad x = e^z. \quad (8.18)$$

Our third application of the generalized chain rule uses Equation (8.11) to evaluate $D_z^n x^\alpha$ whenever $x$ is a function of $z$ and $\alpha$ is a real or complex number. Substitute $f(x) = x^\alpha$ in (8.11) to obtain

$$
\begin{aligned}
D_z^n x^\alpha &= \sum_{k=0}^n D_x^k x^\alpha \frac{(-1)^k}{k!} \sum_{j=0}^k (-1)^j \binom{k}{j} x^{k-j} D_z^n x^j \\
&= \sum_{k=0}^n (-1)^k \binom{\alpha}{k} \sum_{j=0}^k (-1)^j \binom{k}{j} x^{\alpha-j} D_z^n x^j. \quad (8.19)
\end{aligned}
$$

By interchanging the order of summation, Equation (8.19) becomes

$$
\begin{aligned}
D_z^n x^\alpha &= \sum_{j=0}^n (-1)^j x^{\alpha-j} D_z^n x^j \sum_{k=j}^n (-1)^k \binom{\alpha}{k} \binom{k}{j} \\
&= \sum_{j=0}^n (-1)^j x^{\alpha-j} \binom{\alpha}{j} D_z^n x^j \sum_{k=j}^n (-1)^k \binom{\alpha-j}{k-j}
\end{aligned}
$$

$$= \sum_{j=0}^{n}(-1)^j x^{\alpha-j}\binom{\alpha}{j} D_z^n x^j \sum_{k=0}^{n-j}(-1)^{k+j}\binom{\alpha-j}{k}$$

$$= \sum_{j=0}^{n}(-1)^{n-j} x^{\alpha-j}\binom{\alpha}{j}\binom{\alpha-j-1}{n-j} D_z^n x^j, \qquad \text{by Eq. (1.30)}$$

$$= \sum_{j=0}^{n} x^{\alpha-j}\binom{\alpha}{j}\binom{-\alpha+n}{n-j} D_z^n x^j. \tag{8.20}$$

We continue to manipulate Equation (8.20) by applying the $-1$-Transformation and obtaining

$$D_z^n x^\alpha = \sum_{j=0}^{n}(-1)^j x^{\alpha-j}\binom{-\alpha+j-1}{j}\binom{-\alpha+n}{n-j} D_z^n x^j. \tag{8.21}$$

Since

$$\binom{-\alpha+j-1}{j}\binom{-\alpha+n}{n-j} = \frac{-\alpha}{-\alpha+j}\binom{-\alpha+n}{n}\binom{n}{j}, \tag{8.22}$$

Equation (8.21) is equivalent to

$$D_z^n x^\alpha = -\alpha\binom{-\alpha+n}{n}\sum_{j=0}^{n}(-1)^j\binom{n}{j}\frac{x^{\alpha-j}}{-\alpha+j} D_z^n x^j, \tag{8.23}$$

an $n^{th}$ difference formula for $D_z^n x^\alpha$.

Some books let $\alpha \to -\alpha$ and write Equations (8.20) and (8.23) as

$$D_z^n x^{-\alpha} = \sum_{j=0}^{n} x^{-\alpha-j}\binom{-\alpha}{j}\binom{\alpha+n}{n-j} D_z^n x^j \tag{8.24}$$

$$D_z^n x^{-\alpha} = \alpha\binom{\alpha+n}{n}\sum_{j=0}^{n}(-1)^j\binom{n}{j}\frac{x^{-\alpha-j}}{\alpha+j} D_z^n x^j.$$

Observe that Equation (8.24) has the symmetrical form

$$x^\alpha D_z^n x^{-\alpha} = \sum_{j=0}^{n}\binom{-\alpha}{j}\binom{n+\alpha}{n-j} x^{-j} D_z^n x^j.$$

Our fourth application of Equation (8.11) involves finding a formula for $D_x^n\left(\frac{1}{f(x)}\right)$. Since

$$D_x^k\left(\frac{1}{x}\right) = \frac{(-1)^k k!}{x^{k+1}}, \tag{8.25}$$

Equation (8.11) becomes

$$D_z^n \left(\frac{1}{x}\right) = \sum_{k=0}^{n} D_x^k \left(\frac{1}{x}\right) \frac{(-1)^k}{k!} \sum_{j=0}^{k} (-1)^j \binom{k}{j} x^{k-j} D_z^n x^j$$

$$= \sum_{k=0}^{n} \sum_{j=0}^{k} (-1)^j \binom{k}{j} \frac{1}{x^{j+1}} D_z^n x^j$$

$$= \sum_{j=0}^{n} \frac{(-1)^j}{x^{j+1}} D_z^n x^j \sum_{k=j}^{n} \binom{k}{j}$$

$$= \sum_{j=0}^{n} (-1)^j \binom{n+1}{j+1} \frac{1}{x^{j+1}} D_z^n x^j, \qquad \text{by Eq. (3.12). (8.26)}$$

Equation (8.26) is Equation (8.23) with $\alpha = -1$.

Since $x$ is a function of $z$, namely $x = \phi(z)$, we can write Equation (8.26) as

$$D_z^n \left(\frac{1}{\phi(z)}\right) = \sum_{j=0}^{n} (-1)^j \binom{n+1}{j+1} \frac{1}{(\phi(z))^{j+1}} D_z^n (\phi(z))^j. \qquad (8.27)$$

Equation (8.27) is our promised formula for the derivative of a reciprocal function. Just let $z \to x$ and $\phi(z) \to f(x)$ to obtain

$$D_x^n \left(\frac{1}{f(x)}\right) = \sum_{j=0}^{n} (-1)^j \binom{n+1}{j+1} \frac{1}{(f(x))^{j+1}} D_x^n (f(x))^j. \qquad (8.28)$$

We turn to a generalization of Equation (8.26). Let $f(x) = \frac{u}{x}$ where $u$ is a function of $x$ and $x$ is a function of $z$. Equation (8.11) implies

$$D_z^n \left(\frac{u}{x}\right) = \sum_{k=0}^{n} D_x^k \left(\frac{u}{x}\right) \frac{(-1)^k}{k!} \sum_{j=0}^{k} (-1) \binom{k}{j} x^{k-j} D_z^n x^j. \qquad (8.29)$$

The Leibniz rule and Equation (8.25) imply that

$$D_x^k \left(\frac{u}{x}\right) = \sum_{r=0}^{k} \binom{k}{r} D_x^{k-r} u D_x^r \left(\frac{1}{x}\right) = \sum_{r=0}^{k} \binom{k}{r} \frac{(-1)^r r!}{x^{r+1}} D_x^{k-r} u. \qquad (8.30)$$

Equation (8.29) is equivalent to

$$D_z^n \left(\frac{u}{x}\right) = \sum_{k=0}^{n} \sum_{r=0}^{k} \binom{k}{r} \frac{(-1)^r r!}{x^{r+1}} D_x^{k-r} u \sum_{j=0}^{k} (-1)^j \binom{k}{j} x^{k-j} D_z^n x^j. \qquad (8.31)$$

There is another way to compute $D_z^n \left( \frac{u}{x} \right)$. Instead of starting with Hoppe's formula, we start with the Leibniz rule and notice that

$$D_z^n \left( \frac{u}{x} \right) = \sum_{k=0}^{n} \binom{n}{k} D_z^{n-k} u D_z^k (x^{-1}).$$

We compute $D_z^k(x^{-1})$ from Equation (8.26), and rewrite the previous line as

$$D_z^n \left( \frac{u}{x} \right) = \sum_{k=0}^{n} \binom{n}{k} D_z^{n-k} u \sum_{j=0}^{k} (-1)^j \binom{k+1}{j+1} \frac{1}{x^{j+1}} D_z^k x^j. \qquad (8.32)$$

Equation (8.32) is simpler than Equation (8.31). This example shows how changing the order in which we apply identities will affect the outcome.

Some books rewrite Equation (8.32) as

$$D_x^n \left( \frac{u}{v} \right) = \sum_{k=0}^{n} \binom{n}{k} D_x^{n-k} u \sum_{j=0}^{k} (-1)^j \binom{k+1}{j+1} \frac{1}{v^{j+1}} D_x^k v^j, \qquad (8.33)$$

where $u$ and $v$ are functions of $x$. The left side of Equation (8.33) mimics the structure of the Leibniz rule.

Our final application of the generalized chain rule is due to G. H. Halphen [1879-1880].

**Theorem 8.2.** *(due to G. H. Halphen) Assume $f$ and $\phi$ are $n$-times differentiable functions of $x$. Define $\phi^{(k)} \left( \frac{1}{x} \right) = D_t^k \phi(t)|_{t=\frac{1}{x}}$. Then*

$$D_x^n \left( f(x) \phi \left( \frac{1}{x} \right) \right) = \sum_{k=0}^{n} (-1)^k \binom{n}{k} \frac{1}{x^k} \phi^{(k)} \left( \frac{1}{x} \right) D_x^{n-k} \left( \frac{f(x)}{x^k} \right). \qquad (8.34)$$

**Proof:** The structure of the left side of Equation (8.34) involves the $n^{th}$ derivative of a product. The Leibniz rule implies that

$$D_x^n \left( f(x) \phi \left( \frac{1}{x} \right) \right) = \sum_{k=0}^{n} \binom{n}{k} D_x^{n-k} f(x) D_x^k \phi \left( \frac{1}{x} \right). \qquad (8.35)$$

We want to evaluate $D_x^k \phi \left( \frac{1}{x} \right)$. Define $t = \frac{1}{x}$ and use Hoppe's formula to find that

$$D_x^k \phi\left(\frac{1}{x}\right) = D_x^n \phi(t) = \sum_{p=0}^{k} D_t^p \phi(t) \frac{(-1)^p}{p!} \sum_{j=0}^{p} (-1)^j \binom{p}{j} t^{p-j} D_x^k t^j$$

$$= \sum_{p=0}^{k} D_t^p \phi(t) \frac{(-1)^p}{p!} \sum_{j=0}^{p} (-1)^j \binom{p}{j} x^{j-p} D_x^k x^{-j}$$

$$= \sum_{p=0}^{k} \frac{(-1)^p}{p!} \phi^{(p)}\left(\frac{1}{x}\right) \sum_{j=0}^{p} (-1)^j \binom{p}{j} x^{j-p} \binom{-j}{k} k! x^{-j-k}$$

$$= k! \sum_{p=0}^{k} \frac{(-1)^p}{p!} x^{-p-k} \phi^{(p)}\left(\frac{1}{x}\right) \sum_{j=0}^{p} (-1)^j \binom{p}{j}\binom{-j}{k}$$

$$= k! \sum_{p=0}^{k} \frac{(-1)^k}{p!} \binom{k-1}{k-p} x^{-p-k} \phi^{(p)}\left(\frac{1}{x}\right), \qquad (8.36)$$

where the final equality follows from Equation (6.38). We now substitute
Equation (8.36) into (8.35). Note that we have interchanged the order of
summation after making our substitution:

$$D_x^n\left(f(x)\phi\left(\frac{1}{x}\right)\right) = \qquad (8.37)$$

$$\sum_{p=0}^{n} \phi^{(p)}\left(\frac{1}{x}\right) \frac{x^{-p}}{p!} \sum_{k=p}^{n} (-1)^k \binom{n}{k} k! \binom{k-1}{k-p} x^{-k} D_x^{n-k} f(x) =$$

$$\sum_{p=0}^{n} \phi^{(p)}\left(\frac{1}{x}\right) x^{-p} \binom{n}{p} \sum_{k=p}^{n} (-1)^k \binom{n-p}{k-p}\binom{k-1}{k-p} (k-p)! x^{-k} D_x^{n-k} f(x).$$

It remains to show the inner sum on the final line of Equation (8.37) equals
$(-1)^p D_x^{n-p}\left(\frac{f(x)}{x^p}\right)$. We once again apply the Leibniz rule and observe that

$$D_x^{n-p}\left(\frac{f(x)}{x^p}\right) = \sum_{k=0}^{n-p} \binom{n-p}{k} D_x^{n-p-k} f(x) D_x^k x^{-p}$$

$$= \sum_{k=p}^{n} \binom{n-p}{k-p} D_x^{n-k} f(x) D_x^{k-p} x^{-p}$$

$$= \sum_{k=p}^{n} \binom{n-p}{k-p} D_x^{n-k} f(x) \binom{-p}{k-p} (k-p)! x^{-k}$$

$$= (-1)^p \sum_{k=p}^{n} (-1)^k \binom{n-p}{k-p}\binom{k-1}{k-p} (k-p)! x^{-k} D_x^{n-k} f(x).$$

Thus we can write Equation (8.37) as

$$D_x^n\left(f(x)\phi\left(\frac{1}{x}\right)\right) = \sum_{p=0}^{n}(-1)^p\binom{n}{p}\frac{1}{x^p}\phi^{(p)}\left(\frac{1}{x}\right)D_x^{n-p}\left(\frac{f(x)}{x^p}\right),$$

which is Halphen's formula.                                              □

Halphen developed Equation (8.34) to facilitate his investigation of the $n^{th}$ derivative of $x^m e^{1/x}$.

# Chapter 9

# Stirling Numbers of the Second Kind $S(n, k)$

We leave the realm of binomial identities and focus on two special combinatorial number sequences, Stirling numbers of the first and second kind. We begin our exploration with $\{S(n, k)\}_{k=0}^{n}$, where $S(n, k)$ is a **Stirling number of the second kind**. To develop a combinatorial meaning for $S(n, k)$ we need a definition. Let $S$ be a nonempty set. A **set partition** of $S$ is a collection of pairwise disjoint nonempty subsets of $S$ whose union is $S$. For example let $S = \{a, b, c\}$. There are 5 set partitions of $S$, namely $S$ itself, the set partition consisting of three subsets, i.e. $\{\{a\}, \{b\}, \{c\}\}$, and the three set partitions consisting of 2 subsets, i.e. $\{\{a\}, \{b, c\}\}$, $\{\{b\}, \{a, c\}\}$, and $\{\{c\}, \{a, b\}\}$. Let $n$ be a positive integer. The quantity $S(n, k)$ counts the set partitions of $[n] = \{1, 2, 3, \ldots, n\}$ which consists of exactly $k$ subsets or parts. By definition $S(n, k) = 0$ if $k = 0$ or $k > n$. For technical reasons we define $S(0, 0) = 1$. Table 9.1 provides numerical values for $S(n, k)$ where $0 \leq k \leq n$ and $0 \leq n \leq 7$.

| | $k = 0$ | $k - 1$ | $k - 2$ | $k - 3$ | $k = 4$ | $k = 5$ | $k = 6$ | $k = 7$ |
|---|---|---|---|---|---|---|---|---|
| $n = 0$ | 1 | | | | | | | |
| $n = 1$ | | 1 | | | | | | |
| $n = 2$ | | 1 | 1 | | | | | |
| $n = 3$ | | 1 | 2 | 1 | | | | |
| $n = 4$ | | 1 | 7 | 6 | 1 | | | |
| $n = 5$ | | 1 | 15 | 25 | 10 | 1 | | |
| $n = 6$ | | 1 | 31 | 90 | 65 | 15 | 1 | |
| $n = 7$ | | 1 | 63 | 301 | 350 | 140 | 21 | 1 |

Table 9.1: Table of values for Stirling numbers of the second kind, $S(n, k)$, with $0 \leq k \leq n$

The Stirling numbers of the second kind satisfy

$$S(n+1,k) = kS(n,k) + S(n,k-1), \qquad 0 \le k-1 \le n. \tag{9.1}$$

To prove Recurrence (9.1) observe that the set partitions of $[n+1]$ with exactly $k$ subsets fall into one of two categories; those which contain the subset $\{n+1\}$ and those which do not. If the set partition contains $\{n+1\}$, delete this subset and obtain a one-to-one correspondence with a set partition of $[n]$ consisting of $k-1$ parts. If $\{n+1\}$ is not a subset of the set partition of $[n+1]$, this set partition came by inserting $n+1$ into one of the $k$ subsets of a set partition of $[n]$.

Recurrence (9.1) provides closed forms for $S(n, n-r)$. Start with Equation (9.1), multiply by $k!$, then let $k \to \alpha$ and $n \to n-1$.

$$\alpha! S(n,\alpha) = \alpha \alpha! S(n-1,\alpha) + \alpha(\alpha-1)! S(n-1,\alpha-1). \tag{9.2}$$

Take Equation (9.2) and let $n \to n-1$, $\alpha \to \alpha-1$ to obtain

$$(\alpha-1)! S(n-1,\alpha-1) = (\alpha-1)(\alpha-1)! S(n-2,\alpha-1)$$
$$+ (\alpha-1)(\alpha-2)! S(n-2,\alpha-2). \tag{9.3}$$

Substitute Equation (9.3) into (9.2). This gives us

$$\alpha! S(n,\alpha) = \alpha \alpha! S(n-1,\alpha) + \alpha(\alpha-1)(\alpha-1)! S(n-2,\alpha-1)$$
$$+ \alpha(\alpha-1)(\alpha-2)! S(n-2,\alpha-2). \tag{9.4}$$

Now repeat this process. Take Equation (9.2), let $n \to n-2$, $\alpha \to \alpha-2$, and obtain

$$(\alpha-2)! S(n-2,\alpha-2) = (\alpha-2)(\alpha-2)! S(n-3,\alpha-2)$$
$$+ (\alpha-2)(\alpha-3)! S(n-2,\alpha-2). \tag{9.5}$$

Substitute (9.5) into Equation (9.4). This gives us

$$\alpha! S(n,\alpha) = \alpha \alpha! S(n-1,\alpha) + \alpha(\alpha-1)(\alpha-1)! S(n-2,\alpha-1)$$
$$+ \alpha(\alpha-1)(\alpha-2)(\alpha-2)! S(n-3,\alpha-2)$$
$$+ \alpha(\alpha-1)(\alpha-2)(\alpha-3)! S(n-3,\alpha-3). \tag{9.6}$$

After $r$ iterations we discover

$$\alpha! S(n,\alpha) = \sum_{k=0}^{r} \binom{\alpha}{k+1} (k+1)! (\alpha-k)! S(n-1-k,\alpha-k)$$
$$+ \binom{\alpha}{r+1} (r+1)! (\alpha-r-1)! S(n-r-1,\alpha-r-1). \tag{9.7}$$

In Equation (9.7) let $r = \alpha - 1$ to obtain

$$\alpha! S(n,\alpha) = \sum_{k=0}^{\alpha-1} \binom{\alpha}{k+1}(k+1)!(\alpha-k)!S(n-1-k,\alpha-k)+\alpha!S(n-r-1,0).$$
(9.8)

As long as $n - r - 1 \neq 0$, $S(n - r - 1, 0) = 0$, and Equation (9.8) becomes

$$\alpha! S(n,\alpha) = \sum_{k=0}^{\alpha-1} \binom{\alpha}{k+1}(k+1)!(\alpha-k)!S(n-1-k,\alpha-k)$$

$$= \sum_{k=1}^{\alpha} \binom{\alpha}{k}k!(\alpha-k+1)!S(n-k,\alpha-k+1). \qquad (9.9)$$

Take Equation (9.9), divide by $\alpha$, and simplify to obtain

$$S(n,\alpha) = \sum_{k=1}^{\alpha}(\alpha-k+1)S(n-k,\alpha-k+1), \qquad \alpha \geq 1, \qquad n \neq \alpha. \;\; (9.10)$$

Equation (9.10), when combined with closed forms for $\sum_{k=0}^{n-r} k^p$, recursively provides polynomials for $S(n, n - r)$ whenever $r$ is a positive integer. To calculate $S(n, n - 1)$ use Equation (9.10) with $\alpha = n - 1$ and observe that

$$S(n,n-1) = \sum_{k=1}^{n-1}(n-k)S(n-k,n-k) = \sum_{k=1}^{n-1}(n-k)$$

$$= n(n-1) - \frac{n(n-1)}{2} = \binom{n}{2}.$$

To obtain the polynomial for $S(n, n-2)$ use Equation (9.10) with $\alpha = n-2$, along with $S(n, n - 1) = \binom{n}{2}$, and calculate

$$S(n,n-2) = \sum_{k=1}^{n-2}(n-k-1)S(n-k,n-k-1)$$

$$= \sum_{k=1}^{n-2}(n-k-1)\binom{n-k}{2}$$

$$= \sum_{k=1}^{n-2}\frac{(n-k-1)^2(n-k)}{2} = \frac{1}{2}\sum_{k=1}^{n-2}k^2(k+1)$$

$$= \frac{1}{2}\left[\sum_{k=1}^{n-2}k^3 + \sum_{k=1}^{n-2}k^2\right]$$

$$= \frac{1}{2}\left[\frac{(n-2)^2(n-1)^2}{4} + \frac{(n-2)(n-1)(2n-3)}{6}\right]$$

$$= \frac{n(n-1)(n-2)(3n-5)}{24}.$$

Continuing this recursive procedure shows that

$$S(n, n-3) = \frac{n(n-1)(n-2)^2(n-3)^2}{48}$$

$$S(n, n-4) = \frac{n(n-1)(n-2)(n-3)(n-4)(15n^3 - 150n^2 + 485n - 502)}{5760}$$

$$S(n, n-5) = \frac{n(n-1)(n-2)(n-3)(n-4)^2(n-5)^2(3n^2 - 23n + 38)}{11520}.$$

Recurrence (9.1) also implies that

$$x^n = \sum_{j=0}^{n} \binom{x}{j} j! S(n, j) \tag{9.11}$$

whenever $n$ is a nonnegative integer. To prove Equation (9.11) let $\{C(n, j)\}_{j=0}^{n}$ be the coefficient of $\binom{x}{j}$ in the expansion $x^n = \sum_{j=0}^{n} \binom{x}{j} j! C(n, j)$. Notice that $C(0, 0) = 1$ and $C(n, j) = 0$ if $j > n$. By definition we have $x^{n+1} = \sum_{j=0}^{n+1} \binom{x}{j} j! C(n+1, j)$. On the other hand

$$x^{n+1} = x \cdot x^n = x \sum_{j=0}^{n} \binom{x}{j} j! C(n, j)$$

$$= \sum_{j=0}^{n} (x-j) \binom{x}{j} j! C(n, j) + \sum_{j=0}^{n} j \binom{x}{j} j! C(n, j)$$

$$= \sum_{j=1}^{n+1} (x-j+1) \binom{x}{j-1} (j-1)! C(n, j-1) + \sum_{j=0}^{n} j \binom{x}{j} j! C(n, j)$$

$$= \sum_{j=0}^{n+1} \binom{x}{j} j! C(n, j-1) + \sum_{j=0}^{n+1} j \binom{x}{j} j! C(n, j)$$

$$= \sum_{j=0}^{n+1} [C(n, j-1) + j C(n, j)] j! \binom{x}{j}.$$

Comparing the coefficient of $j!\binom{x}{j}$ implies that $C(n+1, j) = j C(n, j) + C(n, j-1)$. This is precisely the recurrence obeyed by $S(n, j)$. Since $C(0, 0) = S(0, 0) = 1$ and $C(n, j) = 0 = S(n, j)$ for $j > n$, we conclude that $C(n, j) = S(n, j)$ for all nonnegative integers $j$ and $n$.

Equation (9.11) is the called the **basis definition** of $S(n, j)$. This definition provides easy proofs of the following three identities:

$$\sum_{j=0}^{n}(-1)^j j! S(n,j) = (-1)^n \tag{9.12}$$

$$\sum_{j=0}^{n}(-1)^j \binom{x+j-1}{j} j! S(n,j) = (-1)^n x^n \tag{9.13}$$

$$\sum_{j=0}^{n}(-1)^j \binom{2j}{j} \frac{j!}{2^{2j}} S(n,j) = \frac{(-1)^n}{2^n}. \tag{9.14}$$

Equation (9.14) is Equation (9.11) with $x = -\frac{1}{2}$; Equation (9.12) is Equation (9.11) with $x = -1$ while Equation (9.13) is Equation (9.11) with $x \to -x$. If $x = 1$, Equation (9.13) becomes Equation (9.12). This suggests that we take Equation (9.13) and let $x = 2$ to find that

$$\sum_{j=0}^{n}(-1)^j(j+1)j! S(n,j) = \sum_{j=0}^{n}(-1)^j jj! S(n,j) + \sum_{j=0}^{n}(-1)^j j! S(n,j)$$
$$= (-1)^n 2^n. \tag{9.15}$$

Use Equation (9.12) to evaluate the right most sum of Equation (9.15). After simplification we obtain the identity

$$\sum_{j=0}^{n}(-1)^j jj! S(n,j) = (-1)^n \left[2^n - 1\right]. \tag{9.16}$$

We can continue this process, substitute $x = r + 1$ into Equation (9.13), and use the previously computed closed forms for $\sum_{j=0}^{n}(-1)^j j^p j! S(n,j)$ to inductively compute a closed form for $\sum_{j=0}^{n}(-1)^j j^r j! S(n,j)$, whenever $r$ is a nonnegative integer. In particular we find that

$$\sum_{j=0}^{n}(-1)^j j^2 j! S(n,j) = (-1)^{n+1}\left[-2 \cdot 3^n + 3 \cdot 2^n - 1\right]$$

$$\sum_{j=0}^{n}(-1)^j j^3 j! S(n,j) = (-1)^n(6 \cdot 4^n - 1 + 7 \cdot 2^n - 12 \cdot 3^n)$$

$$\sum_{j=0}^{n}(-1)^j j^4 j! S(n,j) = (-1)^{n+1}(-24 \cdot 5^n - 1 + 15 \cdot 2^n - 50 \cdot 3^n + 60 \cdot 4^n)$$

$$\sum_{j=0}^{n}(-1)^j j^5 j! S(n,j) =$$
$$(-1)^n(120 \cdot 6^n - 1 + 31 \cdot 2^n - 180 \cdot 3^n + 390 \cdot 4^n - 360 \cdot 5^n)$$

$$\sum_{j=0}^{n}(-1)^j j^6 j! S(n,j) =$$

$$(-1)^{n+1}(-720 \cdot 7^n - 1 + 63 \cdot 2^n - 602 \cdot 3^n + 2100 \cdot 4^n - 3360 \cdot 5^n + 2520 \cdot 6^n)$$

The preceding results are summarized as

$$\sum_{j=0}^{n}(-1)^j j^r j! S(n,j) = (-1)^{n+r+1}\sum_{j=0}^{r+1}(-1)^j (j-1)! j^n S(r+1,j). \quad (9.17)$$

To prove Equation (9.17), take the right side, write it as
$(-1)^{n+r+1}\sum_{j=0}^{r+1}(-1)^j j! j^{n-1} S(r+1,j)$, set $r+1 = n$, and obtain
$(-1)^{2n}\sum_{j=0}^{n}(-1)^j j! j^r S(n,j)$.

Another property of $S(n,j)$ obtained via Equation (9.11) is

$$\alpha! S(n,\alpha) = \sum_{j=\alpha}^{n}(-1)^{n-j}\binom{j-1}{\alpha-1}j! S(n,j). \quad (9.18)$$

To prove Equation (9.18) start with Equation (9.11) and let $x \to -x$. This gives us

$$(-x)^n = \sum_{j=0}^{n}\binom{-x}{j}j! S(n,j) = \sum_{j=0}^{n}(-1)^j \binom{x+j-1}{j}j! S(n,j), \quad (9.19)$$

where the last equality follows from the $-1$-Transformation. We take Equation (9.19) and use the Vandermonde convolution to obtain

$$x^n = \sum_{j=0}^{n}(-1)^{n-j}j! S(n,j)\sum_{\alpha=0}^{j}\binom{x}{j}\binom{j-1}{j-\alpha}$$

$$= \sum_{\alpha=0}^{n}\binom{x}{\alpha}\sum_{j=\alpha}^{n}(-1)^{n-j}\binom{j-1}{\alpha-1}j! S(n,j). \quad (9.20)$$

Since Equation (9.11) implies that $x^n = \sum_{\alpha=0}^{n}\binom{x}{\alpha}\alpha! S(n,\alpha)$, and this expansion for $x^n$ in terms of $\binom{x}{\alpha}$ is unique, compare the coefficients of $\binom{x}{\alpha}$ in Equation (9.20) with that of Equation (9.11) to obtain Equation (9.18).

## 9.1    Euler's Formula for $S(n,k)$

So far we have not provided an explicit formula for $S(n,j)$. We remedy that by noting that

$$S(n,j) = \frac{(-1)^j}{j!}\sum_{k=0}^{j}(-1)^k\binom{j}{k}k^n = \frac{1}{j!}\sum_{k=0}^{j}(-1)^k\binom{j}{k}(j-k)^n. \quad (9.21)$$

Equation (9.21) is known as Euler's formula for Stirling numbers [Gould, 1987]. We verify Equation (9.21) by defining $f(n,k) = \frac{(-1)^j}{j!} \sum_{k=0}^{j}(-1)^k\binom{j}{k}k^n$ and showing that $f(n+1,k) = kf(n,k)+f(n,k-1)$, $f(1,k) = S(1,k)$ and $f(0,0) = 1 = S(0,0)$.

Euler's formula for $S(n,j)$ is intimately connected with the $n^{th}$ difference operator $\Delta^n_{x,h}$ where $\Delta^n_{x,h}f(x) = \frac{(-1)^n}{h^n} \sum_{k=0}^{n}(-1)^k\binom{n}{k}f(x+kh)$. Let $f(x) = x^p$ for any nonnegative integer $p$ and observe that

$$\Delta^n_{x,1}x^p\big|_{x=0} = (-1)^n \sum_{k=0}^{n}(-1)^k\binom{n}{k}k^p = n!S(p,n). \qquad (9.22)$$

More generally

$$\Delta^n_{x,h}x^p = \sum_{r=0}^{p}\binom{p}{r}x^{p-r}h^{r-n}n!S(r,n), \qquad (9.23)$$

since

$$\Delta^n_{x,h}x^p = \frac{(-1)^n}{h^n} \sum_{k=0}^{n}(-1)^k\binom{n}{k}(x+kh)^p$$

$$= \frac{(-1)^n}{h^n} \sum_{k=0}^{n}(-1)^k\binom{n}{k}\sum_{r=0}^{p}\binom{p}{r}(kh)^r x^{p-r}$$

$$= \frac{(-1)^n}{h^n} \sum_{r=0}^{p}\binom{p}{r}x^{p-r}h^r \sum_{k=0}^{n}(-1)^k\binom{n}{k}k^r$$

$$= \sum_{r=0}^{p}\binom{p}{r}x^{p-r}h^{r-n}n!S(r,n).$$

Equation (9.21) has far reaching consequences. For instance it provides a proof of the alternative recurrence

$$S(n,j) = \sum_{\alpha=0}^{n}(-1)^{\alpha+j}\binom{n}{\alpha}j^{n-\alpha}S(\alpha,j), \qquad (9.24)$$

since

$$j!S(n,j) = \sum_{k=0}^{j}(-1)^k\binom{j}{k}(j-k)^n = \sum_{k=0}^{j}(-1)^k\binom{j}{k}\sum_{\alpha=0}^{n}\binom{n}{\alpha}(-k)^\alpha j^{n-\alpha}$$

$$= \sum_{\alpha=0}^{n}(-1)^\alpha\binom{n}{\alpha}j^{n-\alpha}\sum_{k=0}^{j}(-1)^k\binom{j}{k}k^\alpha$$

$$= \sum_{\alpha=0}^{n}(-1)^{\alpha+j}\binom{n}{\alpha}j^{n-\alpha}j!S(\alpha,j).$$

Equation (9.21) is also used in the verification of

$$\binom{r+\alpha}{r} S(n, r+\alpha) = \sum_{k=0}^{n} \binom{n}{k} S(k, r) S(n-k, \alpha). \qquad (9.25)$$

To prove Equation (9.25) start with the right side and use Equation (9.21) to expand $S(k, r)$ and $S(n-k, \alpha)$. Then interchange the order of summation and simplify the results.

$$\sum_{k=0}^{n} \binom{n}{k} S(k, r) S(n-k, \alpha)$$

$$= \frac{1}{r!\alpha!} \sum_{k=0}^{n} \binom{n}{k} \sum_{i=0}^{r} (-1)^{r-i} \binom{r}{i} i^k \sum_{j=0}^{\alpha} (-1)^{\alpha-j} \binom{\alpha}{j} j^{n-k}$$

$$= \frac{1}{r!\alpha!} \sum_{i=0}^{r} (-1)^{r-i} \binom{r}{i} \sum_{j=0}^{\alpha} (-1)^{\alpha-j} \binom{\alpha}{j} \sum_{k=0}^{n} \binom{n}{k} i^k j^{n-k}$$

$$= \frac{1}{r!\alpha!} \sum_{i=0}^{r} (-1)^{r-i} \binom{r}{i} \sum_{j=0}^{\alpha} (-1)^{\alpha-j} \binom{\alpha}{j} (i+j)^n$$

$$= \frac{1}{r!\alpha!} \sum_{i=0}^{r} (-1)^{r-i} \binom{r}{i} \sum_{k=i}^{\alpha+i} (-1)^{\alpha-k+i} \binom{\alpha}{k-i} k^n$$

$$= \frac{1}{r!\alpha!} \sum_{i=0}^{r} (-1)^{r-i} \binom{r}{i} \sum_{k=0}^{\alpha+r} (-1)^{\alpha-k+i} \binom{\alpha}{k-i} k^n$$

$$= \frac{1}{r!\alpha!} \sum_{k=0}^{\alpha+r} (-1)^{\alpha+r-k} k^n \sum_{i=0}^{r} \binom{r}{i} \binom{\alpha}{k-i}$$

$$= \frac{1}{r!\alpha!} \sum_{k=0}^{\alpha+r} (-1)^{\alpha+r-k} k^n \sum_{i=0}^{k} \binom{r}{i} \binom{\alpha}{k-i}$$

$$= \frac{1}{r!\alpha!} \sum_{k=0}^{\alpha+r} (-1)^{\alpha+r-k} \binom{\alpha+r}{k} k^n, \qquad \text{Vandermonde convolution}$$

$$= \frac{(\alpha+r)!}{r!\alpha!} S(n, \alpha+r), \qquad \text{by Eq. (9.21).}$$

Equation (9.25), when combined with Equations (9.11) and (9.21), provides an evaluation of $\sum_{j=0}^{n} (-1)^j \binom{n}{j} (z-j)^{n+\alpha}$ where $n$ and $\alpha$ are nonnegative integers and $z$ is a complex number. In particular we have

$$\sum_{j=0}^{n}(-1)^{j}\binom{n}{j}(z-j)^{n+\alpha} = \sum_{j=0}^{n}(-1)^{j}\binom{n}{j}(z-n+n-j)^{n+\alpha}$$

$$= \sum_{j=0}^{n}(-1)^{j}\binom{n}{j}\sum_{k=0}^{n+\alpha}\binom{n+\alpha}{k}(z-n)^{k}(n-j)^{n+\alpha-k}$$

$$= \sum_{j=0}^{n}(-1)^{j}\binom{n}{j}\sum_{k=0}^{n+\alpha}\binom{n+\alpha}{k}(n-j)^{n+\alpha-k}\sum_{r=0}^{k}\binom{z-n}{r}r!S(k,r)$$

$$= \sum_{r=0}^{n+\alpha}\binom{z-n}{r}\sum_{k=r}^{n+\alpha}\binom{n+\alpha}{k}r!S(k,r)\sum_{j=0}^{n}(-1)^{j}\binom{n}{j}(n-j)^{n+\alpha-k}. \quad (*)$$

Euler's finite difference theorem implies that $\sum_{j=0}^{n}(-1)^{j}\binom{n}{j}(n-j)^{n+\alpha-k} = 0$ if $k > \alpha$. But $k > \alpha$ if and only if $r > \alpha$. Thus we may assume that $0 \le r \le \alpha$ and write Line $(*)$ as

$$\sum_{j=0}^{n}(-1)^{j}\binom{n}{j}(z-j)^{n+\alpha}$$

$$= \sum_{r=0}^{\alpha}\binom{z-n}{r}\sum_{k=r}^{n+\alpha}\binom{n+\alpha}{k}r!S(k,r)\sum_{j=0}^{n}(-1)^{j}\binom{n}{j}(n-j)^{n+\alpha-k}$$

$$= \sum_{r=0}^{\alpha}\binom{z-n}{r}\sum_{k=r}^{n+\alpha}\binom{n+\alpha}{k}r!S(k,r)\sum_{j=0}^{n}(-1)^{n-j}\binom{n}{j}j^{n+\alpha-k}$$

$$= \sum_{r=0}^{\alpha}\binom{z-n}{r}\sum_{k=r}^{n+\alpha}\binom{n+\alpha}{k}r!S(k,r)n!S(n+\alpha-k,n), \quad \text{by Eq. (9.21)}$$

$$= \sum_{r=0}^{\alpha}\binom{z-n}{r}(r+n)!S(n+\alpha,r+n), \qquad \text{by Eq. (9.25)}.$$

In summary we have shown for complex $z$ and arbitrary nonnegative integers $n$ and $\alpha$ that

$$\sum_{j=0}^{n}(-1)^{j}\binom{n}{j}(z-j)^{n+\alpha} = \sum_{r=0}^{\alpha}\binom{z-n}{r}(n+r)!S(n+\alpha,n+r). \quad (9.26)$$

If $\alpha = 0$, Equation (9.26) becomes $\sum_{j=0}^{n}(-1)^{j}\binom{n}{j}(z-j)^{n} = 1$, a special case of Euler's finite difference theorem.

Equation (9.26) is a generalization of Equation (9.21). We now consider another generalization. Define $S_n(x,r) = \sum_{k=0}^{n}(-1)^{k}\binom{x}{k}k^{r}$. We have seen some special cases of $S_n(x, r)$. If $r = 0$, we obtain Equation (1.30) which we reproduce below:

$$S_n(x, 0) = \sum_{k=0}^{n} (-1)^k \binom{x}{k} = (-1)^n \binom{x-1}{n}. \tag{9.27}$$

We claim that $S_n(x, r)$ for $r \geq 1$ satisfies

$$S_n(x, r) = -x \sum_{j=0}^{r-1} \binom{r-1}{j} S_{n-1}(x - 1, j), \tag{9.28}$$

since

$$
\begin{aligned}
S_n(x, r) &= \sum_{k=0}^{n} (-1)^k \binom{x}{k} k^r = x \sum_{k=0}^{n} (-1)^k \binom{x}{k} \frac{k}{x} k^{r-1} \\
&= x \sum_{k=1}^{n} (-1)^k \binom{x-1}{k-1} k^{r-1} \\
&= x \sum_{k=0}^{n-1} (-1)^{k+1} \binom{x-1}{k} (k+1)^{r-1} \\
&= -x \sum_{k=0}^{n-1} (-1)^k \binom{x-1}{k} \sum_{j=0}^{r-1} \binom{r-1}{j} k^j \\
&= -x \sum_{j=0}^{r-1} \binom{r-1}{j} \sum_{k=0}^{n-1} (-1)^k \binom{x-1}{k} k^j \\
&= -x \sum_{j=0}^{r-1} \binom{r-1}{j} S_{n-1}(x - 1, j).
\end{aligned}
$$

By recursively applying Equation (9.28) we deduce that

$$S_n(x, r) = \sum_{k=0}^{n} (-1)^k \binom{x}{k} k^r = (-1)^n \sum_{j=0}^{r} \binom{x-j-1}{n-j} \binom{x}{j} j! S(r, j). \tag{9.29}$$

To prove Equation (9.29) start with Equation (9.11) and observe that

$$
\begin{aligned}
S_n(x, r) &= \sum_{k=0}^{n} (-1)^k \binom{x}{k} k^r = \sum_{k=0}^{n} (-1)^k \binom{x}{k} \sum_{j=0}^{r} \binom{k}{j} j! S(r, j) \\
&= \sum_{j=0}^{r} j! S(r, j) \sum_{k=j}^{n} (-1)^k \binom{x}{k} \binom{k}{j}
\end{aligned}
$$

$$= \sum_{j=0}^{r} \binom{x}{j} j! S(r, j) \sum_{k=j}^{n} (-1)^k \binom{x-j}{k-j}$$

$$= \sum_{j=0}^{r} (-1)^j \binom{x}{j} j! S(r, j) \sum_{k=0}^{n-j} (-1)^k \binom{x-j}{k}$$

$$= (-1)^n \sum_{j=0}^{r} \binom{x}{j} j! S(r, j) \binom{x-j-1}{n-j}, \qquad \text{by Eq. (9.27)}.$$

If we take Equation (9.29), let $x \to -x - 1$ and apply the $-1$-Transformation, we get

$$\sum_{k=0}^{n} \binom{x+k}{k} k^r = \sum_{j=0}^{r} \binom{x+j}{j} \binom{x+n+1}{n-j} j! S(r, j)$$

$$= \sum_{j=0}^{r} \binom{x+j}{j} \binom{x+n+1}{n-j} \sum_{k=0}^{j} (-1)^k \binom{j}{k} (j-k)^r, \qquad (9.30)$$

where the last equality follows from Equation (9.21).

Here are some special cases of Equation (9.30):

$$\sum_{k=0}^{n} \binom{x+k}{k} = \binom{n+x+1}{n}$$

$$\sum_{k=0}^{n} \binom{n+k}{k} k^r = \sum_{j=0}^{r} \binom{n+j}{j} \binom{2n+1}{n-j} j! S(r, j)$$

$$\sum_{k=0}^{n} \binom{n+k}{k} = \binom{2n+1}{n}, \qquad \sum_{k=0}^{n} \binom{n+k}{k} k = (n+1) \binom{2n+1}{n-1}$$

$$\sum_{k=0}^{n} \binom{n+k}{k} k^2 = (n+1) \binom{2n+1}{n-1} + (n+2)(n+1) \binom{2n+1}{n-2}$$

$$\sum_{k=0}^{n} \binom{n+k}{k} k^3 = (n+1) \binom{2n+1}{n-1}$$

$$+ 3(n+2)(n+1) \binom{2n+1}{n-2} + (n+3)(n+2)(n+1) \binom{2n+1}{n-3}.$$

Because the $-1$-Transformation implies that $\binom{x-j-1}{n-j} = (-1)^{n-j} \binom{n-x}{n-j}$, Equation (9.29) often appears in the literature as

$$\sum_{k=0}^{n} (-1)^k \binom{x}{k} k^r = \sum_{j=0}^{n} (-1)^j \binom{n-x}{n-j} \binom{x}{j} j! S(r, j). \qquad (9.31)$$

Equation (9.31) implies that

$$\sum_{k=0}^{n}(-1)^k\binom{x}{k}(z+yk)^p = \sum_{r=0}^{p}\binom{p}{r}z^{p-r}y^r\sum_{k=0}^{n}(-1)^k\binom{x}{k}k^r$$

$$= \sum_{r=0}^{p}\binom{p}{r}z^{p-r}y^r\sum_{j=0}^{r}(-1)^j\binom{x}{j}\binom{n-x}{n-j}j!S(r,j), \qquad (9.32)$$

whenever $x, y, z$ are complex numbers and $p$ is a nonnegative integer.

**Remark 9.1.** The same techniques used to verify Equation (9.29) show that

$$S_r(x) = \sum_{n=0}^{\infty}\frac{n^r x^n}{n!} = e^x\sum_{j=0}^{r}x^j S(r,j), \qquad r \text{ nonnegative integer,} \quad (9.33)$$

and that

$$S_{r+1}(x) = x\sum_{j=0}^{r}\binom{r}{j}S_j(x). \qquad (9.34)$$

Since $x$ is an arbitrary complex number, we may let $x \to e^{ix} = \cos x + i\sin x$ in Equation (9.33), apply De Moivre's Theorem, and collect the real and imaginary parts to obtain

$$\sum_{k=0}^{\infty}\frac{k^r\cos kx}{k!} =$$

$$e^{\cos x}\left[\cos(\sin x)\sum_{j=0}^{r}\cos(jx)S(r,j) - \sin(\sin x)\sum_{j=0}^{r}\sin(jx)S(r,j)\right] \quad (9.35)$$

$$\sum_{k=0}^{\infty}\frac{k^r\sin kx}{k!} =$$

$$e^{\cos x}\left[\cos(\sin x)\sum_{j=0}^{r}\sin(jx)S(r,j) + \sin(\sin x)\sum_{j=0}^{r}\cos(jx)S(r,j)\right]. \quad (9.36)$$

Both Equations (9.11) and (9.33) generalize to arbitrary polynomials

$f(x) = \sum_{k=0}^{n} a_k x^k$. Equations (9.11) and (9.21) imply that

$$f(x) = \sum_{k=0}^{n} a_k x^k = \sum_{k=0}^{n} a_k \sum_{\alpha=0}^{k} \binom{x}{\alpha} \alpha! S(k, \alpha) = \sum_{\alpha=0}^{n} \binom{x}{\alpha} \alpha! \sum_{k=\alpha}^{n} a_k S(k, \alpha)$$

$$= \sum_{\alpha=0}^{n} \binom{x}{\alpha} \alpha! \sum_{k=0}^{n} a_k S(k, \alpha)$$

$$= \sum_{\alpha=0}^{n} \binom{x}{\alpha} \alpha! \sum_{k=0}^{n} a_k \frac{1}{\alpha!} \sum_{j=0}^{\alpha} (-1)^j \binom{\alpha}{j} (\alpha - j)^k$$

$$= \sum_{\alpha=0}^{n} \binom{x}{\alpha} \sum_{j=0}^{\alpha} (-1)^j \binom{\alpha}{j} \sum_{k=0}^{n} a_k (\alpha - j)^k$$

$$= \sum_{\alpha=0}^{n} \binom{x}{\alpha} \sum_{j=0}^{\alpha} (-1)^j \binom{\alpha}{j} f(\alpha - j).$$

These calculations prove the following polynomial expansion theorem:

**Theorem 9.1.** *If $f(x)$ is a polynomial of degree $n$, then*

$$f(x) = \sum_{k=0}^{n} \binom{x}{k} \sum_{j=0}^{k} (-1)^j \binom{k}{j} f(k - j). \qquad (9.37)$$

If $f(x) = x^n$, Equation (9.37) becomes Equation (9.11). Here are three more examples of Theorem 9.1:

$$\binom{x+n}{n} = \sum_{k=0}^{n} \binom{x}{k} \sum_{j=0}^{k} (-1)^j \binom{k}{j} \binom{k-j+n}{n} \qquad (9.38)$$

$$\binom{mx}{n} = \sum_{k=0}^{n} \binom{x}{k} \sum_{j=0}^{k} (-1)^j \binom{k}{j} \binom{mk - mj}{n} \qquad (9.39)$$

$$\binom{x}{r}^p = \sum_{k=0}^{rp} \binom{x}{k} \sum_{j=0}^{k} (-1)^j \binom{k}{j} \binom{k-j}{r}^p, \qquad rp = n. \quad (9.40)$$

We may also use polynomials to generalize Equation (9.33).

**Theorem 9.2.** *Let $f(x)$ be a polynomial of degree $n$. Then*

$$\sum_{k=0}^{\infty} \frac{x^k}{k!} f(k) = e^x \sum_{k=0}^{n} \frac{x^k}{k!} \sum_{j=0}^{k} (-1)^j \binom{k}{j} f(k - j). \qquad (9.41)$$

If $f(x) = x^n$, Equation (9.41) becomes Equation (9.33).

To prove Equation (9.41) let $f(x) = \sum_{r=0}^{n} a_r x^r$ and observe that

$$\sum_{k=0}^{\infty} \frac{x^k}{k!} f(k) = \sum_{k=0}^{\infty} \frac{x^k}{k!} \sum_{r=0}^{n} a_r k^r$$

$$= \sum_{r=0}^{n} a_r \sum_{k=0}^{\infty} \frac{k^r x^k}{k!} = e^x \sum_{r=0}^{n} a_r \sum_{j=0}^{r} x^j S(r,j), \qquad \text{by Eq. (9.33)}$$

$$= e^x \sum_{j=0}^{r} x^j \sum_{r=j}^{n} a_r S(r,j) = e^x \sum_{j=0}^{r} x^j \sum_{r=0}^{n} a_r S(r,j)$$

$$= e^x \sum_{j=0}^{n} \frac{x^j}{j!} \sum_{r=0}^{n} a_r \sum_{k=0}^{j} (-1)^k \binom{j}{k} (j-k)^r$$

$$= e^x \sum_{j=0}^{n} \frac{x^j}{j!} \sum_{k=0}^{j} (-1)^k \binom{j}{k} \sum_{r=0}^{n} a_r (j-k)^r$$

$$= e^x \sum_{j=0}^{n} \frac{x^j}{j!} \sum_{k=0}^{j} (-1)^k \binom{j}{k} f(j-k).$$

Let $f(x) = \binom{x+n}{n}$. Equation (9.41) combined with Equation (6.35) implies that

$$\sum_{k=0}^{\infty} \binom{n+k}{n} \frac{x^k}{k!} = e^x \sum_{k=0}^{n} \frac{x^k}{k!} \sum_{j=0}^{k} (-1)^j \binom{k}{j} \binom{k-j+n}{n}$$

$$= e^x \sum_{k=0}^{n} (-1)^k \frac{x^k}{k!} \sum_{j=0}^{k} (-1)^j \binom{k}{j} \binom{j+n}{n}$$

$$= e^x \sum_{k=0}^{n} (-1)^k \frac{x^k}{k!} (-1)^k \binom{n}{n-k} = e^x \sum_{k=0}^{n} \binom{n}{k} \frac{x^k}{k!}.$$

## 9.2 Grunert's Operational Formula

So far we have used Stirling numbers of the second kind to obtain expansions for $x^n$, $\sum_{k=0}^{n} (-1)^k \binom{x}{k} k^r$, and $\sum_{n=0}^{\infty} \frac{n^r x^n}{n!}$. We now use them to obtain an expansion for the formal power series $S_n(x) = \sum_{k=0}^{\infty} k^n x^k$. To derive this expansion we need

$$\left( x \frac{d}{dx} \right)^r S = \sum_{j=0}^{r} x^j S(r,j) \frac{d^j S}{dx^j}. \qquad (9.42)$$

Equation (9.42) is known as Grunert's formula [Gould, 1987, p. 89]. Equation (9.42) is proven via induction on $r$. In order to expand $S_n(x)$ observe that

$$\left(x\frac{d}{dx}\right)S_n(x) = \sum_{k=0}^{\infty} k^{n+1}x^k = S_{n+1}(x), \qquad (9.43)$$

and that

$$\left(x\frac{d}{dx}\right)^2 S_n(x) = x\frac{d}{dx}\sum_{k=0}^{\infty} k^{n+1}x^k = \sum_{k=0}^{\infty} k^{n+2}x^k = S_{n+2}(x). \qquad (9.44)$$

By induction on $r$, we see that

$$\left(x\frac{d}{dx}\right)^r S_n(x) = \sum_{k=0}^{\infty} k^{n+r}x^k = S_{n+r}(x). \qquad (9.45)$$

Equation (9.45) implies that

$$S_r(x) = \left(x\frac{d}{dx}\right)^r S_0(x) = \left(x\frac{d}{dx}\right)^r \frac{1}{1-x}, \qquad |x| < 1. \qquad (9.46)$$

Grunert's formula when applied to Equation (9.46) implies that

$$\left(x\frac{d}{dx}\right)^r \frac{1}{1-x} = \sum_{j=0}^{r} S(r,j)x^j \frac{d^j}{dx^j}\left(\frac{1}{1-x}\right)$$

$$= \sum_{j=0}^{r} \frac{(-1)^j}{j!} \sum_{k=0}^{j}(-1)^k \binom{j}{k}k^r x^j \frac{d^j}{dx^j}\left(\frac{1}{1-x}\right), \text{ by Eq. (9.21)}$$

$$= \sum_{j=0}^{r}(-1)^j \sum_{k=0}^{j}(-1)^k \binom{j}{k}k^r \frac{x^j}{(1-x)^{j+1}}.$$

In summary we have shown that for $n \geq 0$ and $|x| < 1$ that

$$\sum_{k=0}^{\infty} k^n x^k = \sum_{j=0}^{n} j!S(n,j)\frac{x^j}{(1-x)^{j+1}}$$

$$= \sum_{j=0}^{n}(-1)^j \sum_{k=0}^{j}(-1)^k \binom{j}{k}k^n \frac{x^j}{(1-x)^{j+1}}$$

$$= \sum_{j=0}^{n}\sum_{k=0}^{j}(-1)^k \binom{j}{k}(j-k)^n \frac{x^j}{(1-x)^{j+1}}. \qquad (9.47)$$

We may also use Grunert's formula to evaluate $\mathcal{S}_{n,r}(x) = \sum_{k=0}^{n} k^r x^k$. Induction on $r$ implies that

$$\mathcal{S}_{n,r}(x) = \left(x\frac{d}{dx}\right)^r \mathcal{S}_{n,0}(x) = \left(x\frac{d}{dx}\right)^r \sum_{k=0}^{n} x^k$$

$$= \left(x\frac{d}{dx}\right)^r \left(\frac{x^{n+1}-1}{x-1}\right). \tag{9.48}$$

We then use Grunert's formula to evaluate the third sum in Equation (9.48):

$$\mathcal{S}_{n,r}(x) = \sum_{j=0}^{r} S(r,j)x^j \sum_{k=0}^{n} \frac{d^j}{dx^j}(x^k)$$

$$= \sum_{j=0}^{r} x^j \frac{(-1)^j}{j!} \sum_{\alpha=0}^{j} (-1)^\alpha \binom{j}{\alpha} \alpha^r \sum_{k=0}^{n} \frac{d^j}{dx^j}(x^k)$$

$$= \sum_{j=0}^{r} x^j \frac{(-1)^j}{j!} \sum_{\alpha=0}^{j} (-1)^\alpha \binom{j}{\alpha} \alpha^r \sum_{k=0}^{n} \frac{k!}{(k-j)!} x^{k-j}.$$

This result is efficiently written as

$$\sum_{k=0}^{n} k^r x^r = \sum_{j=0}^{r} (-1)^j \sum_{\alpha=0}^{j} (-1)^\alpha \binom{j}{\alpha} \alpha^r \sum_{k=0}^{n} \binom{k}{j} x^k$$

$$= \sum_{j=0}^{r} j! S(r,j) \sum_{k=0}^{n} \binom{k}{j} x^k. \tag{9.49}$$

If $x = 1$, Equation (9.49) becomes

$$\sum_{k=0}^{n} k^r = \sum_{j=0}^{r} j! S(r,j) \sum_{k=0}^{n} \binom{k}{j} = \sum_{j=0}^{r} j! S(r,j) \binom{n+1}{j+1}, \tag{9.50}$$

where the last equality follows from the hockey stick identity [Gould, 1987, Identity (1.52)].

We can take Equation (9.50) and apply Equation (9.18) to obtain

$$\sum_{k=0}^{n} k^r = \sum_{\alpha=0}^{r} \binom{n+1}{\alpha+1} \sum_{j=\alpha}^{r} (-1)^{r-j} \binom{j-1}{j-\alpha} j! S(r,j)$$

$$= \sum_{j=0}^{r} (-1)^{r-j} j! S(r,j) \sum_{\alpha=0}^{j} \binom{n+1}{\alpha+1}\binom{j-1}{j-\alpha}$$

$$= \sum_{j=0}^{r} (-1)^{r-j} j! S(r,j) \sum_{\alpha=1}^{j+1} \binom{n+1}{\alpha}\binom{j-1}{j+1-\alpha}.$$

We would like to extend the range of summation to include $\alpha = 0$.

$$\sum_{j=0}^{r}(-1)^{r-j}j!S(r,j)\sum_{\alpha=0}^{j+1}\binom{n+1}{\alpha}\binom{j-1}{j+1-\alpha}$$

$$=\sum_{j=0}^{r}(-1)^{r-j}j!S(r,j)\left[\binom{j-1}{j+1}+\sum_{\alpha=1}^{j+1}\binom{n+1}{\alpha}\binom{j-1}{j+1-\alpha}\right].$$

If $j > 0$, $\binom{j-1}{j+1} = 0$. If $j = 0$, the first term in the previous line becomes $(-1)^{r}S(r,0)\binom{n+1}{1}$ and this vanishes whenever $r > 0$ since $S(r,0) = 0$. Therefore, if $r \geq 1$

$$\sum_{k=0}^{n}k^{r}=\sum_{j=0}^{r}(-1)^{r-j}j!S(r,j)\sum_{\alpha=0}^{j+1}\binom{n+1}{\alpha}\binom{j-1}{j+1-\alpha}$$

$$=\sum_{j=0}^{r}(-1)^{r-j}j!S(r,j)\binom{n+j}{j+1},$$

and we have proven the transformation

$$\sum_{j=0}^{p}\binom{n+1}{j+1}j!S(p,j)=\sum_{j=0}^{p}(-1)^{p-j}\binom{n+j}{j+1}j!S(p,j),\qquad p\geq 1,\ n\geq 0.$$

$$(9.51)$$

If $x = -1$, Equation (9.49) implies that

$$\sum_{k=0}^{n}(-1)^{k}k^{r}=\sum_{j=0}^{r}j!S(r,j)\sum_{k=j}^{n}(-1)^{k}\binom{k}{j}=\left(x\frac{d}{dx}\right)^{r}\sum_{k=0}^{n}x^{k}\Big|_{x=-1}$$

$$=\left(x\frac{d}{dx}\right)^{r}\left(\frac{x^{n+1}-1}{x-1}\right)\Big|_{x=-1},\qquad(9.52)$$

and there is no simple closed form for $\sum_{k=j}^{n}(-1)^{k}\binom{k}{j}$. We list five specific instances of Equation (9.52):

$$\sum_{k=0}^{n}(-1)^{k}=\frac{1+(-1)^{n}}{2}=\left\lfloor\frac{n}{2}\right\rfloor-\left\lfloor\frac{n-1}{2}\right\rfloor$$

$$\sum_{k=0}^{n}(-1)^{k}k=\frac{(-1)^{n}(2n+1)-1}{4}=(-1)^{n}\left\lfloor\frac{n+1}{2}\right\rfloor$$

$$\sum_{k=0}^{n}(-1)^{k}k^{2}=\frac{(-1)^{n}(n^{2}+n)}{2}$$

$$\sum_{k=0}^{n}(-1)^k k^3 = \frac{(-1)^n(4n^3 + 6n^2 - 1) + 1}{8} = \frac{(-1)^n(2n^3 + 3n^2)}{4} + \frac{1 - (-1)^n}{8}$$

$$\sum_{k=0}^{n}(-1)^k k^4 = \frac{(-1)^n(n^4 + 2n^3 - n)}{2}.$$

By using the bisection formula, we can derive an alternate expansion for $\sum_{k=0}^{n}(-1)^k k^p$ whenever $p$ is a nonnegative integer. The bisection formula implies that $\sum_{k=0}^{n} f(k) + \sum_{k=0}^{n}(-1)^k f(k) = 2\sum_{k=0}^{\lfloor \frac{n}{2} \rfloor} f(2k)$. Set $f(k) = k^p$ to obtain

$$\sum_{k=0}^{n} k^p + \sum_{k=0}^{n}(-1)^k k^p = 2\sum_{k=0}^{\lfloor \frac{n}{2} \rfloor}(2k)^p = 2^{p+1}\sum_{k=0}^{\lfloor \frac{n}{2} \rfloor} k^p. \qquad (9.53)$$

Equation (9.50) shows that $\sum_{k=0}^{n} k^p = \sum_{j=0}^{p}\binom{n+1}{j+1} j! S(p,j)$. Put this result into Equation (9.53) to obtain

$$\sum_{k=0}^{n}(-1)^{k+1} k^p = \sum_{k=0}^{n} k^p - 2^{p+1}\sum_{k=0}^{\lfloor \frac{n}{2} \rfloor} k^p$$

$$= \sum_{j=0}^{p}\left[\binom{n+1}{j+1} - 2^{p+1}\binom{\lfloor \frac{n}{2} \rfloor + 1}{j+1}\right] j! S(p,j). \qquad (9.54)$$

Grunert's formula also provides an evaluation of $\mathcal{S}_n(x,r) = \sum_{k=0}^{n}\binom{n}{k} k^r x^k$. If $x = -1$, $\mathcal{S}_n(-1,r) = \sum_{k=0}^{n}(-1)^k\binom{n}{k} k^r$ is evaluated by either Euler's finite difference theorem or by Equation (9.21). By induction on $p$ we readily show that

$$\left(x\frac{d}{dx}\right)^p \mathcal{S}_n(x,r) = \mathcal{S}_n(x, p+r). \qquad (9.55)$$

From Equation (9.55) we deduce that

$$\mathcal{S}_n(x,r) = \left(x\frac{d}{dx}\right)^r \mathcal{S}_n(x,0)$$

$$= \left(x\frac{d}{dx}\right)^r \sum_{k=0}^{n}\binom{n}{k} x^k = \left(x\frac{d}{dx}\right)^r (1+x)^n. \qquad (9.56)$$

Apply Grunert's formula to the right expression in Equation (9.56) to obtain

$$\left(x\frac{d}{dx}\right)^r(1+x)^n = \sum_{j=0}^r S(r,j)x^j\frac{d^j}{dx^j}(1+x)^n$$

$$= \sum_{j=0}^r S(r,j)x^j\frac{n!}{(n-j)!}(1+x)^{n-j}$$

$$= \sum_{j=0}^r(-1)^j\sum_{k=0}^j(-1)^k\binom{j}{k}k^r\binom{n}{j}x^j(1+x)^{n-j}.$$

We summarize the results of these calculations as

$$\sum_{k=0}^n\binom{n}{k}k^r x^k = (1+x)^n\sum_{j=0}^r j!\binom{n}{j}S(r,j)\frac{x^j}{(1+x)^j}$$

$$= (1+x)^n\sum_{j=0}^r(-1)^j\binom{n}{j}\frac{x^j}{(1+x)^j}\sum_{k=0}^j(-1)^k\binom{j}{k}k^r. \qquad (9.57)$$

If $x=1$, Equation (9.57) becomes

$$\sum_{k=0}^n\binom{n}{k}k^r = 2^n\sum_{j=0}^r j!\binom{n}{j}S(r,j)\frac{1}{2^j}$$

$$= 2^n\sum_{j=0}^r(-1)^j\binom{n}{j}\frac{1}{2^j}\sum_{k=0}^j(-1)^k\binom{j}{k}k^r. \qquad (9.58)$$

## 9.3  Expansions of $\left(\frac{e^x-1}{x}\right)^n$

We now turn our attention to an expansion that provides a generating function representing Stirling numbers of the second kind. It begins with a basic expansion of $(e^x-1)^n$. The binomial theorem implies that

$$(e^x-1)^n = \sum_{k=0}^n(-1)^{n-k}\binom{n}{k}e^{kx} = \sum_{k=0}^n(-1)^{n-k}\binom{n}{k}\sum_{j=0}^\infty\frac{(kx)^j}{j!}$$

$$= \sum_{j=0}^\infty\frac{x^j}{j!}\sum_{k=0}^n(-1)^{n-k}\binom{n}{k}k^j = \sum_{j=0}^\infty\frac{x^j}{j!}n!S(j,n)$$

$$= \sum_{j=n}^\infty\frac{x^j}{j!}n!S(j,n) = \sum_{j=0}^\infty\frac{x^{j+n}}{(j+n)!}n!S(j+n,n).$$

By rewriting the final line we discover that

$$\left(\frac{e^x-1}{x}\right)^n = \sum_{j=0}^{\infty} \frac{x^j}{(j+n)!} n! S(j+n,n), \qquad (9.59)$$

or equivalently that

$$D_x^n \left(\frac{e^x-1}{x}\right)^j \Bigg|_{x=0} = \frac{1}{\binom{n+j}{j}} S(n+j,j). \qquad (9.60)$$

We may also use Equation (9.59) to compute derivatives of the reciprocal function $\left(\frac{x}{e^x-1}\right)^n$. Recall Equation (8.24), namely

$$D_x^n z^{-j} = \sum_{j=0}^{n} (-1)^j \binom{\alpha+j-1}{j} \binom{\alpha+n}{n-j} \frac{1}{z^{\alpha+j}} D_x^n z^j. \qquad (9.61)$$

Since

$$\binom{\alpha+n}{n-j}\binom{\alpha+j-1}{j} = \frac{\alpha}{\alpha+j}\binom{\alpha+n}{n}\binom{n}{j},$$

we may rewrite Equation (9.61) as

$$D_x^n z^{-\alpha} = \alpha\binom{\alpha+n}{n} \sum_{j=0}^{n} (-1)^j \binom{n}{j} \frac{1}{\alpha+j} \frac{1}{z^{\alpha+j}} D_x^n z^j. \qquad (9.62)$$

Take Equation (9.62) and let $z = \frac{e^x-1}{x}$ to discover that

$$D_x^n \left(\frac{x}{e^x-1}\right)^{\alpha} \Bigg|_{x=0}$$

$$= \alpha\binom{\alpha+n}{n} \sum_{j=0}^{n} (-1)^j \binom{n}{j} \frac{1}{\alpha+j} \left(\frac{x}{e^x-1}\right)^{\alpha+j} D_x^n \left(\frac{e^x-1}{x}\right)^j \Bigg|_{x=0}.$$

Since $\lim_{x\to 0} \left(\frac{x}{e^x-1}\right)^{\alpha+j} = 1$, we may rewrite the previous line as

$$D_x^n \left(\frac{x}{e^x-1}\right)^{\alpha} \Bigg|_{x=0} = \alpha\binom{\alpha+n}{n} \sum_{j=0}^{n} (-1)^j \binom{n}{j} \frac{1}{\alpha+j} D_x^n \left(\frac{e^x-1}{x}\right)^j \Bigg|_{x=0}$$

$$= \alpha\binom{\alpha+n}{n} \sum_{j=0}^{n} (-1)^j \binom{n}{j} \frac{1}{(\alpha+j)\binom{n+j}{j}} S(n+j,j),$$

where the final equality follows from Equation (9.60).

## 9.4 Bell Numbers

We now investigate the sequence associated with the row sums of Table 9.1. This sum, denoted $B(n)$, is called the $n^{th}$ Bell number in honor of E. T. Bell [Gould and Glatzer, 1979] . Table 9.2 lists $\{B(n)\}_{n=0}^{10}$.

| $n$ | 0 | 1 | 2 | 3 | 4 | 5 | 6 | 7 | 8 | 9 | 10 |
|------|---|---|---|---|----|----|-----|-----|------|-------|--------|
| $B(n)$ | 1 | 1 | 2 | 5 | 15 | 52 | 203 | 877 | 4140 | 21147 | 115975 |

Table 9.2: The first eleven Bell numbers

Since $B(n) = \sum_{k=0}^{n} S(n,k)$, we may interpret $B(n)$ as the number of set partitions of $[n] = \{1, 2, \ldots, n\}$. By carefully applying this combinatorial interpretation we can show that

$$B(n+1) = \sum_{k=0}^{n} \binom{n}{k} B(n-k). \tag{9.63}$$

The left side of (9.63) counts the set partitions of $[n+1] = \{1, 2, \ldots, n+1\}$. We claim the right side sum also counts the set partitions of $[n+1]$. Let $P$ be a set partition of $[n+1]$ . One subset of $P$ *must* contain $n+1$. Fix $k$, where $0 \le k \le n$, and let $k$ count the number of elements from $[n]$ which are in part containing $n+1$. There are $\binom{n}{k}$ ways to select $k$ elements from $[n]$. To finish constructing $P$ we observe that the remaining $n-k$ elements of $[n]$ must form a set partition for a set of size $n-k$. Since there are $B(n-k)$ ways to form such set partitions, the rule of products implies there are $\binom{n}{k} B(n-k)$ possible ways to construct $P$. By varying $k$ for $0 \le k \le n$, we obtain all set partitions of $[n+1]$.

Equation (9.63) often appears in the literature as

$$B(n+1) = \sum_{k=0}^{n} \binom{n}{k} B(k). \tag{9.64}$$

Equation (9.64) is easily inverted through

**Theorem 9.3.**

$$f(n) = \sum_{k=0}^{n} (-1)^{n-k} \binom{n}{k} g(k) \text{ if and only if } g(n) = \sum_{k=0}^{n} \binom{n}{k} f(k). \tag{9.65}$$

Applying Theorem 9.3 to Equation (9.64) gives the identity

$$\sum_{k=0}^{n} (-1)^{n-k} \binom{n}{k} B(k+1) = \Delta_{j,1}^{n} B(j)|_{j=1} = B(n). \tag{9.66}$$

Clearly $\sum_{k=0}^{n}(-1)^{n-k}\binom{n}{k}B(k+1) = \sum_{k=0}^{n}(-1)^{k}\binom{n}{k}B(n-k+1)$. The principle of inclusion/exclusion [Brualdi, 2014, Sec. 6.1] provides a combinatorial interpretation for $\sum_{k=0}^{n}(-1)^{k}\binom{n}{k}B(n-k+1)$. Let $1 \leq i \leq n$. A set partition of $[n+1]$ has property $P_{i,i+1}$ if it contains the part which includes $\{i, i+1\}$. Then $\sum_{k=0}^{n}(-1)^{k}\binom{n}{k}B(n-k+1)$ counts the number of set partitions of $[n+1]$ which have none of the properties. Define $\mathcal{B}_1(n)$ to be the set partitions of $[n+1]$ such that none of the subsets contain elements whose difference is one. By definition $|\mathcal{B}_1(n)| = B_1(n)$. Equation (9.66) implies that $B_1(n) = B(n)$. For example $B_1(3) = B(3) = 5$ since

$$\mathcal{B}_1(3) =$$

$$\{\{\{1,3\},\{2\},\{4\}\},\{\{1,4\},\{2\},\{3\}\},\{\{1\},\{2,4\},\{3\}\}\}$$

$$\cup \{\{\{1,3\},\{2,4\}\},\{\{1\},\{2\},\{3\},\{4\}\}\}\},$$

while the set partitions of $[3]$ are

$$\{\{\{1,2,3\}\},\{\{1,2\},\{3\}\},\{\{1,3\},\{2\}\},\{\{1\},\{2,3\}\},\{\{1\},\{2\},\{3\}\}\}.$$

To describe the bijection between $\mathcal{B}_1(n)$ and the set partitions of $[n]$ we need some additional notation [Zeilberger (n.d.)]. Let $S_1(n,k)$ count the number of set partitions of $[n]$ with $k$ subsets such that no subset has two elements whose difference is one. An adjustment of the proof of Equation (9.1) implies that

$$S_1(n,k) = S_1(n-1,k-1) + (k-1)S_1(n-1,k). \qquad (9.67)$$

The first summand on the right side corresponds to adjoining $\{\{n\}\}$ while the second summand corresponds to placing $n$ in one of the previous $k$ subsets. The factor of $k-1$ reflects the fact that $n$ cannot be placed in the subset which contains $n-1$. Equation (9.1) with $n \to n-2$ and $k \to k-1$ becomes

$$S(n-1,k-1) = S(n-2,k-2) + (k-1)S(n-2,k-1). \qquad (9.68)$$

Both Equations (9.67) and (9.68) have the structure $A(n,k) = A(n-1,k-1) + (k-1)A(n-1,k)$. Thus we deduce that $S_1(n,k)$ and $S(n-1,k-1)$ obey the same recurrence. Since $S_1(1,1) = S(0,0) = 1$ and $S_1(2,1) = S(1,0) = 0$, we conclude that $S_1(n,k) = S(n-1,k-1)$.

We are now ready to describe the bijection between $S_1(n+1,k+1)$ and $S(n,k)$. Take a set partition of $[n]$ with $k$ subsets. Let $1 \leq i \leq n-1$. At Step $i$ we remove $n-i+1$ and keep track of which subset we removed it from via the largest remaining element in that subset. We do this until we reach

$\{\{1\}\}$. Next send $\{\{1\}\}$ to $\{\{1\}, \{2\}\}$ and work backwards through the steps inserting $n - i + 2$ into the subset which contains the largest element associated with $n - i + 1$. The result will be a set partition of $n + 1$ with $k + 1$ subsets, none of which contain adjacent elements. To best understand this bijection we demonstrate why $S_1(5,3) = S(4,2)$ in Figure 9.1.

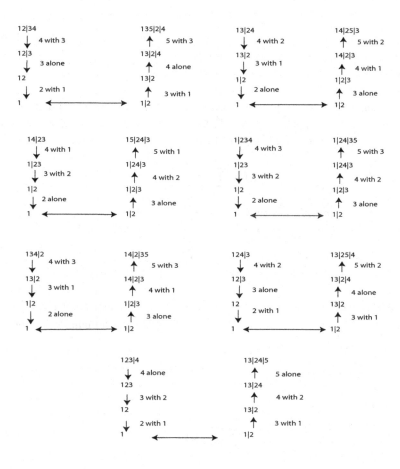

Figure 9.1: The bijection between $S_1(5,3) = S(4,2)$

The bijection sends

$$\{\{1,2\},\{3,4\}\} \text{ to } \{\{1,3,5\},\{2\},\{4\}\},$$
$$\{\{1,3\},\{2,4\}\} \text{ to } \{\{1,4\},\{2,5\},\{3\}\},$$
$$\{\{1,4\},\{2,3\}\} \text{ to } \{\{1,5\},\{2,4\},\{3\}\},$$
$$\{\{1\},\{2,3,4\}\} \text{ to } \{\{1\},\{2,4\},\{3,5\}\},$$
$$\{\{1,3,4\},\{2\}\} \text{ to } \{\{1,4\},\{2\},\{3,5\}\},$$
$$\{\{1,2,4\},\{3\}\} \text{ to } \{\{1,3\},\{2,5\},\{4\}\}.$$

We leave it to the reader to rigorously prove this algorithm maps each set partition of $[n]$ with $k$ parts to a unique set partition of $[n+1]$ with $k+1$ parts, none of which contains adjacent elements, and vice versa.

Equation (9.64) also provides an exponential generating function for $\{B(n)\}_{n=0}^{\infty}$. Let $f(x) = \sum_{n=0}^{\infty} \frac{x^n}{n!} B(n)$. For our calculations it suffices to consider $f(x)$ in the context of formal power series. Then

$$f'(x) = \sum_{n=1}^{\infty} \frac{n x^{n-1}}{n!} B(n) = \sum_{n=1}^{\infty} \frac{x^{n-1}}{(n-1)!} B(n) = \sum_{n=0}^{\infty} \frac{x^n}{n!} B(n+1)$$

$$= \sum_{n=0}^{\infty} \frac{x^n}{n!} \sum_{k=0}^{n} \binom{n}{k} B(k) = \sum_{k=0}^{\infty} \frac{B(k)}{k!} \sum_{n=k}^{\infty} \frac{x^n}{(n-k)!}$$

$$= \sum_{k=0}^{\infty} \frac{x^k}{k!} B(k) \sum_{n=0}^{\infty} \frac{x^n}{n!} = e^x \sum_{k=0}^{\infty} \frac{x^k}{k!} B(k) = e^x f(x).$$

These calculations imply that $\frac{f'(t)}{f(t)} = e^t$, or equivalently that $D_t \ln f(t) = e^t$. Since $f(0) = B(0) = 1$, $\ln f(0) = 0$. The fundamental theorem of calculus implies that

$$\int_0^x D_t \ln f(t)\, dt = \ln f(x) - \ln f(0) = \ln f(x) = \int_0^x e^t\, dt = e^x - 1.$$

Therefore

$$f(x) = \sum_{n=0}^{\infty} \frac{x^n}{n!} B(n) = e^{e^x - 1}. \tag{9.69}$$

We mention that Equation (9.69) is a consequence of Equation (9.59). When deriving Equation (9.59) we showed that

$$\frac{(e^x - 1)^k}{k!} = \sum_{n=0}^{\infty} \frac{x^n}{n!} S(n,k). \tag{9.70}$$

Since $B(n) = \sum_{k=0}^{n} S(n,k)$, we have

$$f(x) = \sum_{n=0}^{\infty} \frac{x^n}{n!} B(n) = \sum_{n=0}^{\infty} \frac{x^n}{n!} \sum_{k=0}^{n} S(n,k) = \sum_{n=0}^{\infty} \frac{x^n}{n!} \sum_{k=0}^{\infty} S(n,k)$$

$$= \sum_{k=0}^{\infty} \sum_{n=0}^{\infty} \frac{x^n}{n!} S(n,k) = \sum_{k=0}^{\infty} \frac{(e^x - 1)^k}{k!} = e^{e^x - 1}.$$

We may also use Equation (9.33) with $x = 1$ to obtain

$$B(n) = \frac{1}{e} \sum_{k=0}^{\infty} \frac{k^n}{k!}, \qquad n \geq 1. \tag{9.71}$$

Equation (9.71) is known as Dobinski's formula [Dobinski,1877]. Dobinski derived Equation (9.33) and analyzed the special case of $x = -1$, namely

$$\sum_{k=0}^{n} (-1)^k S(n,k) = e \sum_{k=0}^{\infty} (-1)^k \frac{k^n}{k!}, \qquad n \geq 1. \tag{9.72}$$

Define $d_0 = 1$ and $d_n = \sum_{k=0}^{n} (-1)^k S(n,k)$ for $n \geq 1$. The sequence $\{d_n\}_{n=0}^{\infty} = \{1, -1, 0, 1, 1, -2, -9, -9, 50, 267, 413, \ldots\}$ is known as the Rao-Uppuluri Carpenter sequence [Rao Uppuluri and Carpenter, 1969] and is Sequence A00587 in [*The Online Encyclopedia of Integer Sequences*, 2015]. The techniques used to derive the exponential generating function associated with the Bell numbers yield an exponential generating function for the Rao-Uppuluri sequence. In particular we have

$$\sum_{n=0}^{\infty} \frac{x^n}{n!} d_n = \sum_{n=0}^{\infty} \frac{x^n}{n!} \sum_{k=0}^{n} (-1)^k S(n,k) = \sum_{k=0}^{\infty} (-1)^k \sum_{n=0}^{\infty} \frac{x^n}{n!} S(n,k)$$

$$= \sum_{k=0}^{\infty} \frac{(1 - e^x)^k}{k!} = e^{1 - e^x} = \frac{1}{e^{e^x - 1}}. \tag{9.73}$$

Thus the generating function for the Rao-Uppuluri sequence is reciprocal to the generating function of the Bell numbers. For this reason $\{d_n\}_{n=0}^{\infty}$ are also known as the complementary Bell numbers.

To discover a recurrence formula for $\{d_n\}_{n=0}^{\infty}$ apply the $D_x$ operator to Equation (9.73). On one hand we have

$$D_x \sum_{n=0}^{\infty} d_n \frac{x^n}{n!} = \sum_{n=1}^{\infty} \frac{x^{n-1}}{(n-1)!} d_n = \sum_{n=0}^{\infty} \frac{x^n}{n!} d_{n+1}. \tag{9.74}$$

Since $\sum_{n=0}^{\infty} \frac{x^n}{n!} d_n = e^{1-e^x}$, we also know that

$$D_x \sum_{n=0}^{\infty} d_n \frac{x^n}{n!} = D_x e^{1-e^x} = -e^x e^{1-e^x}$$

$$= -\sum_{n=0}^{\infty} d_n \frac{x^n}{n!} \sum_{k=0}^{\infty} \frac{x^k}{k!}$$

$$= -\sum_{n=0}^{\infty} x^n \sum_{k=0}^{n} \frac{d_k}{k!} \frac{1}{(n-k)!}$$

$$= -\sum_{n=0}^{\infty} \frac{x^n}{n!} \sum_{k=0}^{n} \binom{n}{k} d_k. \qquad (9.75)$$

Comparing the coefficients of $x^n$ in Equations (9.74) and (9.75) yields

$$-d_{n+1} = \sum_{k=0}^{n} \binom{n}{k} d_k. \qquad (9.76)$$

Equation (9.76) corresponds to (9.64). Apply Theorem 9.3 to Recurrence (9.76) to obtain

$$(-1)^{n+1} d_n = \sum_{j=0}^{n} (-1)^j \binom{n}{j} d_{j+1}, \qquad (9.77)$$

which is the complement to Equation (9.66).

We should mention that both the generating function for the Bell numbers and the Rao-Uppuluri numbers can be directly obtained from Equation (9.33). Define $G_n(x) = \sum_{k=0}^{n} S(n,k) x^k$. Equation (9.33) may be rewritten as

$$G_n(x) = e^{-x} \sum_{k=0}^{\infty} \frac{x^k k^n}{k!}. \qquad (9.78)$$

Now

$$\sum_{n=0}^{\infty} \frac{t^n}{n!} G_n(x) = e^{-x} \sum_{n=0}^{\infty} \frac{t^n}{n!} \sum_{k=0}^{\infty} \frac{x^k k^n}{k!} = e^{-x} \sum_{k=0}^{\infty} \frac{x^k}{k!} \sum_{n=0}^{\infty} \frac{(tk)^n}{n!}$$

$$= e^{-x} \sum_{k=0}^{\infty} \frac{x^k}{k!} e^{tk} = e^{-x} e^{xe^t}.$$

If $x = 1$, the preceding line becomes Equation (9.69), while if $x = -1$ it becomes Equation (9.73).

# Chapter 10

# Eulerian Numbers

Equation (9.11) for Stirling numbers of the second kind states that $x^n = \sum_{k=0}^{n} \binom{x}{k} k! S(n, k)$. We called this equation the basis definition for $S(n, k)$ since $\left\{ \binom{x}{k} \right\}_{k=0}^{n}$ is a basis for the vector space $\mathcal{P}_n(x)$, the set of polynomials in $x$ of degree less than or equal to $n$. There are many basis for $\mathcal{P}_n(x)$. An obvious basis is $\{x^m\}_{m=0}^{n}$. Another basis is $\mathcal{B} \left\{ \binom{x+k-1}{n} \right\}_{k=0}^{n}$. Each element of $\mathcal{B}$ is itself a polynomial of degree $n$ in $x$. This is the basis Euler used when defining $x^n$ as

$$x^n = \sum_{k=0}^{n} \binom{x+k-1}{n} A_{n,k}, \tag{10.1}$$

where $\{A_{n,k}\}_{k=0}^{n}$ is a collection of nonnegative integers known as **Eulerian numbers** [Carlitz, 1959]. We require that $A_{n,k} = 0$ if $n$ is negative or $k > n$.

There is a relationship between Eulerian numbers and Stirling numbers of the second kind. Take Equation (10.1) and apply the Vandermonde convolution:

$$x^n = \sum_{j=0}^{n} \binom{x+j-1}{n} A_{n,j} = \sum_{j=0}^{n} \sum_{\alpha=0}^{n} \binom{x}{\alpha} \binom{j-1}{n-\alpha} A_{n,j}$$

$$= \sum_{\alpha=0}^{n} \binom{x}{\alpha} \sum_{j=0}^{n} \binom{j-1}{n-\alpha} A_{n,j}.$$

Since $x^n = \sum_{\alpha=0}^{n} \binom{x}{\alpha} \alpha! S(n, \alpha)$, comparing the coefficients of $\binom{x}{\alpha}$ implies that

$$\frac{1}{\alpha!} \sum_{j=0}^{n} \binom{j-1}{n-\alpha} A_{n,j} = S(n, \alpha). \tag{10.2}$$

Equation (10.1) also provides a recurrence for $A_{n+1,j}$. By definition $x^{n+1} = \sum_{j=0}^{n+1} \binom{x+j-1}{n+1} A_{n+1,j}$. However

$$x^{n+1} = x \cdot x^n = x \sum_{j=0}^{n} \binom{x+j-1}{n} A_{n,j}$$

$$= \sum_{j=0}^{n} \left[ \frac{(x+j)(n+1-j)}{n+1} \binom{x+j-1}{n} + \frac{x+j-1-n}{n+1} \binom{x+j-1}{n} j \right] A_{n,j}$$

$$= \sum_{j=0}^{n} \binom{x+j}{n+1} (n+1-j) A_{n,j} + \sum_{j=0}^{n} \binom{x+j-1}{n+1} j A_{n,j}$$

$$= \sum_{j=0}^{n+1} [(n+2-j) A_{n,j-1} + j A_{n,j}].$$

Thus we have shown that

$$A_{n+1,j} = (n+2-j) A_{n,j-1} + j A_{n,j}, \qquad \text{where } x^n = \sum_{k=0}^{n} \binom{x+k-1}{n} A_{n,k}.$$
$$\tag{10.3}$$

By applying Equation (10.3) we will show that

$$A_{n,j} = \sum_{k=0}^{j} (-1)^k \binom{n+1}{k} (j-k)^n. \tag{10.4}$$

Define $\mathcal{A}_{n,j} = \sum_{k=0}^{j} (-1)^k \binom{n+1}{k} (j-k)^n$. Observe that $\mathcal{A}_{n,j} = 0$ if $j > n$. We claim that $\mathcal{A}_{n+1,j} = (n+2-j) \mathcal{A}_{n,j-1} + j \mathcal{A}_{n,j}$ since

$$\mathcal{A}_{n+1,j} = \sum_{k=0}^{j} (-1)^k \binom{n+2}{k} (j-k)^{n+1}$$

$$= \sum_{k=0}^{j} (-1)^k (j-k)^n \left[ -k \binom{n+2}{k} + j \binom{n+2}{k} \right]$$

$$= \sum_{k=0}^{j} (-1)^k (j-k)^n \left[ -(n+2) \binom{n+1}{k-1} + j \binom{n+2}{k} \right]$$

$$= \sum_{k=0}^{j} (-1)^k (j-k)^n \left[ -(n+2) \binom{n+1}{k-1} + j \left[ \binom{n+1}{k-1} + \binom{n+1}{k} \right] \right]$$

$$= \sum_{k=0}^{j} (-1)^k (j-k)^n \left[ -(n+2-j) \binom{n+1}{k-1} + j \binom{n+1}{k} \right]$$

$$= (n+2-j) \sum_{k=1}^{j} (-1)^{k-1} \binom{n+1}{k-1} (j-k)^n$$

$$+ j \sum_{k=0}^{j} (-1)^k \binom{n+1}{k} (j-k)^n$$

$$= (n+2-j) \sum_{k=0}^{j-1} (-1)^k \binom{n+1}{k} (j-1-k)^n$$

$$+ j \sum_{k=0}^{j} (-1)^k \binom{n+1}{k} (j-k)^n$$

$$= (n+2-j)\mathcal{A}_{n,j-1} + j\mathcal{A}_{n,j}.$$

Since $A_{0,0} = 1 = \mathcal{A}_{0,0}$ and both sequences obey the same recurrence relation, we conclude that $A_{n,k} = \mathcal{A}_{n,k}$.

Equation (10.4) is important for its computational aspects and is the basis for the data of Table 10.1.

| | $k=0$ | $k=1$ | $k=2$ | $k=3$ | $k=4$ | $k=5$ | $k=6$ | $k=7$ |
|---|---|---|---|---|---|---|---|---|
| $n=0$ | 1 | | | | | | | |
| $n=1$ | | 1 | | | | | | |
| $n=2$ | | 1 | 1 | | | | | |
| $n=3$ | | 1 | 4 | 1 | | | | |
| $n=4$ | | 1 | 11 | 11 | 1 | | | |
| $n=5$ | | 1 | 26 | 66 | 26 | 1 | | |
| $n=6$ | | 1 | 57 | 302 | 302 | 57 | 1 | |
| $n=7$ | | 1 | 120 | 1191 | 2416 | 1191 | 120 | 1 |

Table 10.1: Table of values for Eulerian numbers, $A_{n,k}$, with $0 \le k \le n$

By analyzing the rows of Table 10.1 we conjecture that

$$A_{n,j} = A_{n,n-j+1}, \qquad n \ge 1. \tag{10.5}$$

To verify Equation (10.5) first rewrite Equation (10.4) as

$$A_{n,j} = (-1)^j \sum_{k=0}^{j} (-1)^k \binom{n+1}{j-k} k^n. \tag{10.6}$$

Then

$$A_{n,n-j+1} = (-1)^{n-j+1} \sum_{k=0}^{n-j+1} (-1)^k \binom{n+1}{n-j+1-k} k^n$$

$$= (-1)^{n-j+1} \sum_{k=0}^{n-j+1} (-1)^k \binom{n+1}{j+k} k^n$$

$$= (-1)^{n-j+1} \sum_{k=j}^{n+1} (-1)^{k-j} \binom{n+1}{k} (k-j)^n$$

$$= -\sum_{k=j}^{n+1} (-1)^k \binom{n+1}{k} (j-k)^n.$$

If we combine this calculation with Equation (10.4), we discover that

$$A_{n,j} - A_{n,n-j+1} = \sum_{k=0}^{j} (-1)^k \binom{n+1}{k} (j-k)^n + \sum_{k=j}^{n+1} (-1)^k \binom{n+1}{k} (j-k)^n$$

$$= \sum_{k=0}^{n+1} (-1)^k \binom{n+1}{k} (j-k)^n + (-1)^j \binom{n+1}{j} (j-j)^n.$$

Observe that the summation is the $(n+1)^{st}$ difference of $(j-k)^n$, which is a polynomial in $k$ of degree $n$. By Euler's finite difference theorem, this sum must be zero. Furthermore if $n \geq 1$, then $(-1)^j \binom{n+1}{j} (j-j)^n$ is also zero. This analysis implies that the preceding line is equivalent to

$$A_{n,j} - A_{n-j+1} = \begin{cases} 0, & n \geq 1 \\ (-1)^j, & n = 0, \quad j = 0, 1. \end{cases}$$

Take another look at Table 10.1 and add up the entries in each row. Observe that

$$\sum_{j=0}^{n} A_{n,j} = n!. \tag{10.7}$$

Equation (10.7) is a consequence of Equation (10.2). Take Equation (10.2) and let $\alpha = n$ to obtain $S(n,n) = 1 = \frac{1}{n!} \sum_{j=0}^{n} A_{n,j}$.

## 10.1 Functional Expansions Involving Eulerian Numbers

In Chapter 9 we used Stirling numbers of the second kind in a variety of series expansions. For example we showed that

$$\sum_{k=0}^{\infty} k^n x^k = \sum_{j=0}^{n} j! S(n,j) \frac{x^j}{(1-x)^{j+1}}. \tag{10.8}$$

See Equation (9.47). We will now show that

$$\sum_{k=0}^{\infty} k^n x^k = \sum_{j=0}^{n} A_{n,j} \frac{x^j}{(1-x)^{n+1}}, \qquad n \geq 1. \tag{10.9}$$

Equation (10.9) is a consequence of (10.8), (10.2), and (10.5) since

$$\sum_{k=0}^{\infty} k^n x^k = \sum_{\alpha=0}^{n} \alpha! S(n,\alpha) \frac{x^\alpha}{(1-x)^{\alpha+1}} = \sum_{\alpha=0}^{n} \sum_{j=0}^{n} \binom{j-1}{n-\alpha} A_{n,j} \frac{x^\alpha}{(1-x)^{\alpha+1}}$$

$$= \sum_{\alpha=0}^{n} \sum_{j=n-\alpha}^{n} \binom{j-1}{n-\alpha} A_{n,j} \frac{x^\alpha}{(1-x)^{\alpha+1}}, \qquad n \geq 1$$

$$= \sum_{\alpha=0}^{n} \sum_{j=\alpha}^{n} \binom{j-1}{\alpha} A_{n,j} \frac{x^{n-\alpha}}{(1-x)^{n-\alpha+1}}$$

$$= \sum_{j=0}^{n} A_{n,j} \frac{x^n}{(1-x)^{n+1}} \sum_{\alpha=0}^{j} \binom{j-1}{\alpha} \left(\frac{1-x}{x}\right)^\alpha$$

$$= \sum_{j=1}^{n} A_{n,j} \frac{x^n}{(1-x)^{n+1}} \sum_{\alpha=0}^{j-1} \binom{j-1}{\alpha} \left(\frac{1}{x}-1\right)^\alpha, \qquad \text{since } A_{n,0} = 0$$

$$= \sum_{j=1}^{n} \frac{x^{n-j+1}}{(1-x)^{n+1}} A_{n,j} = \sum_{j=1}^{n} \frac{x^j}{(1-x)^{n+1}} A_{n,n-j+1}$$

$$= \sum_{j=0}^{n} \frac{x^j}{(1-x)^{n+1}} A_{n,j}.$$

Equation (10.9) is valid if $n = 0$ since $\sum_{k=0}^{\infty} x^k = \frac{1}{1-x}$. This observation allows us to combine Equations (10.8) and (10.9) as

$$(1-x) \sum_{k=0}^{\infty} k^n x^k = \sum_{j=0}^{n} j! S(n,j) \left(\frac{x}{1-x}\right)^j = \sum_{j=0}^{n} A_{n,j} \frac{x^j}{(1-x)^n}. \tag{10.10}$$

Equation (10.10) provides an inversion of Equation (10.2). In particular

$$\sum_{\alpha=0}^{n} A_{n,\alpha} x^\alpha = (1-x)^n \sum_{\alpha=0}^{n} \alpha! S(n,\alpha) \frac{x^\alpha}{(1-x)^\alpha}$$

$$= \sum_{\alpha=0}^{n} \alpha! S(n,\alpha) x^\alpha (1-x)^{n-\alpha}$$

$$= \sum_{\alpha=0}^{n} \alpha! S(n,\alpha) x^\alpha \sum_{j=0}^{n-\alpha} \binom{n-\alpha}{j} (-1)^{n-\alpha-j} x^{n-\alpha-j}$$

$$= \sum_{j=0}^{n} x^{n-j} \sum_{\alpha=0}^{n} (-1)^{n-\alpha-j} \binom{n-\alpha}{j} \alpha! S(n,\alpha)$$

$$= \sum_{j=0}^{n} (-1)^j x^j \sum_{\alpha=0}^{n} (-1)^{\alpha} \binom{n-\alpha}{n-j} \alpha! S(n,\alpha)$$

$$= \sum_{\alpha=0}^{n} (-1)^{\alpha} x^{\alpha} \sum_{j=0}^{n} (-1)^j \binom{n-j}{n-\alpha} j! S(n,j).$$

By comparing the coefficients of $x^{\alpha}$ we find that

$$A_{n,k} = (-1)^k \sum_{j=0}^{n} (-1)^j \binom{n-j}{n-k} j! S(n,j). \tag{10.11}$$

Another application of Equation (10.9) provides a formal two variable exponential generating function for $\{A_{n,k}\}_{n,k=0}^{\infty}$. We claim that

$$\sum_{n=0}^{\infty} \sum_{j=0}^{\infty} A_{n,j} \frac{x^n y^j}{n!} = \frac{1-y}{1 - y e^{(1-y)x}}. \tag{10.12}$$

To verify Equation (10.12) observe that

$$\frac{1-y}{1 - y e^{(1-y)x}} = (1-y) \sum_{k=0}^{\infty} (y e^{(1-y)x})^k = (1-y) \sum_{k=0}^{\infty} y^k e^{(1-y)xk}$$

$$= (1-y) \sum_{k=0}^{\infty} y^k \sum_{n=0}^{\infty} \frac{((1-y)xk)^n}{n!}$$

$$= \sum_{n=0}^{\infty} \frac{x^n}{n!} (1-y)^{n+1} \sum_{k=0}^{\infty} k^n y^k$$

$$= \sum_{n=0}^{\infty} \frac{x^n}{n!} (1-y)^{n+1} \sum_{j=0}^{n} A_{n,j} \frac{y^j}{(1-y)^{n+1}}$$

$$= \sum_{n=0}^{\infty} \sum_{j=0}^{\infty} A_{n,j} \frac{x^n y^j}{n!}.$$

## 10.2   Combinatorial Interpretation of $A(n,m)$

Just as the Bell numbers and Stirling numbers of the second kind have combinatorial interpretations so too do Eulerian numbers. This interpretation involves permutations of $[n] = \{1, 2, \ldots, n\}$. For a fixed $n$ we visualize a permutation as an ordered $n$-tuple $(a_1, a_2, \ldots a_n)$ where $a_i \in \{1, 2, \ldots, n\}$ and $a_i = a_j$ if and only if $i = j$. For example the permutations of $\{1, 2, 3\}$ are $\{\{1, 2, 3\}, \{1, 3, 2\}, \{2, 1, 3\}, \{3, 1, 2\}, \{2, 3, 1\}, \{3, 2, 1\}\}$. The definition ensures there are $n!$ different permutations.

Let $P$ be a permutation of $\{1, 2, \ldots, n\}$, represented as an $n$-tuple $(a_1, a_2, \ldots a_n)$. There are $n - 1$ adjacent pairs of entries in this $n$-tuple. Denote a typical adjacent pair by $a_i, a_{i+1}$ where $1 \leq i \leq n - 1$. Such an adjacency is an *ascent* if $a_i < a_{i+1}$. It is a *descent* if $a_i > a_{i+1}$. Let $p$ be the number of ascents in $P$ and $q$ be the number of descents. Notice that $p + q = n - 1$. Define $E(n, m)$ to be the number of permutations of $\{1, 2, \ldots, n\}$ with precisely $m$ ascents where $0 \leq m \leq n - 1$. It is not hard to calculate by hand that $E(1, 0) = 1$, $E(2, 0) = 1$, $E(2, 1) = 1$, $E(3, 0) = 1$, $E(3, 1) = 4$, and $E(3, 2) = 1$. [1] By comparing these values with the values in rows 2 through 4 of Table 10.1 we conjecture that $A_{n,m} = E(n, m - 1)$ whenever $n \geq 1$ and $1 \leq m \leq n$. We verify this conjecture by demonstrating that

$$E(n, m) = (n - m)E(n - 1, m - 1) + (m + 1)E(n - 1, m). \qquad (10.13)$$

The left side of Equation (10.13) counts the permutations of $[n]$ which have $m$ ascents. We describe how to build such permutations from the permutations of $[n-1]$ by appending $n$. There are two types of permutations of $[n - 1]$ which yield a permutation of $[n]$ with exactly $m$ ascents. The first type has the $m$ ascents. To such a permutation we either put $n$ at the beginning or $n$ between one of the $m$ ascents. This reasoning justifies the summand of $(m + 1)E(n - 1, m)$. The second type of permutation has only $m - 1$ ascents and consequently $n - m - 1$ descents. For this type of permutation we place the $m$ either at the end of the permutation or insert it between one of the descents. This reasoning justifies the summand of $(n - m)E(n - 1, m - 1)$. Figure 10.1 demonstrates this construction for $E(4, 2) = 2E(3, 1) + 3E(3, 2)$.

---

[1] Wikipedia, *Eulerian number*, http://en.wikipedia.org/wiki/Eulerian_number

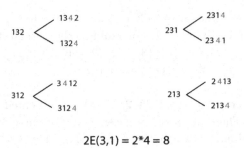

$$2E(3,1) = 2*4 = 8$$

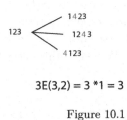

$$3E(3,2) = 3 *1 = 3$$

Figure 10.1

We claim that Equation (10.13) is equivalent to Equation (10.3). Start with Equation (10.3) and substitute the hypothesis of $A_{n,j} = E(n, j - 1)$:

$$E(n+1, j-1) = (n+2-j)E(n, j-2)E(n, j-2) + jE(n, j-1). \quad (10.14)$$

Take Equation (10.14) let $n \to n-1$ and $j \to j+1$ to obtain

$$E(n, j) = (n - j)E(n - 1, j - 1) + (j + 1)E(n - 1, j),$$

which is precisely Equation (10.13). Hence $\{A_{n,j}\}_{n=1}^{\infty}$ with $1 \leq j \leq n$ and $\{E(n, j - 1)\}_{n=1}^{\infty}$ with $1 \leq j \leq n$ obey the same recurrence and share the same initial conditions. Therefore we conclude that $E(n, j) = A_{n,j+1}$ for $0 \leq j \leq n - 1$.

# Chapter 11

# Worpitzky Numbers

We begin this chapter by comparing the explicit formula for Stirling numbers of the second kind with that of Eulerian numbers. Equation (9.21) states

$$j! S(n, j) = \sum_{k=0}^{j} (-1)^k \binom{j}{k} (j - k)^n, \tag{11.1}$$

while Equation (10.4) states

$$A_{n,j} = \sum_{k=0}^{j} (-1)^k \binom{n+1}{k} (j - k)^n. \tag{11.2}$$

Take a careful look at the structure of the sums in Equations (11.1) and (11.2). They have the form $\sum_{k=0}^{\square} (-1)^k \binom{\triangle}{k} (\square - k)^{\Diamond}$, where $\square$, $\triangle$, and $\Diamond$ are three independent nonnegative integers. The nineteenth century mathematician Julius Worpitzky was the first person who recognized this structural similarity. Worpitzky [1883] defined a class of numbers which included $j! S(n, j)$ and $A_{n,j}$. Let $j$, $m$, and $n$ be nonnegative integers. The Worpitzky number $B_{j,m}^n$ is defined to be

$$B_{j,m}^n = \sum_{k=0}^{j} (-1)^k \binom{m}{k} (j - k)^n. \tag{11.3}$$

By definition $B_{j,j}^n = j! S(n, j)$ and $B_{j,n+1}^n = A_{n,j}$.

Since Worpitzky numbers are generalizations of Eulerian numbers, we are able to reinterpret many of the properties discussed in the preceding

147

chapter. We start by generalizing Equation (10.5). By definition we have

$$B_{m-k+1,m+1}^n = \sum_{j=0}^{m-k+1} (-1)^j \binom{m+1}{j}(m-k+1-j)^n$$

$$= \sum_{j=k}^{m+1} (-1)^{j-k} \binom{m+1}{j-k}(m+1-j)^n$$

$$= \sum_{j=k}^{m+1} (-1)^{m+1-j} \binom{m+1}{m+1-j}(j-k)^n$$

$$= (-1)^{m+n+1} \sum_{j=k}^{m+1} (-1)^j \binom{m+1}{j}(k-j)^n.$$

These calculations imply that

$$B_{k,m+1}^n - (-1)^{m+n} B_{m-k+1,m+1}^n$$

$$= \sum_{j=0}^{k} (-1)^j \binom{m+1}{j}(k-j)^n + \sum_{j=k}^{m+1} (-1)^j \binom{m+1}{j}(k-j)^n$$

$$= \sum_{j=0}^{m+1} (-1)^j \binom{m+1}{j}(k-j)^n + (-1)^k \binom{m+1}{k}(k-k)^n.$$

By Euler's finite difference theorem, the series in the preceding line is zero whenever $m \geq n$. The additional term is also zero if $n \geq 1$. Therefore we have shown

$$B_{k,m+1}^n - (-1)^{m+n} B_{m-k+1,m+1}^n = \begin{cases} 0 & m \geq n \geq 1 \\ (-1)^k \binom{m+1}{k} & m \geq n, \ n = 0. \end{cases}$$

$$(11.4)$$

If $m = n$, Equation (11.4) becomes Equation (10.5).

We next generalize Equation (10.2). If $k$ is a nonnegative integer such that $1 \leq k \leq j$, then $\binom{k-1}{j} = 0$. Next observe that $B_{0,q}^n = \sum_{k=0}^{0} (-1)^k \binom{q}{k}(0-k)^n = \binom{q}{0}(0-0)^n$. This quantity is guaranteed to be zero whenever $n \geq 1$. With these facts in mind we are ready to evaluate $\sum_{k=0}^{m+1} \binom{k-1}{m-\alpha} B_{k,m+1}^n$ for

$m \geq n \geq 1$. In particular we have

$$\sum_{k=0}^{m+1} \binom{k-1}{m-\alpha} B_{k,m+1}^n = \sum_{k=m-\alpha}^{m+1} \binom{k-1}{m-\alpha} B_{k,m+1}^n$$

$$= \sum_{k=0}^{\alpha+1} \binom{k+m-\alpha-1}{m-\alpha} B_{k+m-\alpha,m+1}^n$$

$$= \sum_{k=0}^{\alpha+1} \binom{k+m-\alpha-1}{k-1} B_{k+m-\alpha,m+1}^n$$

$$= \sum_{k=0}^{\alpha+1} (-1)^{k-1} \binom{\alpha-m-1}{k-1} B_{k+m-\alpha,m+1}^n$$

$$= \sum_{k=0}^{\alpha+1} (-1)^{k+\alpha} \binom{\alpha-m-1}{\alpha-k} B_{m-k+1,m+1}^n$$

$$= (-1)^{\alpha+m+n} \sum_{k=0}^{\alpha+1} (-1)^k \binom{\alpha-m-1}{\alpha-k} B_{k,m+1}^n$$

$$= (-1)^{\alpha+m+n} \sum_{k=0}^{\alpha+1} (-1)^k \binom{\alpha-m-1}{\alpha-k} \sum_{j=0}^{k} (-1)^j \binom{m+1}{j} (k-j)^n$$

$$= (-1)^{\alpha+m+n} \sum_{k=0}^{\alpha+1} (-1)^k \binom{\alpha-m-1}{\alpha-k} \sum_{j=0}^{k} (-1)^{k-j} \binom{m+1}{k-j} j^n$$

$$= (-1)^{\alpha+m+n} \sum_{j=0}^{\alpha+1} (-1)^j j^n \sum_{k=j}^{\alpha+1} \binom{\alpha-m-1}{\alpha-k} \binom{m+1}{k-j}$$

$$= (-1)^{\alpha+m+n} \sum_{j=0}^{\alpha+1} (-1)^j j^n \sum_{k=0}^{\alpha-j+1} \binom{\alpha-m-1}{\alpha-k-j} \binom{m+1}{k}$$

$$= (-1)^{\alpha+m+n} \sum_{j=0}^{\alpha+1} (-1)^j j^n \sum_{k=0}^{\alpha-j} \binom{\alpha-m-1}{\alpha-j-k} \binom{m+1}{k}$$

$$= (-1)^{\alpha+m+n} \sum_{j=0}^{\alpha+1} (-1)^j j^n \binom{\alpha}{\alpha-j} = (-1)^{\alpha+m+n} \sum_{j=0}^{\alpha} (-1)^j \binom{\alpha}{j} j^n$$

$$= (-1)^{m+n} \alpha! S(n,\alpha).$$

We conclude that

$$\sum_{k=0}^{m+1} \binom{k-1}{m-\alpha} B_{k,m+1}^n = (-1)^{m+n} \alpha! S(n,\alpha), \qquad m \geq n \geq \alpha \geq 0. \quad (11.5)$$

Our calculations have only verified Equation (11.5) for the case of $m \geq n \geq 1$. We leave it to the reader to verify the case of $n = 0$.

If we assume that $m \geq n \geq 1$, we may use Equation (11.4) to transform Equation (11.5) into the identity

$$\alpha! S(n, \alpha) = \sum_{k=0}^{m+1} \binom{m-k}{m-\alpha} B_{k,m+1}^n \qquad m \geq n \geq 1. \qquad (11.6)$$

Equation (11.5) writes a Stirling number of the second kind in terms of a Worpitzky number. We now invert this formula to obtain

$$B_{k,m+1}^n = (-1)^k \sum_{\alpha=0}^m (-1)^\alpha \alpha! S(n, \alpha) \binom{m-\alpha}{m-k}, \qquad m \geq n \geq 0. \qquad (11.7)$$

To prove Equation (11.7) apply Equation (11.5) and observe that

$$(-1)^k \sum_{\alpha=0}^m (-1)^\alpha \alpha! S(n, \alpha) \binom{m-\alpha}{m-k}$$

$$= (-1)^k \sum_{\alpha=0}^m (-1)^\alpha \binom{m-\alpha}{m-k} (-1)^{m+n} \sum_{j=0}^{m+1} \binom{j-1}{m-\alpha} B_{j,m+1}^n$$

$$= (-1)^{k+m+n} \sum_{j=0}^{m+1} B_{j,m+1}^n \sum_{\alpha=0}^m (-1)^\alpha \binom{m-\alpha}{m-k} \binom{j-1}{m-\alpha}$$

$$= (-1)^{k+n} \sum_{j=0}^{m+1} B_{j,m+1}^n \sum_{\alpha=0}^m (-1)^\alpha \binom{j-1}{\alpha} \binom{\alpha}{m-k}$$

$$= (-1)^{k+n} \sum_{j=0}^{m+1} B_{j,m+1}^n (-1)^{j-1} \binom{0}{j-1-(m-k)}, \qquad \text{by Eq. (6.35)}$$

$$= (-1)^{m+n} B_{m-k+1,m+1}^n = B_{k,m+1}^n, \qquad \text{by Eq. (11.4)}.$$

An important application of Equation (11.5) proves a formula attributed to N. Nielsen [1923]. Nielsen showed that

$$x^n = (-1)^{m+n} \sum_{k=0}^{m+1} \binom{x+k-1}{m} B_{k,m+1}^n, \qquad m \geq n \geq 0. \qquad (11.8)$$

If $m = n$, Equation (11.8) becomes Equation (10.1).

To prove Equation (11.8) start with Equation (9.11) and observe that

$$(-1)^{m+n}x^n = (-1)^{m+n}\sum_{j=0}^{n}\binom{x}{j}j!S(n,j)$$

$$= (-1)^{m+n}\sum_{j=0}^{m}\binom{x}{j}j!S(n,j)$$

$$= \sum_{j=0}^{m}\binom{x}{j}\sum_{k=0}^{m+1}\binom{k-1}{m-j}B_{k,m+1}^n$$

$$= \sum_{k=0}^{m+1}B_{k,m+1}^n\sum_{j=0}^{m}\binom{x}{j}\binom{k-1}{m-j}$$

$$= \sum_{k=0}^{m+1}\binom{x+k-1}{m}B_{k,m+1}^n.$$

Nielsen's formula for $x^n$ provides a two term recurrence for the Worpitzky numbers. Take Equation (11.8), replace $m$ with $m-1$, and multiply by $x$ to obtain

$$(-1)^{m+n-1}x^{n+1} = \sum_{k=0}^{m}x\binom{x+k-1}{m-1}B_{k,m}^n, \qquad m-1\ge n\ge 0. \quad (11.9)$$

An application of the committee/chair identity shows that

$$x\binom{x+k-1}{m-1} = \binom{x+k}{m}(m-k) + \binom{x+k-1}{m}k, \qquad m\ge 1. \quad (11.10)$$

If we substitute Equation (11.10) into the right side of Equation (11.9), we discover that

$$(-1)^{m+n-1}x^{n+1} = \sum_{k=0}^{m}\binom{x+k}{m}(m-k)B_{k,m}^n + \sum_{k=0}^{m}\binom{x+k-1}{m}kB_{k,m}^n$$

$$= \sum_{k=1}^{m+1}\binom{x+k-1}{m}(m-k+1)B_{k-1,m}^n + \sum_{k=0}^{m}\binom{x+k-1}{m}kB_{k,m}^n$$

$$= \sum_{k=0}^{m+1}\binom{x+k-1}{m}\left[(m-k+1)B_{k-1,m}^n + kB_{k,m}^n\right].$$

Since

$$(-1)^{m+n-1}x^{n+1} = \sum_{k=0}^{m+1}\binom{x+k-1}{m}B_{k,m+1}^{n+1}, \qquad m\ge n+1\ge 1, \quad (11.11)$$

comparing the coefficients of $\binom{x+k-1}{m}$ implies that

$$B_{k,m+1}^{n+1} = (m-k+1)B_{k-1,m}^n + kB_{k,m}^n, \qquad m\ge n+1\ge 1. \quad (11.12)$$

If $m = n+1$, Equation (11.12) becomes Equation (10.3).

## 11.1    Polynomial Expansions from Nielsen's Formula

Equation (11.8) is useful for expanding polynomials. Let $f(x) = \sum_{k=0}^{m} a_k x^k$. By Equation (11.8) we have

$$
f(x) = \sum_{k=0}^{m} a_k x^k = \sum_{k=0}^{m} a_k (-1)^{m+k} \sum_{\alpha=0}^{m+1} \binom{x+\alpha-1}{m} B_{\alpha,m+1}^{k}
$$

$$
= (-1)^m \sum_{k=0}^{m} (-1)^k a_k \sum_{\alpha=0}^{m+1} \binom{x+\alpha-1}{m} \sum_{j=0}^{\alpha} (-1)^j \binom{m+1}{j} (\alpha-j)^k
$$

$$
= (-1)^m \sum_{\alpha=0}^{m+1} \binom{x+\alpha-1}{m} \sum_{j=0}^{\alpha} (-1)^j \binom{m+1}{j} \sum_{k=0}^{m} (-1)^k (\alpha-j)^k a_k
$$

$$
= (-1)^m \sum_{\alpha=0}^{m+1} \binom{x+\alpha-1}{m} \sum_{j=0}^{\alpha} (-1)^j \binom{m+1}{j} f(j-\alpha).
$$

Since $f(x)$ is a polynomial of degree $m$, Euler's finite difference theorem tells us that $\sum_{j=0}^{m+1} (-1)^j \binom{m+1}{j} f(j-\alpha) = 0$. Therefore we may truncate the outer sum in the previous line at $m$ to obtain

$$
f(x) = (-1)^m \sum_{\alpha=0}^{m} \binom{x+\alpha-1}{m} \sum_{j=0}^{\alpha} (-1)^j \binom{m+1}{j} f(j-\alpha).
$$

We summarize the results of these calculation in the following theorem:

**Theorem 11.1.** *(Nielsen's Polynomial Expansion: Version 1) Let $f(x) = \sum_{k=0}^{m} a_k x^k$ be a polynomial in $x$ of degree $m$. Then*

$$
f(x) = (-1)^m \sum_{\alpha=0}^{m} \binom{x+\alpha-1}{m} \sum_{j=0}^{\alpha} (-1)^j \binom{m+1}{j} f(j-\alpha). \qquad (11.13)
$$

Theorem 11.1 should be contrasted with Theorem 9.1. It is instructive to look at special cases of Theorem 11.1. For our first example take $f(x) = \binom{x}{r}^p$, where $r$ and $p$ are positive integers such that $rp \leq m$. Equation (11.13) becomes

$$
\binom{x}{r}^p = (-1)^m \sum_{\alpha=0}^{m} \binom{x+\alpha-1}{m} \sum_{j=0}^{\alpha} (-1)^j \binom{m+1}{j} \binom{j-\alpha}{r}^p
$$

$$
= (-1)^{m+rp} \sum_{\alpha=0}^{m} \binom{x+\alpha-1}{m} \sum_{j=0}^{\alpha} (-1)^j \binom{m+1}{j} \binom{\alpha-j+r-1}{r}^p. \qquad (11.14)
$$

If $m = rp$, Equation (11.14) becomes

$$
\binom{x}{r}^p = \sum_{\alpha=0}^{rp} \binom{x+\alpha-1}{rp} \sum_{j=0}^{\alpha} (-1)^j \binom{rp+1}{j} \binom{\alpha-j+r-1}{r}^p. \qquad (11.15)
$$

Since $\binom{r-1}{r} = 0$, we may start the outer sum at $\alpha = 1$. We further claim that we need only sum $\alpha$ as far as $rp - r + 1$. Thus Equation (11.15) is equivalent to

$$\binom{x}{r}^p = \sum_{\alpha=1}^{rp-r+1} \binom{x+\alpha-1}{rp} \sum_{j=0}^{\alpha} (-1)^j \binom{rp+1}{j} \binom{\alpha-j+r-1}{r}^p.$$

(11.16)

To justify the range of summation, fix positive integers $r$ and $p$ and rewrite Equation (11.15) as

$$\binom{x}{r}^p = \sum_{\alpha=0}^{rp} \binom{x+\alpha-1}{rp} C_\alpha, \quad C_\alpha = \sum_{j=0}^{\alpha} (-1)^j \binom{rp+1}{j} \binom{\alpha-j+r-1}{r}^p.$$

(11.17)

Technically we should index the inner summation as $C_{\alpha,r,p}$, but for purposes of exposition we suppress the fixed quantities $r$ and $p$. Notice that

$$C_0 = \sum_{j=0}^{0} (-1)^j \binom{rp+1}{j} \binom{-j+r-1}{r}^p = \binom{r-1}{r}^p = 0.$$

If $\alpha$ is a positive integer the range of summation for $C_\alpha$ need only go to $\alpha - 1$. By definition

$$C_{rp-r-\alpha+2} = \sum_{j=0}^{rp-r-\alpha+1} (-1)^j \binom{rp+1}{j} \binom{rp-\alpha-j+1}{r}^p$$

$$= \sum_{j=\alpha+r}^{rp+1} (-1)^{j-\alpha-r} \binom{rp+1}{j-\alpha-r} \binom{rp-j+r+1}{r}^p$$

$$= \sum_{j=\alpha+r}^{rp+1} (-1)^{rp+1-j} \binom{rp+1}{rp+1-j} \binom{j-\alpha}{r}^p$$

$$= -\sum_{j=\alpha+r}^{rp+1} (-1)^j \binom{rp+1}{j} \binom{\alpha-j+r-1}{r}^p.$$

Since $r$ and $p$ are positive integers, $\binom{\alpha-j+r-1}{r}^p$ is zero whenever $0 \le \alpha - j + r - 1 \le r - 1$, or whenever $\alpha \le j \le \alpha + r - 1$. The preceding line becomes

$$C_{rp-r-\alpha+2} = -\sum_{j=\alpha}^{rp+1} (-1)^j \binom{rp+1}{j} \binom{\alpha-j+r-1}{r}^p.$$

Combine this result with $C_\alpha$ to obtain

$$C_\alpha - C_{rp-r-\alpha+2} = \sum_{j=0}^{rp+1} (-1)^j \binom{rp+1}{j} \binom{\alpha-j+r-1}{r}^p = 0,$$

since we have an $(rp+1)^{th}$ difference of a polynomial of degree $rp$. Therefore we conclude that

$$C_\alpha = C_{rp-r-\alpha+2}, \qquad r, p \geq 1. \tag{11.18}$$

By definition $C_\alpha$ vanishes if $\alpha$ is negative. Equation (11.18) shows this occurs when $\alpha > rp - r + 2$. Furthermore $C_0 = C_{rp-r-\alpha+2} = 0$. This analysis allows us to rewrite Equation (11.17) as

$$\binom{x}{r}^p = \sum_{\alpha=0}^{rp} \binom{x+\alpha-1}{rp} C_\alpha = \sum_{\alpha=0}^{rp-r+1} \binom{x+\alpha-1}{rp} C_\alpha,$$

which is precisely Equation (11.16).

For our second example of Theorem 11.1 take $f(x) = \binom{x+n}{n}$, with $n \leq m$, and obtain

$$\binom{x+n}{n} = (-1)^m \sum_{\alpha=0}^{m} \binom{x+\alpha-1}{m} \sum_{j=0}^{\alpha} (-1)^j \binom{m+1}{j} \binom{j-\alpha+n}{n}.$$
$$\tag{11.19}$$

Equation (11.19) often appears in the literature as

$$\binom{x}{n} = (-1)^n \sum_{\alpha=0}^{n} \binom{x-n+\alpha-1}{n} \sum_{j=0}^{\alpha} (-1)^j \binom{n+1}{j} \binom{j-\alpha+n}{n}.$$
$$\tag{11.20}$$

Equation (11.20) should be contrasted with the expansion of $\binom{x}{n}$ provided via Equation (9.38), namely

$$\binom{x}{n} = \sum_{\alpha=0}^{n} \binom{x-n}{\alpha} \sum_{j=0}^{\alpha} (-1)^j \binom{\alpha}{j} \binom{\alpha-j+n}{n}. \tag{11.21}$$

To find yet another expansion of $\binom{x}{n}$ let $f(x) = \binom{px}{n}$ with $n \leq m$. Equation (11.13) becomes

$$\binom{px}{n} = (-1)^m \sum_{\alpha=0}^{m} \binom{x+\alpha-1}{m} \sum_{j=0}^{\alpha} (-1)^j \binom{m+1}{j} \binom{pj-p\alpha}{n}. \tag{11.22}$$

Take $p = 1$ and $n = m$ to obtain

$$\binom{x}{n} = (-1)^n \sum_{\alpha=0}^{n} \binom{x+\alpha-1}{n} \sum_{j=0}^{\alpha} (-1)^j \binom{n+1}{j} \binom{j-\alpha}{n}. \tag{11.23}$$

Equation (11.23) complements Equation (9.39) with $m = 1$.

We next use Theorem 11.1 to prove the following corollary:

**Corollary 11.1.** *Let* $f(x) = \sum_{k=0}^{m} a_k x^k$. *Then*

$$\sum_{k=0}^{n} (-1)^k \binom{n}{k} f(k) = (-1)^{m+n} \sum_{\alpha=0}^{m} \binom{\alpha-1}{m-n} K_\alpha, \qquad (11.24)$$

*where*

$$K_\alpha = \sum_{j=0}^{\alpha} (-1)^j \binom{m+1}{j} f(j-\alpha).$$

**Proof:** Theorem 11.1 states that $f(k) = (-1)^m \sum_{\alpha=0}^{m} \binom{k+\alpha-1}{m} K_\alpha$. Take this equation, multiply by $(-1)^k \binom{n}{k}$, and sum over $k$ to obtain

$$\sum_{k=0}^{n} (-1)^k \binom{n}{k} f(k) = \sum_{k=0}^{n} (-1)^k \binom{n}{k} (-1)^m \sum_{\alpha=0}^{m} \binom{k+\alpha-1}{m} K_\alpha$$

$$= (-1)^m \sum_{\alpha=0}^{m} K_\alpha \sum_{k=0}^{n} (-1)^k \binom{n}{k} \binom{\alpha-1+k}{m}.$$

Next recall Equation (6.35) which states

$$\sum_{k=0}^{n} (-1)^k \binom{n}{k} \binom{x+k}{j} = (-1)^n \binom{x}{j-n}.$$

Apply this identity to the preceding line and obtain

$$\sum_{k=0}^{n} (-1)^k \binom{n}{k} f(k) = (-1)^m \sum_{\alpha=0}^{m} K_\alpha (-1)^n \binom{\alpha-1}{m-n}. \qquad \square$$

Let us take Equation (11.24) and explore what happens when $f(x) = \binom{x}{r}^p$, where $r$ and $p$ are positive integers such that $rp \leq m$. We initially obtain

$$\sum_{k=0}^{n} (-1)^k \binom{n}{k} \binom{k}{r}^p = (-1)^{m+n} \sum_{\alpha=0}^{m} \binom{\alpha-1}{m-n} \sum_{j=0}^{\alpha} (-1)^j \binom{m+1}{j} \binom{j-\alpha}{r}^p$$

$$= (-1)^{rp+m+n} \sum_{\alpha=0}^{m} \binom{\alpha-1}{m-n} \sum_{j=0}^{\alpha} (-1)^j \binom{m+1}{j} \binom{\alpha-j+r-1}{r}^p. \qquad (11.25)$$

Take Equation (11.25) and let $m = n$. This gives us

$$\sum_{k=0}^{n} (-1)^k \binom{n}{k} \binom{k}{r}^p = (-1)^{rp} \sum_{\alpha=0}^{n} \sum_{j=0}^{\alpha} (-1)^j \binom{n+1}{j} \binom{\alpha-j+r-1}{r}^p, \quad rp \leq n.$$

$$(11.26)$$

Observe that $\binom{k}{r}^p$ is a polynomial in $k$ of degree $rp$. If $rp < n$, Euler's finite difference theorem tells us $\sum_{k=0}^{n}(-1)^k\binom{n}{k}\binom{k}{r}^p = 0$. Therefore we may assume $rp = n$ and apply Euler's finite difference theorem to obtain

$$\sum_{k=0}^{n}(-1)^k\binom{n}{k}\binom{k}{r}^p = \sum_{k=0}^{rp}(-1)^k\binom{rp}{k}\binom{k}{r}^p = (-1)^{rp}\frac{(rp)!}{r!^p}, \qquad (11.27)$$

since the coefficient of $x^{rp}$ in $\binom{x}{r}^p$ is $\frac{1}{r!^p}$.

Combine Equations (11.26) and (11.27) to obtain the identity

$$\sum_{\alpha=0}^{rp}\sum_{j=0}^{\alpha}(-1)^j\binom{rp+1}{j}\binom{\alpha-j+r-1}{r}^p = \frac{(rp)!}{r!^p}. \qquad (11.28)$$

The inner sum on the left side of Equation (11.28) is precisely the $C_\alpha$ defined at Equation (11.17). If $r$ and $p$ are positive integers we may apply the reasoning for deriving Equation (11.16) and conclude that

$$\sum_{\alpha=0}^{rp-r+1}\sum_{j=0}^{\alpha}(-1)^j\binom{rp+1}{j}\binom{\alpha-j+r-1}{r}^p = \frac{(rp)!}{r!^p}, \qquad r,p \geq 1.$$

$$(11.29)$$

## 11.2  Nielsen's Expansion with Taylor's Theorem

We now turn our attention to combining Nielsen's formula (11.8) with Taylor's theorem. Assume $f(x)$ is a polynomial of degree $n$ in $x$, i.e. $f(x) = \sum_{k=0}^{n} a_k x^k$. Furthermore assume $m \geq n$. Recall that $f^{(k)}(a)$ means $\frac{d^k}{dx^k}f(x)_{|x=a}$. By Taylor's theorem we have

$$
\begin{aligned}
f(x+y) &= \sum_{k=0}^{n}\frac{x^k}{k!}f^{(k)}(y) = \sum_{k=0}^{n}\frac{f^{(k)}(y)}{k!}(-1)^{m+k}\sum_{\alpha=0}^{m+1}\binom{x+\alpha-1}{m}B_{\alpha,m+1}^k \\
&= \sum_{k=0}^{n}\frac{f^{(k)}(y)}{k!}(-1)^{m+k}\sum_{\alpha=0}^{m+1}\binom{x+\alpha-1}{m}\sum_{j=0}^{\alpha}(-1)^j\binom{m+1}{j}(\alpha-j)^k \\
&= (-1)^m\sum_{\alpha=0}^{m+1}\binom{x+\alpha-1}{m}\sum_{j=0}^{\alpha}(-1)^j\binom{m+1}{j}\sum_{k=0}^{n}\frac{(j-\alpha)^k}{k!}f^{(k)}(y) \\
&= (-1)^m\sum_{\alpha=0}^{m+1}\binom{x+\alpha-1}{m}\sum_{j=0}^{\alpha}(-1)^j\binom{m+1}{j}f(j-\alpha+y).
\end{aligned}
$$

In summary we have proven the following generalization of Theorem 11.1:

**Theorem 11.2.** *(Nielsen's Polynomial Expansion: Version 2) Let $f(x) = \sum_{k=0}^{n} a_k x^k$ be a polynomial in $x$ of degree $n$. Then*

$$f(x+y) = (-1)^m \sum_{\alpha=0}^{m+1} \binom{x+\alpha-1}{m} \sum_{j=0}^{\alpha}(-1)^j \binom{m+1}{j} f(j-\alpha+y), \quad m \geq n.$$

$$(11.30)$$

If $y = 0$, Equation (11.30) becomes Equation (11.13).

In a similar manner we may use Taylor's theorem, along with Equation (9.11), to generalize Theorem 9.1. Once again assume $f(x)$ is a polynomial in $x$ of degree $n$. Then

$$\begin{aligned}
f(x+y) &= \sum_{k=0}^{n} \frac{x^k}{k!} f^{(k)}(y) = \sum_{k=0}^{n} \frac{f^{(k)}(y)}{k!} \sum_{\alpha=0}^{k} \binom{x}{\alpha} \alpha! S(k,\alpha) \\
&= \sum_{\alpha=0}^{n} \binom{x}{\alpha} \alpha! \sum_{k=0}^{n} \frac{f^{(k)}(y)}{k!} S(k,\alpha) \\
&= \sum_{\alpha=0}^{n} \binom{x}{\alpha} \sum_{k=0}^{n} \frac{f^{(k)}(y)}{k!} \sum_{j=0}^{\alpha}(-1)^j \binom{\alpha}{j}(\alpha-j)^k \\
&= \sum_{\alpha=0}^{n} \binom{x}{\alpha} \sum_{j=0}^{\alpha}(-1)^j \binom{\alpha}{j} \sum_{k=0}^{n} \frac{(\alpha-j)^k}{k!} f^{(k)}(y) \\
&= \sum_{\alpha=0}^{n} \binom{x}{\alpha} \sum_{j=0}^{\alpha}(-1)^j \binom{\alpha}{j} f(\alpha-j+y).
\end{aligned}$$

We record the result of these calculations as

**Theorem 11.3.** *(Nielsen's Polynomial Expansion: Version 3) Let $f(x) = \sum_{k=0}^{n} a_k x^k$ be a polynomial in $x$ of degree $n$. Then*

$$f(x+y) = \sum_{\alpha=0}^{n} \binom{x}{\alpha} \sum_{j=0}^{\alpha}(-1)^j \binom{\alpha}{j} f(\alpha-j+y). \qquad (11.31)$$

If $y = 0$, Equation (11.31) becomes Equation (9.37).

We now spend some time looking at particular examples of Theorems 11.2 and 11.3. First let $f(x) = \binom{x}{r}^p$ with $r$ and $p$ fixed positive integers. If

$rp \leq m$, Theorem 11.2 implies that

$$\binom{x+y}{r}^p = (-1)^m \sum_{\alpha=0}^{m+1} \binom{x+\alpha-1}{m} \sum_{j=0}^{\alpha} (-1)^j \binom{m+1}{j} \binom{j-\alpha+y}{r}^p$$

$$= (-1)^{rp+m} \sum_{\alpha=0}^{m+1} \binom{x+\alpha-1}{m} \sum_{j=0}^{\alpha} (-1)^j \binom{m+1}{j} \binom{\alpha-j-y+r-1}{r}^p. \quad (11.32)$$

If $y = 0$, Equation (11.32) becomes Equation (11.14).

In Equation (11.32) let $m = rp$ and $y \to -y$ to obtain

$$\binom{x-y}{r}^p = \sum_{\alpha=0}^{rp+1} \binom{x+\alpha-1}{rp} \sum_{j=0}^{\alpha} (-1)^j \binom{rp+1}{j} \binom{\alpha-j+y+r-1}{r}^p.$$

$$(11.33)$$

Take Equation (11.33), let $x = 1$, and observe that

$$\binom{1-y}{r}^p = \sum_{\alpha=0}^{rp+1} \binom{\alpha}{rp} \sum_{j=0}^{\alpha} (-1)^j \binom{rp+1}{j} \binom{\alpha-j+y+r-1}{r}^p$$

$$= \sum_{\alpha=rp}^{rp+1} \binom{\alpha}{rp} \sum_{j=0}^{\alpha} (-1)^j \binom{rp+1}{j} \binom{\alpha-j+y+r-1}{r}^p$$

$$= \sum_{j=0}^{rp} (-1)^j \binom{rp+1}{j} \binom{rp-j+y+r-1}{r}^p$$

$$+ \binom{rp+1}{rp} \sum_{j=0}^{rp+1} (-1)^j \binom{rp+1}{j} \binom{rp-j+y+r}{r}^p$$

$$= \sum_{j=0}^{rp} (-1)^j \binom{rp+1}{j} \binom{rp-j+y+r-1}{r}^p + 0.$$

A remark about the last equality. Notice that $\sum_{j=0}^{rp+1} (-1)^j \binom{rp+1}{j} \binom{rp-j+y+r}{r}^p$ is the $(rp+1)^{st}$ difference of $\binom{rp-j+y+r}{r}^p$, a polynomial in $j$ of degree $rp$. Since the degree of the polynomial is less than the difference being calculated, Euler's finite difference theorem tells us this sum must be zero. We summarize these calculations as

$$\sum_{j=0}^{rp} (-1)^j \binom{rp+1}{j} \binom{rp-j+y+r-1}{r}^p = \binom{1-y}{r}^p. \quad (11.34)$$

For our second example of Theorem 11.2 let $f(x) = \binom{px}{n}$. Assume $m \geq n$.

Equation (11.30) becomes

$$
\binom{px + py}{n} = (-1)^m \sum_{\alpha=0}^{m+1} \binom{x + \alpha - 1}{m} \sum_{j=0}^{\alpha} (-1)^j \binom{m+1}{j} \binom{pj - p\alpha + py}{n}
$$

$$
= (-1)^{m+n} \sum_{\alpha=0}^{m+1} \binom{x + \alpha - 1}{m}
$$

$$
\cdot \sum_{j=0}^{\alpha} (-1)^j \binom{m+1}{j} \binom{p\alpha - pj - py + n - 1}{n}. \tag{11.35}
$$

If $y = 0$, Equation (11.35) becomes Equation (11.22).

Equation (11.35) is a fertile source of binomial identities. Let $m = n$ and $x = 1$ to obtain

$$
\binom{py + p}{n} = \sum_{\alpha=0}^{n+1} \binom{\alpha}{n} \sum_{j=0}^{\alpha} (-1)^j \binom{n+1}{j} \binom{p\alpha - pj - py + n - 1}{n}
$$

$$
= \sum_{\alpha=n}^{n+1} \binom{\alpha}{n} \sum_{j=0}^{\alpha} (-1)^j \binom{n+1}{j} \binom{p\alpha - pj - py + n - 1}{n}
$$

$$
= \sum_{j=0}^{n} (-1)^j \binom{n+1}{j} \binom{(p+1)n - pj - py - 1}{n}
$$

$$
+ (n+1) \sum_{j=0}^{n+1} (-1)^j \binom{n+1}{j} \binom{(p+1)n + p - pj - py - 1}{n}
$$

$$
= \sum_{j=0}^{n} (-1)^j \binom{n+1}{j} \binom{(p+1)n - pj - py - 1}{n} + 0.
$$

The final equality made use of the fact that we have an $(n+1)^{st}$ difference of a polynomial in $j$ of degree $n$. These calculations prove

$$
\sum_{j=0}^{n} (-1)^j \binom{n+1}{j} \binom{(p+1)n - pj - py - 1}{n} = \binom{py + p}{n}. \tag{11.36}
$$

Take Equation (11.36) and let $py = n - r - 1$ to obtain

$$
\sum_{j=0}^{n} (-1)^j \binom{n+1}{j} \binom{pn - pj + r}{n} = \binom{n - r + p - 1}{n}. \tag{11.37}
$$

In Equation (11.37) set $r = 0$ to obtain

$$
\sum_{j=0}^{n} (-1)^j \binom{n+1}{j} \binom{pn - pj}{n} = \binom{n + p - 1}{n}. \tag{11.38}
$$

Observe that $\binom{pn-pj}{n} = 0$ whenever $n \geq 1$ and $0 \leq p(n-j) < n$, i.e. whenever $\frac{(p-1)n}{p} < j \leq n$. Hence we may truncate the sum in Equation (11.38) at $\left\lfloor \frac{(p-1)n}{p} \right\rfloor$ to obtain

$$\sum_{j=0}^{\left\lfloor \frac{(p-1)n}{p} \right\rfloor} (-1)^j \binom{n+1}{j} \binom{pn-pj}{n} = \binom{n+p-1}{n}. \tag{11.39}$$

The special case of $p = 2$ is attributed to B. C. Wong [1930].

Now take Equation (11.37) and set $r = 1$ to obtain

$$\sum_{j=0}^{n} (-1)^j \binom{n+1}{j} \binom{p(n-j)+1}{n} = \binom{n+p-2}{n}. \tag{11.40}$$

Take $n \to n-1$ in Equation (11.40) and conclude that

$$\sum_{j=0}^{n-1} (-1)^j \binom{n}{j} \binom{p(n-j)+1-p}{n-1} = \binom{n+p-3}{n-1}, \qquad n \geq 1. \tag{11.41}$$

Observe that $\binom{p(n-j)+1-p}{n-1} = 0$ whenever $0 \leq p(n-j)+1-p < n-1$, or whenever $\frac{pn-p-n+2}{p} < j \leq \frac{pn-p+1}{p}$. If $p \geq 2$, then $\frac{pn-p-n+2}{p} < \frac{pn-n}{p} < j \leq \frac{pn-p+1}{p}$ ensures that $\binom{p(n-j)+1-p}{n-1} = 0$. This last inequality implies that we may rewrite Equation (11.41) as

$$\sum_{j=0}^{\left\lfloor \frac{(p-1)n}{p} \right\rfloor} (-1)^j \binom{n}{j} \binom{p(n-j)+1-p}{n-1} = \binom{n+p-3}{n-1}, \qquad n \geq 1, \ p \geq 2. \tag{11.42}$$

The special case of $p = 2$ is also attributed to B. C. Wong [1931].

Finally take Equation (11.37) let $r = -1$ and obtain

$$\sum_{j=0}^{n} (-1)^j \binom{n+1}{j} \binom{pn-pj-1}{n} = \binom{n+p}{n}. \tag{11.43}$$

Assume $n \geq 1$. Analysis similar to that done for Equations (11.39) and (11.42) implies that $\binom{pn-pj-1}{n} = 0$ whenever $0 \leq pn - pj - 1 < n$, or whenever $\frac{(p-1)n-1}{p} < j \leq \frac{pn-1}{p}$. Therefore the only nonzero summands in the left side of Equation (11.43) occur when $0 \leq j \leq \left\lfloor \frac{(p-1)n}{p} \right\rfloor$ and $j = n$. This means we may rewrite Equation (11.43) as

$$\sum_{j=0}^{\left\lfloor \frac{(p-1)n}{p} \right\rfloor} (-1)^j \binom{n+1}{j} \binom{pn-pj-1}{n} = \begin{cases} 1, & n = 0 \\ \binom{n+p}{n} - (n+1), & n \geq 1. \end{cases} \tag{11.44}$$

We end this section by comparing the expansion of $\binom{px}{n}$ provided by Theorem 11.2 with the expansion provided by Theorem 11.3. Equation (11.31) implies that

$$\binom{px + py}{n} = \sum_{\alpha=0}^{n} \binom{x}{\alpha} \sum_{j=0}^{\alpha} (-1)^j \binom{\alpha}{j} \binom{p\alpha - pj + py}{n}$$

$$= (-1)^n \sum_{\alpha=0}^{n} \binom{x}{\alpha} \sum_{j=0}^{\alpha} (-1)^j \binom{\alpha}{j} \binom{pj - p\alpha - py + n - 1}{n}. \quad (11.45)$$

Equation (11.45) corresponds to Equation (11.35). To obtain the equation corresponding to (11.36) take Equation (11.45) and set $x = 1$. This gives us

$$\binom{py + p}{n} = (-1)^n \sum_{\alpha=0}^{1} \binom{1}{\alpha} \sum_{j=0}^{\alpha} (-1)^j \binom{\alpha}{j} \binom{pj - p\alpha - py + n - 1}{n}, \qquad n \geq 1$$

$$= (-1)^n \left[ \binom{-py + n - 1}{n} + \sum_{j=0}^{1} (-1)^j \binom{1}{j} \binom{pj - p - py + n - 1}{n} \right]$$

$$= (-1)^n \left[ \binom{-py + n - 1}{n} + \binom{-p - py + n - 1}{n} - \binom{-py + n - 1}{n} \right]$$

$$= (-1)^n \binom{-p - py + n - 1}{n}.$$

Hence, when $x = 1$, Equation (11.45) reduces to $\binom{py+p}{n} = (-1)^n \binom{-p-py+n-1}{n}$, a result which is directly obtained from the $-1$-Transformation.

## 11.3 Nielsen Numbers

We have spent most of this chapter working with the Worpitzky polynomial expansion for $x^n$ discovered by Nielsen. When discovering Equation (11.8) Nielsen used a generalization of the Worpitzky numbers [Nielsen, 1923] known as the Nielsen numbers. This generalization, denoted $\mathcal{B}_p^{m,n}(\alpha)$ for $p, m$, and $n$ nonnegative integers and $\alpha$ a complex number, is defined by the sum

$$\mathcal{B}_p^{m,n}(\alpha) = \sum_{k=0}^{p} (-1)^k \binom{m+1}{k} (p + \alpha - k)^n. \quad (11.46)$$

Notice that

$$\mathcal{B}_p^{m,n}(0) = \sum_{k=0}^{p} (-1)^k \binom{m+1}{k} (p - k)^n = B_{p,m+1}^n. \quad (11.47)$$

Equation (11.47) demonstrates the obvious relationship between Worpitzky numbers and Nielsen numbers. The binomial theorem provides another way to describe $\mathcal{B}_p^{m,n}(\alpha)$ in terms of Worpizky numbers. Start with Equation (11.46) and observe that

$$
\begin{aligned}
\mathcal{B}_p^{m,n}(\alpha) &= \sum_{k=0}^{p}(-1)^k \binom{m+1}{k}(p+\alpha-k)^n \\
&= \sum_{k=0}^{p}(-1)^k \binom{m+1}{k} \sum_{j=0}^{n}\binom{n}{j}(p-k)^j \alpha^{n-j} \\
&= \sum_{j=0}^{n}\binom{n}{j}\alpha^{n-j}\sum_{k=0}^{p}(-1)^k \binom{m+1}{k}(p-k)^j \\
&= \sum_{j=0}^{n}\binom{n}{j}\alpha^{n-j}B_{p,m+1}^j.
\end{aligned}
$$

In summary we have shown that

$$
\mathcal{B}_p^{m,n}(\alpha) = \sum_{j=0}^{n}\binom{n}{j}\alpha^{n-j}B_{p,m+1}^j. \tag{11.48}
$$

If we combine Equation (11.48) with Equation (11.8) we discover that

$$
\begin{aligned}
(-1)^{m+n}(x-\alpha)^n &= (-1)^{m+n}\sum_{j=0}^{n}\binom{n}{j}\alpha^{n-j}(-1)^{n-j}x^j \\
&= \sum_{j=0}^{n}\binom{n}{j}\alpha^{n-j}\sum_{k=0}^{m+1}\binom{x+k-1}{m}B_{k,m+1}^j \\
&= \sum_{k=0}^{m+1}\binom{x+k-1}{m}\sum_{j=0}^{n}\binom{n}{j}\alpha^{n-j}B_{k,n+1}^j \\
&= \sum_{k=0}^{m+1}\binom{x+k-1}{m}\mathcal{B}_k^{m,n}(\alpha).
\end{aligned}
$$

These calculations show that

$$
(-1)^{m+n}(x-\alpha)^n = \sum_{k=0}^{m+1}\binom{x+k-1}{m}\mathcal{B}_k^{m,n}(\alpha), \qquad m \geq n \geq 0. \tag{11.49}
$$

Equation (11.49) appears as Relation (10) in [Nörlund, 1924, p. 28]. If $\alpha = 0$, Equation (11.49) reduces to Equation (11.8).

If we set $-\alpha = y$, Equation (11.49) becomes

$$(-1)^n(x+y)^n = (-1)^m \sum_{\alpha=0}^{m+1} \binom{x+\alpha-1}{m} \mathcal{B}_\alpha^{m,n}(-y)$$

$$= (-1)^{m+n} \sum_{\alpha=0}^{m+1} \binom{x+\alpha-1}{m} \sum_{j=0}^{\alpha} (-1)^j \binom{m+1}{j} (j+y-\alpha)^n,$$

a result which readily follows from Theorem 11.2.

We highly recommend the reader peruse Nielsen's seminal work *Traité élémentaire des nombres de Bernoulli* [Nielsen, 1923]. It is a treasure trove of many fascinating polynomial expansions including

$$f(x+y) = x\binom{\frac{x}{\alpha}+m}{m} \sum_{k=0}^{m} \frac{(-1)^k}{x+\alpha k} \binom{m}{k} f(y-k\alpha), \tag{11.50}$$

where $f(x)$ is a polynomial of degree at most $m$ [Nielsen, 1923, Relation (7)].

One should compare Nielsen's proof of Equation (11.50), which is based on the Lagrange interpolation formula, with our justification via Melzak's theorem. By Melzak's theorem we have

$$y\binom{y+m}{m} \sum_{k=0}^{m} (-1)^k \binom{m}{k} \frac{f(z-k)}{y+k} = f(z+y), \qquad f(x) = \sum_{i=0}^{m} a_i x^i. \tag{11.51}$$

In Equation (11.51) let $y = \frac{x}{\alpha}$. Furthermore define $\phi(\alpha x) = \sum_{i=0}^{m} b_i \alpha^i x^i = f(x)$. Equation (11.51) becomes

$$x\binom{\frac{x}{\alpha}+m}{m} \sum_{k=0}^{m} (-1)^k \binom{m}{k} \frac{\phi(\alpha z - \alpha k)}{x+\alpha k} = \phi(\alpha z + x). \tag{11.52}$$

If we set $y = \alpha z$, Equation (11.52) becomes (11.50).

# Chapter 12

# Stirling Numbers of the First Kind $s(n, k)$

We next explore a collection of combinatorial numbers which complement the Stirling numbers of the second kind, the Stirling numbers of the first kind, denoted $s(n, k)$. Perhaps the easiest way to define $s(n, k)$ is by inverting $x^n = \sum_{k=0}^{n} \binom{x}{k} k! S(n, k)$. Both $\{x^k\}_{k=0}^{n}$ and $\left\{\binom{x}{k} k!\right\}_{k=0}^{n}$ are bases for the vector space of polynomials in $x$ of degree $n$. Equation (9.11) writes $x^n$ in terms of the basis $\left\{\binom{x}{k} k!\right\}_{k=0}^{n}$. We now want to write $\binom{x}{n} n!$ in terms of the basis $\{x_k\}_{k=0}^{n}$. We do so by using $\{s(n, k)\}_{k=0}^{n}$ and say that

$$n!\binom{x}{n} = \sum_{k=0}^{n} s(n, k) x^k. \tag{12.1}$$

Equation (12.1) implies that $s(n, k) = 0$ if $k > n$ or $k < 0$. Table 12.1 provides the values of $\{s(n, k)\}_{k=0}^{n}$ for $0 \le n \le 7$.

|         | $k=0$ | $k=1$ | $k=2$ | $k=3$ | $k=4$ | $k=5$ | $k=6$ | $k=7$ |
|---------|-------|-------|-------|-------|-------|-------|-------|-------|
| $n=0$   | 1     |       |       |       |       |       |       |       |
| $n=1$   |       | 1     |       |       |       |       |       |       |
| $n=2$   |       | -1    | 1     |       |       |       |       |       |
| $n=3$   |       | 2     | -3    | 1     |       |       |       |       |
| $n=4$   |       | -6    | 11    | -6    | 1     |       |       |       |
| $n=5$   |       | 24    | -50   | 35    | -10   | 1     |       |       |
| $n=6$   |       | -120  | 274   | -225  | 85    | -15   | 1     |       |
| $n=7$   |       | 720   | -1764 | 1624  | -735  | 175   | -21   | 1     |

Table 12.1: Stirling numbers of the first kind, $s(n, k)$, with $0 \le k \le n$ and $0 \le n \le 7$

It is not hard to show that $(n+1)!\binom{x}{n+1} = n!\binom{x}{n}x - n!\binom{x}{n}n$. This identity when combined with Equation (12.1) provides a two term recurrence for

$s(n+1, k)$. By Definition (12.1) we have $(n+1)! \binom{x}{n+1} = \sum_{k=0}^{n+1} s(n+1, k) x^k$. On the other hand

$$(n+1)! \binom{x}{n+1} = n! \binom{x}{n} x - n! \binom{x}{n} n = x \sum_{k=0}^{n} s(n, k) x^k - n \sum_{k=0}^{n} s(n, k) x^k$$

$$= \sum_{k=0}^{n+1} [s(n, k-1) - n s(n, k)] x^k,$$

since $s(n, -1) = s(n, n+1) = 0$. Comparing the coefficients of $x^k$ gives us

$$s(n+1, k) = s(n, k-1) - n s(n, k). \qquad (12.2)$$

Recurrence (12.2) provides a combinatorial interpretation for $(-1)^{n-k} s(n, k)$ in terms of permutations. The factor of $(-1)^{n-k}$ ensures that all the entries in Table 12.1 are positive integers. Let $n$ be a positive integer. A permutation of $[n] = \{1, 2, \dots, n\}$ is a map $\alpha : [n] \to [n]$ which is one-to-one and onto. There are three ways to represent a permutation. The first representation involves a $2 \times n$ array, where the first row is $1\,2 \dots n$ while the second row is the image $\alpha(1)\,\alpha(2) \dots \alpha(n)$. The second representation for $\alpha$ is just the second row of the $2 \times n$ array. The third representation is the cycle notation of $\alpha$. A cycle of a permutation $\alpha$ is a nonempty ordered subset of $[n]$ given by $(a_1 a_2 \dots a_j)$ where $a_{i+1} = \alpha(a_i)$ for $1 \leq i \leq j-1$ and $\alpha(a_j) = a_1$. Because $\alpha$ is one-to-one and onto it can be decomposed into $k$ disjoint cycles where $1 \leq k \leq n$. The cycle representation is found by vertically tracing left to right through the $2 \times n$ array. We demonstrate the three representations for the $3!$ permutations of $[3]$ in Table 12.2.

Let $\begin{bmatrix} n \\ k \end{bmatrix}$ enumerate the permutations of $[n]$ with exactly $k$ cycles. Clearly $\begin{bmatrix} n \\ k \end{bmatrix} = 0$ if $k > n$ or $k$ is negative. Furthermore define $\begin{bmatrix} 0 \\ 0 \end{bmatrix} = 1$. We show that

$$\begin{bmatrix} n+1 \\ k \end{bmatrix} = \begin{bmatrix} n \\ k-1 \end{bmatrix} + n \begin{bmatrix} n \\ k \end{bmatrix}. \qquad (12.3)$$

The left side of Equation (12.3) counts the permutations of $[n+1]$ with precisely $k$ cycles. We claim that the right side of Equation (12.3) also counts this quantity. Let $\alpha$ be a permutation of $[n+1]$ with precisely $k$ cycles. Either $\alpha$ has a cycle of the form $(n+1)$ or $n+1$ is in a cycle of length greater than 1, i.e. $n+1$ is part of a cycle which contains at least two elements. If $(n + 1)$ is an independent cycle the rest of $\alpha$ is a permutation of $[n]$ with $k - 1$ cycles. Such permutations are enumerated by $\left[\begin{array}{c} n \\ k-1 \end{array}\right]$. Now suppose that $n + 1$ is not isolated. Then it must belong to one of the cycles of $\alpha_2$, where $\alpha_2$ is a permutation of $[n]$ with $k$ cycles. Represent $\alpha_2$ as $(a_{11}a_{12}\ldots a_{1j_1})(a_{21}a_{22}\ldots a_{2j_2})\ldots(a_{k1}a_{k2}\ldots a_{kj_k})$ where $\sum_{i=1}^{k} j_k = n$ Working from left to right there are $n$ ways to insert $n + 1$ into these $k$ cycles, namely by placing it immediately to the right one of the $a_{ij_i}$ digits in the cycle structure. The rule of products implies there are $n\left[\begin{array}{c} n \\ k \end{array}\right]$ such permutations of $[n + 1]$ which have $k$ cycles, none of which have the form $(n + 1)$. By applying the rule of sums we obtain the right side of Equation (12.3).

| array representation | row representation | cycle representation |
|---|---|---|
| 1 2 3<br>1 2 3 | 123 | (1)(2)(3) |
| 1 2 3<br>1 3 2 | 132 | (1)(23) |
| 1 2 3<br>2 1 3 | 213 | (12)(3) |
| 1 2 3<br>2 3 1 | 231 | (123) |
| 1 2 3<br>3 1 2 | 312 | (132) |
| 1 2 3<br>3 2 1 | 321 | (13)(2) |

Table 12.2: The permutations of [3]

We claim that $\left[\begin{array}{c} n \\ k \end{array}\right] = (-1)^{n-k}s(n, k)$. We leave it to the reader to verify that $(-1)^{n-k}s(n, k) = \left[\begin{array}{c} n \\ k \end{array}\right]$ for $0 \le k \le n$ and $1 \le n \le 3$. Take Equation (12.2) and multiply by $(-1)^{n+1-k}$ to obtain

$$(-1)^{n+1-k}s(n + 1, k) = (-1)^{n+1-k}s(n, k - 1) + (-1)^{n-k}ns(n, k). \quad (12.4)$$

Equation (12.4) is precisely Equation (12.3) with the substitution $\left[ \begin{matrix} n \\ k \end{matrix} \right] = (-1)^{n-k} s(n, k)$. In other words $\left\{ \left[ \begin{matrix} n \\ k \end{matrix} \right] \right\}_{n=0}^{\infty}$ and $\{(-1)^{n-k} s(n, k)\}_{n=0}^{\infty}$ obey the same recurrence relation and share the same initial conditions. We use the equivalence of the recurrence relations to conclude that $\left[ \begin{matrix} n \\ k \end{matrix} \right] = (-1)^{n-k} s(n, k)$ for all nonnegative integers $n$ and $k$.

## 12.1  Properties of $s(n, k)$

We are in a position to use Equation (12.1) to derive some basic properties of $s(n, k)$. We first show that

$$\sum_{n=0}^{\infty} \frac{z^n}{n!} s(n, k) = \frac{(\ln(1 + z))^k}{k!}. \tag{12.5}$$

To prove Equation (12.5) start with $(1 + z)^{\alpha}$, use the binomial theorem, and observe that

$$(1 + z)^{\alpha} = \sum_{n=0}^{\infty} \binom{\alpha}{n} z^n = \sum_{n=0}^{\infty} \frac{z^n}{n!} \sum_{k=0}^{n} s(n, k) \alpha^k = \sum_{k=0}^{\infty} \alpha^k \sum_{n=0}^{\infty} \frac{z^n}{n!} s(n, k). \tag{12.6}$$

On the other hand the Taylor series expansion for $e^x$ implies that

$$(1 + z)^{\alpha} = e^{\alpha \ln(1+z)} = \sum_{k=0}^{\infty} \frac{(\ln(1 + z))^k}{k!} \alpha^k. \tag{12.7}$$

Comparing the coefficient of $\alpha^k$ in Equation (12.6) with that of $\alpha^k$ in Equation (12.7) gives Equation (12.5).

Taking Equation (12.5), letting $z \to -z$, and multiplying by $(-1)^k$ gives

$$\sum_{n=0}^{\infty} \frac{z^n}{n!} \left[ \begin{matrix} n \\ k \end{matrix} \right] = (-1)^k \frac{(\ln(1 - z))^k}{k!}. \tag{12.8}$$

We now develop a derivative operator formula for $s(n, k)$. There is no explicit formula for $s(n, k)$ corresponding to Euler's formula for $S(n, k)$ given by Equation (9.21). But we can use the Taylor expansion for $\binom{x}{n}$ to describe $s(n, k)$. Taylor's theorem implies that

$$n! \binom{x}{n} = n! \sum_{k=0}^{n} D_x^k \binom{x}{n} \bigg|_{x=0} \frac{x^k}{k!}.$$

Since $n!\binom{x}{n} = \sum_{k=0}^{n} s(n,k)x^k$, coefficient comparison shows that

$$\frac{n!}{k!} D_x^k \binom{x}{n}\Bigg|_{x=0} = s(n,k). \tag{12.9}$$

We next concentrate on deriving a convolution formula involving Stirling numbers of the first kind. In particular we will show that

$$\sum_{\alpha=0}^{n} \binom{n}{\alpha} s(\alpha,j)s(n-\alpha,k) = s(n,k+j)\binom{k+j}{j}. \tag{12.10}$$

Our proof of Equation (12.10) involves two expansions of $\sum_{\alpha=0}^{n} \binom{x}{\alpha}\binom{y}{n-\alpha}$. By the Vandermonde convolution we have

$$\sum_{\alpha=0}^{n} \binom{x}{\alpha}\binom{y}{n-\alpha} = \binom{x+y}{n} = \frac{1}{n!}\sum_{k=0}^{n} s(n,k)(x+y)^k$$

$$= \frac{1}{n!}\sum_{k=0}^{n} s(n,k)\sum_{j=0}^{k}\binom{k}{j}x^j y^{k-j}$$

$$= \sum_{j=0}^{n} x^j \sum_{k=j}^{n} \frac{s(n,k)}{n!}\binom{k}{j}y^{k-j}. \tag{12.11}$$

On the other hand Equation (12.1) implies that

$$\sum_{\alpha=0}^{n} \binom{x}{\alpha}\binom{y}{n-\alpha} = \sum_{j=0}^{n} x^j \sum_{\alpha=j}^{n} \sum_{k=0}^{n-\alpha} \frac{s(\alpha,j)}{\alpha!}\frac{s(n-\alpha,k)}{(n-\alpha)!}y^k. \tag{12.12}$$

Comparing the coefficient of $x^j$ in Equation (12.11) with that in Equation (12.12) gives us

$$\sum_{k=0}^{n-j} y^k \sum_{\alpha=j}^{n} \frac{s(\alpha,j)}{\alpha!}\frac{s(n-\alpha,k)}{(n-\alpha)!} = \sum_{k=j}^{n} \frac{s(n,k)}{n!}\binom{k}{j}y^{k-j}$$

$$= \sum_{k=0}^{n-j} \frac{s(n,k+j)}{n!}\binom{k+j}{j}y^k. \tag{12.13}$$

If we compare the coefficients of $y^k$ in the first and third sums of Equation (12.13), and multiply by $n!$, we obtain Equation (12.10).

Equation (12.10) is often written with $\gamma = k + j$, namely

$$\sum_{\alpha=0}^{n} \binom{n}{\alpha} s(\alpha,j)s(n-\alpha,\gamma-j) = s(n,\gamma)\binom{\gamma}{j}. \tag{12.14}$$

Besides satisfying the convolution formula of Equation (12.10) Stirling numbers of the first kind also satisfy the transformation

$$\sum_{k=0}^{n}(-1)^k\frac{s(k,\alpha)}{k!} = \frac{(-1)^n}{n!}\sum_{j=\alpha}^{n}(-1)^{j-\alpha}\binom{j}{\alpha}s(n,j). \qquad (12.15)$$

To prove Equation (12.15) recall Equation (1.30), namely

$$\sum_{k=0}^{n}(-1)^k\binom{x}{k} = (-1)^n\binom{x-1}{n}. \qquad (12.16)$$

If we use Equation (12.1) to expand $\binom{x}{k}$ and $\binom{x-1}{n}$ in Equation (12.16) we find that

$$\sum_{k=0}^{n}(-1)^k\sum_{j=0}^{k}\frac{s(k,j)}{k!}x^j = (-1)^n\sum_{j=0}^{n}\frac{s(n,j)}{n!}(x-1)^j.$$

By interchanging the order of summation and applying the binomial theorem we see that

$$\sum_{j=0}^{n}x^j\sum_{k=j}^{n}(-1)^k\frac{s(k,j)}{k!} = (-1)^n\sum_{j=0}^{n}\frac{s(n,j)}{n!}\sum_{\alpha=0}^{j}(-1)^{j-\alpha}\binom{j}{\alpha}x^\alpha$$

$$= (-1)^n\sum_{\alpha=0}^{n}x^\alpha\sum_{j=\alpha}^{n}(-1)^{j-\alpha}\binom{j}{\alpha}\frac{s(n,j)}{n!}.$$

Equation (12.15) follows from coefficient comparison.

By applying a combinatorial argument, it is easy to show that $s(n,1) = (-1)^{n-1}(n-1)!$. Equation (12.15) with $\alpha = 1$ then becomes

$$\sum_{k=1}^{n}\frac{1}{k} = \frac{(-1)^n}{n!}\sum_{j=1}^{n}(-1)^j js(n,j), \qquad (12.17)$$

an identity which writes the harmonic sum in terms of Stirling numbers of the first kind.

## 12.2 Orthogonality Relationships for Stirling Numbers

Let us reexamine Equation (9.11) in terms of Equation (12.1). Equation (9.11) says $x^n = \sum_{k=0}^{n}\binom{x}{k}k!S(n,k)$. If we use Equation (12.1) to expand

$\binom{x}{k}$ we find that

$$
x^n = \sum_{k=0}^{n} \binom{x}{k} k! S(n,k) = \sum_{k=0}^{n} S(n,k) \sum_{\alpha=0}^{k} s(k,\alpha) x^\alpha
$$

$$
= \sum_{\alpha=0}^{n} x^\alpha \sum_{k=\alpha}^{n} S(n,k) s(k,\alpha)
$$

$$
= \sum_{k=0}^{n} x^k \sum_{\alpha=k}^{n} S(n,\alpha) s(\alpha,k).
$$

The coefficient of $x^k$ must vanish unless $k = n$. From this analysis we conclude that

$$
\sum_{\alpha=0}^{n} S(n,\alpha) s(\alpha,k) = \binom{0}{n-k}. \tag{12.18}
$$

If we reverse the order of this calculation, namely take Equation (12.1) and expand $x^n$ in terms of Equation (9.11), we find that

$$
\binom{x}{n} = \sum_{\alpha=0}^{n} \frac{s(n,\alpha)}{n!} x^\alpha = \sum_{\alpha=0}^{n} \frac{s(n,\alpha)}{n!} \sum_{k=0}^{\alpha} \binom{x}{k} k! S(\alpha,k)
$$

$$
= \sum_{k=0}^{n} \binom{x}{k} \sum_{\alpha=k}^{n} \frac{k!}{n!} s(n,\alpha) S(\alpha,k).
$$

Unless $k = n$ the coefficient of $\binom{x}{k}$ must vanish. In summary we find that

$$
\sum_{\alpha=0}^{n} s(n,\alpha) S(\alpha,k) = \binom{0}{n-k}. \tag{12.19}
$$

Equations (12.18) and (12.19) are known as the *orthogonality relations for Stirling numbers*. By using these orthogonality relations we can prove the following inversion theorems.

**Theorem 12.1.** *Let $b_\alpha$ and $a_k$ be a collection of constants independent of $n$. Then*

$$
a_n = \sum_{\alpha=0}^{n} S(n,\alpha) b_\alpha \quad \text{if and only if} \quad b_n = \sum_{k=0}^{n} s(n,k) a_k. \tag{12.20}
$$

**Theorem 12.2.** *Let $b_j$ and $a_k$ be a collection of constants which are independent of $n$. Assume $\alpha$ is nonnegative integer such that $\alpha \geq n$. Then*

$$
a_n = \sum_{j=0}^{\alpha} S(j,n) b_j \quad \text{if and only if} \quad b_n = \sum_{k=0}^{\alpha} s(k,n) a_k. \tag{12.21}
$$

A more complex orthogonality relationship between $s(n,k)$ and $S(n,j)$ is given by

$$\sum_{\alpha=0}^{n} s(r,\alpha)S(n+\alpha-1,k) = \begin{cases} 0, & k < r \\ r^{n-1}, & k = r. \end{cases} \tag{12.22}$$

Equation (12.22) is attributed to Frank Olson [1963] and implicitly implies that $n \geq 1$. Our proof of Equation (12.22) depends on Equation (12.19). By Equation (9.21) we have

$$k!S(n+\alpha,k) = (-1)^k \sum_{j=0}^{k} (-1)^j \binom{k}{j} j^{n+\alpha}. \tag{12.23}$$

On the other hand Equation (9.11) implies that

$$j^\alpha = \sum_{\gamma=0}^{\alpha} \binom{j}{\gamma} \gamma! S(\alpha,\gamma). \tag{12.24}$$

Combine Equations (12.23) and (12.24) to obtain

$$k!S(n+\alpha,k) = (-1)^k \sum_{\gamma=0}^{\alpha} \gamma! S(\alpha,\gamma) \sum_{j=0}^{k} (-1)^j \binom{k}{j} \binom{j}{\gamma} j^n. \tag{12.25}$$

Then

$$\frac{k!}{r!} \sum_{\alpha=0}^{n} s(r,\alpha)S(n+\alpha-1,k)$$

$$= \frac{1}{r!} \sum_{\alpha=0}^{n} s(r,\alpha)(-1)^k \sum_{\gamma=0}^{\alpha} \gamma! S(\alpha,\gamma) \sum_{j=0}^{k} (-1)^j \binom{k}{j} \binom{j}{\gamma} j^{n-1}$$

$$= \frac{(-1)^k}{r!} \sum_{\gamma=0}^{n} \gamma! \sum_{j=0}^{k} (-1)^j \binom{k}{j} \binom{j}{\gamma} j^{n-1} \sum_{\alpha=\gamma}^{n} s(r,\alpha)S(\alpha,\gamma)$$

$$= \frac{(-1)^k}{r!} \sum_{\gamma=0}^{n} \gamma! \sum_{j=0}^{k} (-1)^j \binom{k}{j} \binom{j}{\gamma} j^{n-1} \binom{0}{r-\gamma}$$

$$= (-1)^k \sum_{j=0}^{k} (-1)^j \binom{k}{j} \binom{j}{r} j^{n-1}$$

$$= (-1)^k \sum_{j=r}^{k} (-1)^j \binom{k}{r} \binom{k-j}{j-r} j^{n-1}.$$

Since $\sum_{j=r}^{k}(-1)^j\binom{k}{r}\binom{k-j}{j-r}j^{n-1} = 0$ if $k < r$, we conclude that $\sum_{\alpha=0}^{n} s(r,\alpha)S(n+\alpha-1,k) = 0$ for $k < r$. Take the preceding line and let $k = r$ to obtain

$$\sum_{\alpha=0}^{n} s(r,\alpha)S(n+\alpha-1,k) = (-1)^r \sum_{j=r}^{r}(-1)^j\binom{r-j}{j-r}j^{n-1} = r^{n-1},$$

which is the second part of Olson's identity.

## 12.3    Functional Expansions Involving $s(n, k)$

Since $\binom{x}{n}n! = \sum_{k=0}^{n} s(n,k)x^n$ Stirling numbers of the first kind are a natural tool for expanding polynomials in $x$. We look at two particular examples which will prove useful for the development of the number theoretic definitions for Stirling numbers presented in Chapter 14. The first example is an expansion of $\prod_{k=0}^{n-1}(1+kx)$. Since $\prod_{k=0}^{n-1}(1+kx) = x^n n!\binom{\frac{1}{x}+n-1}{n}$ = $(-1)^n n! x^n\binom{-\frac{1}{x}}{n}$, we use Equation (12.1) to obtain

$$\prod_{k=0}^{n-1}(1+kx) = (-1)^n n! x^n \binom{-\frac{1}{x}}{n} = (-1)^n x^n \sum_{j=0}^{n}(-1)^j s(n,j)x^{-j}$$

$$= \sum_{j=0}^{n}(-1)^j s(n,n-j)x^j. \tag{12.26}$$

We may also use Stirling numbers of the first kind to expand $\binom{x+n}{n}$. The Vandermonde convolution when combined with Equation (12.1) implies that

$$\binom{x+n}{n} = \sum_{j=0}^{n}\binom{n}{n-j}\binom{x}{j} = \sum_{\alpha=0}^{n} x^\alpha \sum_{j=0}^{n}\binom{n}{j}\frac{s(j,\alpha)}{j!}. \tag{12.27}$$

Equation (12.27) is important since the coefficient of $x^\alpha$, namely $q_\alpha^n = \sum_{j=0}^{n}\binom{n}{j}\frac{s(j,\alpha)}{j!}$, provides an alternative recursive formula for $s(n,k)$. To derive this formula observe that

$$\sum_{k=0}^{n}\binom{n}{k}q_\alpha^k = \sum_{k=0}^{n}(-1)^k\binom{n}{k}\sum_{j=0}^{k}\binom{k}{j}\frac{s(j,\alpha)}{j!}$$

$$= \sum_{j=0}^{n}\frac{s(j,\alpha)}{j!}\sum_{k=j}^{n}(-1)^k\binom{n}{k}\binom{k}{j}$$

$$= (-1)^n\frac{s(n,\alpha)}{n!}, \qquad \text{by Eq. (6.25).}$$

From these calculation we deduce that

$$s(n,k) = n! \sum_{j=0}^{n} (-1)^{j+n} \binom{n}{j} \sum_{\alpha=0}^{j} \binom{j}{\alpha} \frac{s(\alpha,k)}{\alpha!}. \tag{12.28}$$

Equation (12.27) should be compared to the reciprocal expansion

$$\frac{1}{\binom{n+x}{n}} = n! \sum_{j=0}^{\infty} \frac{(-1)^j}{x^{j+n}} S(n+j,n). \tag{12.29}$$

To derive Equation (12.29) first observe that

$$\frac{1}{x+k} = \frac{1}{x\left(1+\frac{k}{x}\right)} = \frac{1}{x} \sum_{j=0}^{\infty} (-1)^j \left(\frac{k}{x}\right)^j = \sum_{j=0}^{\infty} (-1)^j \frac{k^j}{x^{j+1}}. \tag{12.30}$$

We combine Equation (12.30) with an application of Melzak's formula. In particular

$$\frac{1}{\binom{n+x}{n}} = \sum_{k=0}^{n} (-1)^k \binom{n}{k} \frac{x}{k+x}, \qquad \text{by Eq. (7.10)}$$

$$= \sum_{k=0}^{n} (-1)^k \binom{n}{k} \sum_{j=0}^{\infty} (-1)^j \frac{k^j}{x^j}$$

$$= \sum_{j=0}^{\infty} (-1)^{j+n} \frac{1}{x^j} (-1)^n \sum_{k=0}^{n} (-1)^k \binom{n}{k} k^j$$

$$= n! \sum_{j=n}^{\infty} (-1)^{j+n} \frac{1}{x^j} S(j,n), \qquad \text{by Eq. (9.21)}$$

$$= n! \sum_{j=0}^{\infty} (-1)^j \frac{1}{x^{j+n}} S(n+j,n).$$

Equation (12.29) provides an expansion for $\prod_{k=1}^{n} (1-kx)^{-1}$. Since $\binom{n+x}{n} = \prod_{k=1}^{n} \frac{x+k}{k} = \prod_{k=1}^{n} \left(1+\frac{x}{k}\right)$, we readily show that

$$\binom{n-\frac{1}{x}}{n} = \frac{(-1)^n}{n! x^n} \prod_{k=1}^{n} (1-kx), \qquad n \geq 1. \tag{12.31}$$

Combine Equation (12.31) with Equation (12.29) to conclude that

$$\frac{1}{\prod_{k=1}^{n}(1-kx)} = \sum_{j=0}^{\infty} S(n+j,n) x^j. \tag{12.32}$$

## 12.4   Derivative Expansions Involving $s(n, k)$

In the last section we used Equation (12.1) to expand various polynomials. In this section Equation (12.1) provides a means of expanding the operator $D_z^n$. Equation (12.1) formally implies that $\binom{D_z}{n} = \sum_{j=0}^{n} \frac{s(n,j)}{n!} D_z^j$. Recall Equation (8.18) which says

$$\binom{D_z}{n} y = \frac{x^n}{n!} D_x^n y, \qquad \text{where } y \text{ is a function of } z \text{ and } x = e^z. \quad (12.33)$$

We combine this result with the symbolic expansion of $\binom{D_z}{n}$ in terms of $s(n, k)$ to obtain

$$\binom{D_z}{n} y = \frac{x^n}{n!} D_x^n y = \sum_{j=0}^{n} \frac{s(n, j)}{n!} D_z^j y,$$

which is equivalent to

$$D_x^n y = x^{-n} \sum_{j=0}^{n} s(n, j) D_z^j y, \qquad x = e^z. \quad (12.34)$$

Since $y$ is a function of $z$ and $z = \ln x$, Equation (12.34) is often written as

$$D_x^n f(\ln x) = x^{-n} \sum_{j=0}^{n} s(n, j) D_z^j f(z), \qquad z = \ln x. \quad (12.35)$$

An important example of Equation (12.35) occurs when $f(z) = z^\alpha$ where $\alpha$ is a nonnegative integer. Equation (12.35) becomes

$$D_x^n (\ln x)^\alpha = x^{-n} \sum_{k=0}^{n} s(n, k) D_z^k z^\alpha = x^{-n} \sum_{k=0}^{\alpha} s(n, k) \frac{\alpha!}{(\alpha - k)!} z^{\alpha-k}$$

$$= \frac{\alpha!}{x^n} \sum_{k=0}^{\alpha} s(n, k) \frac{(\ln x)^{\alpha-k}}{(\alpha - k)!} = \frac{\alpha!}{x^n} \sum_{k=0}^{\alpha} s(n, \alpha - k) \frac{(\ln x)^k}{k!}.$$

In summary we have shown that

$$D_x^n (\ln x)^\alpha = \frac{\alpha!}{x^n} \sum_{k=0}^{\alpha} s(n, \alpha - k) \frac{(\ln x)^k}{k!}. \quad (12.36)$$

If $\alpha = 1$ and $n \geq 1$, Equation (12.36) becomes

$$D_z^n \ln x = (-1)^{n-1} \frac{(n - 1)!}{x^n}, \qquad n \geq 1. \quad (12.37)$$

Another application of Equation (12.36) shows that

$$\sum_{k=0}^{n} S(n, k) x^k D_x^k (\ln x)^n = n!. \quad (12.38)$$

To prove Equation (12.38) observe that

$$\sum_{n=0}^{\alpha} S(\alpha,n)x^n D_x^n(\ln x)^\alpha = \alpha! \sum_{n=0}^{\alpha} S(\alpha,n) \sum_{k=0}^{\alpha} \frac{(\ln x)^k}{k!} s(n,\alpha-k)$$

$$= \alpha! \sum_{k=0}^{\alpha} \frac{(\ln x)^k}{k!} \sum_{n=0}^{\alpha} S(\alpha,n)s(n,\alpha-k)$$

$$= \alpha! \sum_{k=0}^{\alpha} \frac{(\ln x)^k}{k!} \binom{0}{k} = \alpha!.$$

We next discuss the inversion of Equation (12.36). Equation (12.36) expands $D_x^n(\ln x)^\alpha$ in terms of $s(n,\alpha-k)$. We now want to expand $s(n,\alpha)$ in terms of $D_x^n(\ln x)^j$. To derive this inversion first take Equation (12.35) and interchange the roles of $x$ and $z$. This gives us

$$D_z^n f(\ln z) = z^{-n} \sum_{\alpha=0}^{n} s(n,\alpha)D_x^\alpha f(x), \qquad x = \ln z. \qquad (12.39)$$

Recall Hoppe's formula of the generalized chain rule, Equation (8.11), which states

$$D_z^n f(x) = \sum_{\alpha=0}^{n} D_x^\alpha f(x)\frac{(-1)^\alpha}{\alpha!} \sum_{j=0}^{\alpha}(-1)^j \binom{\alpha}{j} x^{\alpha-j} D_z^n x^j. \qquad (12.40)$$

Take Equation (12.40) and let $x = \ln z$ to obtain

$$D_z^n f(\ln z) = \sum_{\alpha=0}^{n} D_x^\alpha f(x)\frac{(-1)^\alpha}{\alpha!} \sum_{j=0}^{\alpha}(-1)^j \binom{\alpha}{j} (\ln z)^{\alpha-j} D_z^n (\ln z)^j. \qquad (12.41)$$

Compare the coefficient of $D_x^\alpha f(x)$ in Equation (12.39) with that of Equation (12.41). Since this coefficient is unique, we set the two variations equal to each other. After simplification we obtain

$$s(n,\alpha) = \frac{(-1)^\alpha z^n}{\alpha!} \sum_{j=0}^{\alpha}(-1)^j \binom{\alpha}{j} (\ln z)^{\alpha-j} D_z^n (\ln z)^j. \qquad (12.42)$$

Equation (12.42) is useful since it provides another differential operator formula for $s(n,\alpha)$. In particular let $z = 1$. The only nonzero summand occurs when $\alpha = j$, and Equation (12.42) reduces to

$$s(n,\alpha) = \frac{1}{\alpha!} D_z^n (\ln z)^\alpha \Big|_{z=1}. \qquad (12.43)$$

Equation (12.43) with $z \to z+1$ is equivalent to

$$s(n,\alpha) = \frac{1}{\alpha!} D_z^n (\ln(z+1))^\alpha \Big|_{z=0}, \qquad (12.44)$$

which is a restatement of Equation (12.5).

# Chapter 13

# Explicit Formulas for $s(n, n - k)$

We now come to a topic in which Henry Gould is a world renowned expert, explicit formulas for $s(n, n - k)$ in terms of $S(n, k)$. Professor Gould develops his formulas through the differentiation of the generating function associated with the Nörlund polynomial $B_k^{(n)}(x)$. Two excellent reference sources for Nörlund polynomials are Chapter 5 of *Calculus of Finite Differences* by Charles Jordan [1957] and the seminal work *Vorlesungen über Differenzenrechnung* by N. E. Nörlund [1924]. Let $n$ be a nonnegative integer. Define the Nörlund polynomial $B_k^{(n)}(x)$ to be the coefficient of $\frac{z^k}{k!}$ in the series expansion of $\frac{z^n e^{xz}}{(e^z - 1)^n}$, i.e.

$$\sum_{k=0}^{\infty} B_k^{(n)}(x) \frac{z^k}{k!} = \frac{z^n e^{xz}}{(e^z - 1)^n}, \qquad |x| < 2\pi. \tag{13.1}$$

We claim that

$$B_k^{(n+1)}(z + 1) = k! D_z^{n-k} \binom{z}{n}. \tag{13.2}$$

When interpreting Equation (13.2) note that a negative exponent in $D_z^{n-k}$ means *integrating* $\binom{z}{n}$ $n - k$ times. Section 13.2 presents Professor Gould's proof of Equation (13.2).

Take Equation (13.2) and let $z = 0$ to obtain

$$B_k^{(n+1)}(1) = k! D_z^{n-k} \binom{z}{n} \Bigg|_{z=0}. \tag{13.3}$$

Since $\binom{z}{n} = \sum_{k=0}^{n} \frac{s(n,k)}{n!} z^k$, we conclude that $\frac{s(n,k)}{n!} = \frac{1}{k!} D_z^k \binom{z}{n}\big|_{z=0}$ and rewrite Equation (13.3) as

$$B_k^{(n+1)}(1) = \frac{1}{\binom{n}{k}} s(n, n - k). \tag{13.4}$$

Recall that $[x^k]f(x)$ denotes the coefficient of $f(x)$ in the formal power series expansion of $f(x)$. Equation (13.1) is equivalent to

$$\frac{B_k^{(n+1)}(1)}{k!} = [x^k]\frac{x^{n+1}e^x}{(e^x-1)^{n+1}} = \frac{1}{k!}D_x^k\left[e^x\left(\frac{x}{e^x-1}\right)^{n+1}\right]_{x=0}.$$

If we combine this result with Equation (13.4) we find that

$$s(n, n-k) = \binom{n}{k}D_x^k\left[e^x\left(\frac{x}{e^x-1}\right)^{n+1}\right]_{x=0}. \qquad (13.5)$$

Equation (13.5) is the basis Professor Gould uses to develop his explicit formulas for $s(n, n-k)$. The formulas depend on various evaluations of $D_x^k\left[e^x\left(\frac{x}{e^x-1}\right)^{n+1}\right]_{x=0}$. By the Leibniz rule we have

$$s(n, n-k) = \binom{n}{k}\sum_{j=0}^{k}\binom{k}{j}D_x^{k-j}e^x D_x^j\left(\frac{x}{e^x-1}\right)^{n+1}\Bigg|_{x=0}$$

$$= \binom{n}{k}\sum_{j=0}^{k}\binom{k}{j}D_x^j\left(\frac{x}{e^x-1}\right)^{n+1}\Bigg|_{x=0}. \qquad (13.6)$$

To evaluate $D_x^j\left(\frac{x}{e^x-1}\right)^{n+1}\Big|_{x=0}$ we turn to the generating function for $S(n+j, n)$ provided by Equation (9.59), namely

$$\left(\frac{e^x-1}{x}\right)^n = n!\sum_{j=0}^{\infty}S(n+j,n)\frac{x^j}{(n+j)!}. \qquad (13.7)$$

Let $f(x) = \left(\frac{e^x-1}{x}\right)^{n+1}$. We have to evaluate $D_x^j\frac{1}{f(x)}\Big|_{x=0}$. To evaluate the derivative of a reciprocal function we use a special case of the generalized chain rule, Equation (8.28), which states

$$D_x^j\left[\frac{1}{f(x)}\right] = \sum_{\alpha=0}^{j}(-1)^\alpha\binom{j+1}{\alpha+1}\frac{1}{(f(x))^{\alpha+1}}D_x^j(f(x))^\alpha. \qquad (13.8)$$

Take Equation (13.8) and let $f(x) = \left(\frac{e^x - 1}{x}\right)^{n+1}$ to obtain

$$D_x^j \left(\frac{x}{e^x - 1}\right)^{n+1}_{x=0}$$

$$= \sum_{\alpha=0}^{j} (-1)^\alpha \binom{j+1}{\alpha+1} \left(\frac{x}{e^x - 1}\right)^{\alpha n + \alpha + n + 1} D_x^j \left(\frac{e^x - 1}{x}\right)^{(n+1)\alpha} \Bigg|_{x=0}$$

$$= \sum_{\alpha=0}^{j} (-1)^\alpha \binom{j+1}{\alpha+1} \left(\frac{x}{e^x - 1}\right)^{\alpha n + \alpha + n + 1}$$

$$* \sum_{k=0}^{\infty} \frac{(\alpha n + \alpha)! S(k + \alpha n + \alpha, \alpha n + \alpha)}{(k + \alpha n + \alpha)!} D_x^j x^k \Bigg|_{x=0}$$

$$= \sum_{\alpha=0}^{j} (-1)^\alpha \binom{j+1}{\alpha+1} \left(\frac{x}{e^x - 1}\right)^{\alpha n + \alpha + n + 1}$$

$$* \sum_{k=j}^{\infty} j! \binom{k}{j} \frac{(\alpha n + \alpha)! S(k + \alpha n + \alpha, \alpha n + \alpha)}{(k + \alpha n + \alpha)!} x^{k-j} \Bigg|_{x=0}$$

$$= \sum_{\alpha=0}^{j} (-1)^\alpha \binom{j+1}{\alpha+1} \sum_{k=j}^{\infty} j! \binom{k}{j} \frac{(\alpha n + \alpha)! S(k + \alpha n + \alpha, \alpha n + \alpha)}{(k + \alpha n + \alpha)!} x^{k-j} \Bigg|_{x=0}$$

$$= \sum_{\alpha=0}^{j} (-1)^\alpha \binom{j+1}{\alpha+1} \frac{j!(\alpha n + \alpha)!}{(j + \alpha n + \alpha)!} S(j + \alpha n + \alpha, \alpha n + \alpha). \tag{13.9}$$

We now combine Equations (13.6) and (13.9) to obtain

$$s(n, n-k) = \binom{n}{k} \sum_{j=0}^{k} \binom{k}{j} \sum_{\alpha=0}^{j} (-1)^\alpha \binom{j+1}{\alpha+1} \frac{S(j + \alpha n + \alpha, \alpha n + \alpha)}{\binom{j + \alpha n + \alpha}{\alpha}}, \tag{13.10}$$

an explicit formula for $s(n, n-k)$ in terms of Stirling numbers of the second kind.

We now derive another double sum formula for $s(n, n-k)$ by applying Equation (8.20) to $D_x^j \left(\frac{x}{e^x - 1}\right)^{n+1}$. Equation (8.20) states

$$D_x^n z^\alpha = \sum_{j=0}^{n} (-1)^{n-j} \binom{\alpha}{j} \binom{\alpha - j - 1}{n - j} z^{\alpha - j} D_x^n z^j. \tag{13.11}$$

Take Equation (13.11) and let $\alpha \to -\alpha$ to obtain

$$D_x^n z^{-\alpha} = \sum_{j=0}^{n} (-1)^{n-j} \binom{-\alpha}{j} \binom{-\alpha - j - 1}{n - j} z^{-\alpha - j} D_x^n z^j$$

$$= \sum_{j=0}^{n} (-1)^j \binom{\alpha + j - 1}{j} \binom{\alpha + n}{n - j} \frac{1}{z^{\alpha + j}} D_x^n z^j. \tag{13.12}$$

It is not hard to show that

$$\binom{\alpha+j-1}{j}\binom{\alpha+n}{n-j} = \frac{\alpha}{\alpha+j}\binom{\alpha+n}{n}\binom{n}{j}. \tag{13.13}$$

We substitute Equation (13.13) into Equation (13.12) to find that

$$D_x^n z^{-\alpha} = \alpha\binom{\alpha+n}{n}\sum_{j=0}^{n}(-1)^j\binom{n}{j}\frac{1}{z^{\alpha+j}(\alpha+j)}D_z^n z^j. \tag{13.14}$$

In Equation (13.14) let $z = \frac{e^x-1}{x}$ and observe that

$$D_x^n\left(\frac{x}{e^x-1}\right)^\alpha\Big|_{x=0}$$

$$= \alpha\binom{\alpha+n}{n}\sum_{j=0}^{n}(-1)^j\binom{n}{j}\frac{1}{\alpha+j}\left(\frac{x}{e^x-1}\right)^{\alpha+j}D_x^n\left(\frac{e^x-1}{x}\right)^j\Big|_{x=0}$$

$$= \alpha\binom{\alpha+n}{n}\sum_{j=0}^{n}(-1)^j\binom{n}{j}\frac{1}{\alpha+j}D_x^n\left(\frac{e^x-1}{x}\right)^j\Big|_{x=0}$$

$$= \alpha\binom{\alpha+n}{n}\sum_{j=0}^{n}(-1)^j\binom{n}{j}\frac{1}{\alpha+j}\sum_{k=0}^{\infty}\frac{j!S(k+j,j)}{(k+j)!}D_x^n x^k\Big|_{x=0}$$

$$= \alpha\binom{\alpha+n}{n}\sum_{j=0}^{n}(-1)^j\binom{n}{j}\frac{1}{\alpha+j}\sum_{k=n}^{\infty}\binom{k}{n}\frac{n!j!}{(k+j)!}S(k+j,j)x^{k-n}\Big|_{x=0}$$

$$= \alpha\binom{\alpha+n}{n}\sum_{j=0}^{n}(-1)^j\binom{n}{j}\frac{1}{(\alpha+j)\binom{n+j}{j}}S(n+j,j). \tag{13.15}$$

We combine Equation (13.15) with Equation (13.6) to obtain

$$s(n,n-k)$$

$$= \binom{n}{k}\sum_{j=0}^{k}\binom{k}{j}(n+1)\binom{n+1+j}{j}\sum_{\alpha=0}^{j}(-1)^\alpha\binom{j}{\alpha}\frac{1}{(n+1+\alpha)\binom{j+\alpha}{j}}S(j+\alpha,\alpha)$$

$$= \binom{n}{k}\sum_{j=0}^{k}\binom{k}{j}\binom{n+j}{j}\sum_{\alpha=0}^{j}(-1)^\alpha\binom{j}{\alpha}\frac{(n+j+1)S(j+\alpha,\alpha)}{(n+\alpha+1)\binom{j+\alpha}{\alpha}}. \tag{13.16}$$

## 13.1 Schläfli's Formula $s(n,n-k)$

Our two expansions for $s(n,n-k)$ in terms of Stirling numbers of the second kind involve double summations. The Swiss mathematician Ludwig Schläfli was able to derive a formula for $s(n,n-k)$ in terms of Stirling numbers of the second kind which only involved a *single* summation [Schläfi, 1852]. By applying formulas used in the derivations of Equations (13.10) and (13.16) we will rederive Schläfli's formula.

To begin note that

$$\frac{n}{n+\alpha}\binom{n+k}{k}\binom{k}{\alpha} = (-1)^\alpha\binom{-n}{\alpha}\binom{k+n}{k-\alpha}. \qquad (13.17)$$

Equation (13.17) allows us to rewrite Equation (13.14) as

$$z^\alpha D_x^n z^{-\alpha} = \sum_{j=0}^{n}\binom{-\alpha}{j}\binom{n+\alpha}{n-j}z^{-j}D_x^n z^j. \qquad (13.18)$$

We next observe that

$$\frac{\binom{-n}{\alpha}}{\binom{k+\alpha}{\alpha}} = (-1)^k\frac{\binom{k-n}{k+\alpha}}{\binom{n-1}{k}}. \qquad (13.19)$$

Combining Equations (13.17) and (13.19) implies that

$$\frac{n\binom{n+k}{k}\binom{k}{\alpha}}{(n+\alpha)\binom{k+\alpha}{\alpha}} = (-1)^{k+\alpha}\frac{\binom{k+n}{k-\alpha}\binom{k-n}{k+\alpha}}{\binom{n-1}{k}}. \qquad (13.20)$$

Take Equation (13.15) let $n \to k$ and $\alpha \to n$ to obtain

$$D_x^k\left(\frac{x}{e^x-1}\right)^n\bigg|_{x=0} = n\binom{n+k}{k}\sum_{\alpha=0}^{k}(-1)^\alpha\binom{k}{\alpha}\frac{S(\alpha+k,\alpha)}{(n+\alpha)\binom{k+\alpha}{\alpha}}. \qquad (13.21)$$

Substitute Equation (13.20) into Equation (13.21) to find that

$$D_x^k\left(\frac{x}{e^x-1}\right)^n\bigg|_{x=0} = \frac{(-1)^k}{\binom{n-1}{k}}\sum_{\alpha=0}^{k}\binom{k+n}{k-\alpha}\binom{k-n}{k+\alpha}S(\alpha+k,\alpha). \qquad (13.22)$$

We now need the following theorem:

**Theorem 13.1.** *Let $k$ be a positive integer and $n \neq 0$. Then*

$$D_x^k\left[e^x\left(\frac{x}{e^x-1}\right)^{n+1}\right]_{x=0} = \frac{n-k}{n}D_x^k\left(\frac{x}{e^x-1}\right)^n\bigg|_{x=0}. \qquad (13.23)$$

**Remark 13.1.** Equation (13.23) often appears in the literature as

$$B_k^{(n+1)}(1) = \frac{n-k}{n}B_k^{(n)}(0). \qquad (13.24)$$

**Proof:** By Taylor's theorem we have

$$
e^z \left( \frac{z}{e^z - 1} \right)^{n+1} = \sum_{k=0}^{\infty} \frac{z^k}{k!} D_x^k \left[ e^x \left( \frac{x}{e^x - 1} \right)^{n+1} \right]_{x=0} . \tag{13.25}
$$

On the other hand,

$$
\sum_{k=0}^{\infty} \frac{z^k}{k!} \frac{n-k}{n} D_x^k \left( \frac{x}{e^x - 1} \right)^n \Bigg|_{x=0}
$$

$$
= \sum_{k=0}^{\infty} \frac{z^k}{k!} D_x^k \left( \frac{x}{e^x - 1} \right)^n \Bigg|_{x=0} - \frac{1}{n} \sum_{k=0}^{\infty} \frac{k z^k}{k!} D_x^k \left( \frac{x}{e^x - 1} \right)^n \Bigg|_{x=0}
$$

$$
= \left( \frac{z}{e^z - 1} \right)^n - \frac{z}{n} \sum_{k=0}^{\infty} \frac{k z^{k-1}}{k!} D_x^k \left( \frac{x}{e^x - 1} \right)^n \Bigg|_{x=0}
$$

$$
= \left( \frac{z}{e^z - 1} \right)^n - \frac{z}{n} D_z \left( \frac{z}{e^z - 1} \right)^n
$$

$$
= \left( \frac{z}{e^z - 1} \right)^n - \left( \frac{z}{e^z - 1} \right)^n \left[ 1 - \frac{z e^z}{e^z - 1} \right]
$$

$$
= \left( \frac{z}{e^z - 1} \right)^n \cdot \frac{z e^z}{e^z - 1} = e^z \left( \frac{z}{e^z - 1} \right)^{n+1} .
$$

Combining this calculation with Equation (13.25) shows that

$$
\sum_{k=0}^{\infty} \frac{z^k}{k!} D_x^k \left[ e^x \left( \frac{x}{e^x - 1} \right)^{n+1} \right]_{x=0} = \sum_{k=0}^{\infty} \frac{z^k}{k!} \frac{n-k}{n} D_x^k \left( \frac{x}{e^x - 1} \right)^n \Bigg|_{x=0} .
$$

Equation (13.23) results by comparing the coefficients $\frac{z^k}{k!}$.  $\square$

Equation (13.5) when combined with Equation (13.23) implies that

$$
s(n, n-k) = \binom{n}{k} D_x^k \left[ e^x \left( \frac{x}{e^x - 1} \right)^{n+1} \right]_{x=0}
$$

$$
= \binom{n-1}{k} D_x^k \left( \frac{x}{e^x - 1} \right)^n \Bigg|_{x=0} . \tag{13.26}
$$

If we go back to Equation (13.1) and let $x = 0$ we find that

$$
\sum_{k=0}^{\infty} B_k^{(n)}(0) \frac{x^k}{k!} = \left( \frac{x}{e^x - 1} \right)^n ,
$$

which in turn implies that $D_x^k \left( \frac{x}{e^x - 1} \right)^n \Big|_{x=0} = B_k^n(0)$. Hence Equation (13.26) may be rewritten as

$$
s(n, n-k) = \binom{n-1}{k} B_k^{(n)}(0), \tag{13.27}
$$

a result which complements Equation (13.4).

We combine Equation (13.26) with Equation (13.9) to obtain

$$s(n, n - k) = \binom{n-1}{k} \sum_{j=0}^{k} (-1)^j \binom{k+1}{j+1} \frac{S(k + jn, jn)}{\binom{k+jn}{k}}, \qquad (13.28)$$

a single summation formula for $s(n, n - k)$ in terms of Stirling numbers of the second kind. The inverse relation to Equation (13.28) is

$$S(n + k, n) = \binom{n+k}{k} \sum_{j=0}^{k} (-1)^j \binom{k+1}{j+1} \frac{s(jn, jn - k)}{\binom{jn-1}{k}}. \qquad (13.29)$$

To prove Equation (13.29) first observe that Equation (13.7) implies

$$S(n + k, n) = \binom{n+k}{k} D_x^k \left( \frac{e^x - 1}{x} \right)^n \Bigg|_{x=0}. \qquad (13.30)$$

Next take Equation (13.30) and combine it with Equation (13.8) to obtain

$$D_x^k \left( \frac{e^x - 1}{x} \right)^n \Bigg|_{x=0} = \sum_{j=0}^{k} (-1)^j \binom{k+1}{j+1} \frac{1}{\left( \frac{x}{e^x-1} \right)^{n(j+1)}} D_x^k \left( \frac{x}{e^x - 1} \right)^{jn} \Bigg|_{x=0}$$

$$= \sum_{j=0}^{k} (-1)^j \binom{k+1}{j+1} \frac{s(jn, jn - k)}{\binom{jn-1}{k}}, \qquad \text{by Eq. (13.26).}$$

An alternative single summation formula for $s(n, n - k)$ is obtained by combining Equation (13.26) with Equation (13.22):

$$s(n, n - k) = (-1)^k \sum_{\alpha=0}^{k} \binom{k+n}{k \quad \alpha} \binom{k-n}{k \mid \alpha} S(\alpha + k, \alpha). \qquad (13.31)$$

Equation (13.31) is Schläfli's formula for $s(n, n - k)$. Since the $-1$-Transformation implies that $\binom{k-n}{k+\alpha} = (-1)^{k+\alpha} \binom{n+\alpha-1}{k+\alpha} = (-1)^{k+\alpha} \binom{n+\alpha-1}{n-k-1}$, and $\binom{k+n}{k-\alpha} = \binom{n+k}{n+\alpha}$, Equation (13.31) is equivalent to

$$s(n, n - k) = \sum_{\alpha=0}^{k} (-1)^\alpha \binom{n+\alpha-1}{n-k-1} \binom{n+k}{n+\alpha} S(\alpha + k, \alpha). \qquad (13.32)$$

We end this section by inverting Equation (13.31). We claim that [Nörlund, 1924, p. 147]

$$\left( \frac{\ln(x+1)}{x} \right)^n = n \sum_{k=0}^{\infty} \frac{x^k}{k!} \frac{B_k^{(n+k)}(0)}{n+k}, \qquad |x| < 1. \qquad (13.33)$$

Equation (13.33) is a consequence of Equation (12.5), namely that

$$\left(\frac{\ln(1+x)}{x}\right)^n = n! \sum_{j=0}^{\infty} \frac{x^j}{(n+j)!} s(j+n,n) = \sum_{j=0}^{\infty} \frac{x^j}{j!} \frac{s(j+n,n)}{\binom{n+j}{j}}. \quad (13.34)$$

Since Equation (13.27) implies that $s(n+j,n) = \binom{n+j-1}{j} B_j^{(n+j)}(0)$, we conclude that the right side of Equation (13.34) is identical to the right side of Equation (13.33). In Equation (13.33) let $n \to -n$ to obtain

$$\left(\frac{x}{\ln(x+1)}\right)^n = \sum_{j=0}^{\infty} \frac{x^j}{j!} \frac{n}{n-j} B_k^{(j-n)}(0). \quad (13.35)$$

We claim that

$$B_k^{(-n)}(0) = \frac{1}{\binom{n+k}{k}} S(n+k,n). \quad (13.36)$$

Equation (13.36) follows from Equations (13.1) and (13.7). Take Equation (13.1) with $x = 0$ and $n \to -n$ to obtain

$$\sum_{j=0}^{\infty} B_j^{(-n)}(0) \frac{x^j}{j!} = \left(\frac{e^x - 1}{x}\right)^n = \sum_{j=0}^{\infty} \frac{n! x^j}{(n+j)!} S(n+j,n).$$

Comparing the coefficients of $x^k$ in the two series gives us Equation (13.36).

Equation (13.36) implies that $B_j^{(j-n)}(0) = \frac{1}{\binom{n}{j}} S(n, n-j)$. If we substitute this result into Equation (13.35) we find that

$$\left(\frac{x}{\ln(x+1)}\right)^n = \sum_{j=0}^{\infty} \frac{x^j}{j!} \frac{n}{(n-j)\binom{n}{j}} S(n, n-j) = \sum_{j=0}^{\infty} \frac{x^j}{j!} \frac{S(n, n-j)}{\binom{n-1}{j}}. \quad (13.37)$$

Taylor's theorem, when combined with Equation (13.37), implies that

$$D_x^k \left(\frac{x}{\ln(x+1)}\right)^n \bigg|_{x=0} = \frac{S(n, n-k)}{\binom{n-1}{k}}. \quad (13.38)$$

To compute $D_x^k \left(\frac{x}{\ln(x+1)}\right)^n \big|_{x=0}$ we use a simple variation of Equation (13.12), namely

$$D_x^k z^{-n} = \sum_{j=0}^{k} \binom{-n}{j} \binom{n+k}{k-j} z^{-n-j} D_x^k z^j. \quad (13.39)$$

Since Equation (13.19) implies that $\binom{-n}{j}\binom{n-1}{k} = (-1)^k \binom{k-n}{k+j}\binom{k+j}{j}$, we may rewrite Equation (13.39) as

$$D_x^k z^{-n} = \frac{(-1)^k}{\binom{n-1}{k}} \sum_{j=0}^{k} \binom{k-n}{k+j} \binom{k+n}{k-j} \binom{k+j}{j} z^{-n-j} D_x^k z^j. \quad (13.40)$$

Take Equation (13.40) and let $z = \frac{\ln(x+1)}{x}$. Since $\lim_{x \to 0} \frac{\ln(x+1)}{x} = 1$, we find that

$$
D_x^k \left( \frac{x}{\ln(x+1)} \right)^n \bigg|_{x=0}
$$

$$
= \frac{(-1)^k}{\binom{n-1}{k}} \sum_{j=0}^{k} \binom{k-n}{k+j} \binom{k+n}{k-j} \binom{k+j}{j} D_x^k \left( \frac{\ln(x+1)}{x} \right)^j \bigg|_{x=0}
$$

$$
= \frac{(-1)^k}{\binom{n-1}{k}} \sum_{j=0}^{k} \binom{k-n}{k+j} \binom{k+n}{k-j} \binom{k+j}{j} \frac{s(k+j,j)}{\binom{k+j}{j}}
$$

$$
= \frac{(-1)^k}{\binom{n-1}{k}} \sum_{j=0}^{k} \binom{k-n}{k+j} \binom{k+n}{k-j} s(k+j,j). \tag{13.41}
$$

If we combine Equation (13.41) with Equation (13.38) we find that

$$
S(n, n-k) = (-1)^k \sum_{j=0}^{k} \binom{k-n}{k+j} \binom{k+n}{k-j} s(k+j,j). \tag{13.42}
$$

Equation (13.42) is the desired inverse relation to Equation (13.31).

## 13.2 Proof of Equation (13.2)

To prove Equation (13.2) define $S_n(x,z) = \sum_{k=0}^{\infty} x^k D_z^{n-k} \binom{z}{n}$. We first show that

$$
S_{n+1}(x, z+1) - S_n(x,z) + \frac{z+1}{n+1} D_z S_n(x,z) - \frac{x}{n+1} D_x S_n(x,z). \tag{13.43}
$$

Next we show that $f_n(x,z) = \frac{x^{n+1} e^{(z+1)x}}{(e^x - 1)^{n+1}}$ also obeys Recurrence (13.43). We finally show that $f_0(x,z) = S_0(x,z)$ and $f_1(x,z) = S_1(x,z)$. From this we conclude that $f_n(x,z) = S_n(x,z)$ for all nonnegative integers $n$. Since Equation (13.1) implies that $f_n(x,z) = \sum_{k=0}^{\infty} B_k^{(n+1)}(z+1) \frac{x^k}{k!}$, we will have completed our proof.

Begin by observing that

$$
S_{n+1}(x, z+1) = \sum_{k=0}^{\infty} x^k D_w^{n+1-k} \binom{z+1}{n+1} \bigg|_{w=z+1}. \tag{13.44}
$$

Now

$$D_w^{m+1}\binom{z+1}{n+1}\Bigg|_{w=z+1} = D_w^m\left[D_w\binom{z+1}{n+1}\right]_{w=z+1}$$

$$= \frac{1}{n+1}D_w^m\left[D_w\left[(z+1)\binom{z}{n}\right]\right]_{w=z+1}$$

$$= \frac{1}{n+1}D_w^m\left[(z+1)D_w\binom{z}{n} + \binom{z}{n}D_{z+1}(z+1)\right]_{w=z+1}$$

$$= \frac{1}{n+1}D_w^m\left[(z+1)D_w\binom{z}{n}\right]_{w=z+1} + \frac{1}{n+1}D_w^m\binom{z}{n}\Bigg|_{w=z+1}$$

$$= \frac{1}{n+1}\left[\sum_{j=0}^m\binom{m}{j}D_w^j(z+1)D_w^{m-j}D_w\binom{z}{n}\right]_{w=z+1} + \frac{1}{n+1}D_w^m\binom{z}{n}\Bigg|_{w=z+1}$$

$$= \frac{1}{n+1}\left[(z+1)D_w^{m+1}\binom{z}{n}\right]_{w=z+1} + \frac{m}{n+1}D_w^m\binom{z}{n}\Bigg|_{w=z+1}$$

$$+ \frac{1}{n+1}D_w^m\binom{z}{n}\Bigg|_{w=z+1}.$$

The last equality made use of the fact that the only nonzero summands occur when $j = 0$ and $j = 1$.

We now recall Equation (8.12), namely $D_z^n f(x)|_{z=a+bx} = \frac{1}{b^n}D_x^n f(x)$. The preceding line becomes

$$D_w^{m+1}\binom{z+1}{n+1}\Bigg|_{w=z+1} = \frac{m+1}{n+1}D_z^m\binom{z}{n} + \frac{z+1}{n+1}D_z^{m+1}\binom{z}{n}. \qquad (13.45)$$

Take Equation (13.45) and let $m \to n - k$ to obtain

$$D_w^{n-k+1}\binom{z+1}{n+1}\Bigg|_{w=z+1} = \frac{n-k+1}{n+1}D_z^{n-k}\binom{z}{n} + \frac{z+1}{n+1}D_z^{n-k+1}\binom{z}{n}.$$

$$(13.46)$$

Substitute Equation (13.46) into Equation (13.44) to obtain

$$S_{n+1}(x, z+1) = \sum_{k=0}^{\infty} x^k \left[ \frac{n-k+1}{n+1} D_z^{n-k}\binom{z}{n} + \frac{z+1}{n+1} D_z^{n-k+1}\binom{z}{n} \right]$$

$$= \sum_{k=0}^{\infty} x^k \left[ \left(1 - \frac{k}{n+1}\right) D_z^{n-k}\binom{z}{n} + \frac{z+1}{n+1} D_z^{n-k+1}\binom{z}{n} \right]$$

$$= \sum_{k=0}^{\infty} x^k D_z^{n-k}\binom{z}{n} - \frac{x}{n+1} \sum_{k=0}^{\infty} k x^{k-1} D_z^{n-k}\binom{z}{n}$$

$$+ \frac{z+1}{n+1} D_z \sum_{k=0}^{\infty} x^k D_z^{n-k}\binom{z}{n}$$

$$= S_n(x, z) - \frac{x}{n+1} D_x S_n(x, z) + \frac{z+1}{n+1} D_z S_n(x, z),$$

which is Recurrence (13.43).

Let $f_n(x, z) = \frac{x^{n+1} e^{(z+1)x}}{(e^x - 1)^{n+1}}$. A computer algebra program such as Maple readily shows that $f_n(x, z)$ obeys Recurrence (13.43), i.e.

$$f_{n+1}(x, z+1) = f_n(x, z) - \frac{x}{n+1} D_x f_n(x, z) + \frac{z+1}{n+1} D_z f_n(x, z).$$

It remains to show that $S_0(x, z) = f_0(x, z)$ and $S_1(x, z) = f_1(x, z)$. By definition

$$S_0(x, z) = \sum_{j=0}^{\infty} x^j D_z^{-j} 1 = \sum_{j=0}^{\infty} x^j \left[ \frac{z^j}{j!} + \sum_{r=0}^{j-1} b_r z^r \right],$$

where $\{b_r\}_{r=0}^{j}$ are a collection of coefficients we do not specifically determine. We observe that a $j^{th}$ anti-derivative of 1 is a polynomial in $z$ of degree $j$ with highest order term $\frac{z^j}{j!}$. By definition $f_0(x, z) = \frac{x e^{(z+1)x}}{e^x - 1}$. We need to expand this function as a formal power series in $x$. Write $\frac{x}{e^x-1} = \sum_{k=0}^{\infty} a_k x^k = 1 + \sum_{k=1}^{\infty} a_k x^k$, where $\{a_k\}_{k=1}^{\infty}$ are undetermined coefficients. We justify $a_0 = 1$ by computing $\lim_{x \to 0} \frac{x}{e^x-1} = 1$. Using this notation we find that

$$f_0(x, z) = \left(\frac{x}{e^x - 1}\right) e^{(z+1)x} = \sum_{k=0}^{\infty} a_k x^k \sum_{k=0}^{\infty} \frac{(z+1)^k}{k!} x^k$$

$$= \sum_{j=0}^{\infty} x^j \sum_{k=0}^{j} a_{j-k} \frac{(z+1)^k}{k!}.$$

The coefficient of $x^j$ is $\sum_{k=0}^{j} a_{j-k} \frac{(z+1)^k}{k!}$, a polynomial in $z$ of degree $j$. It remains to show that the coefficient of $z^j$ is $\frac{1}{j!}$. Since $\{a_j\}_{j=0}^{\infty}$ are independent of $z$, the only way to obtain $z^j$ is from the term with $k = j$, namely $a_0 \frac{(z+1)^j}{j!} = \frac{(z+1)^j}{j!} = \frac{1}{j!} \sum_{i=0}^{j} \binom{j}{i} z^i = \frac{z^j}{j!} + \frac{1}{j!} \sum_{i=0}^{j-1} \binom{j}{i} z^i$. Hence the coefficient of $z^j$ is indeed $\frac{1}{j!}$, which in turn implies that

$$f_0(x,z) = \left( \frac{x}{e^x - 1} \right) e^{(z+1)x} = \sum_{j=0}^{\infty} x^j \sum_{k=0}^{j} a_{j-k} \frac{(z+1)^k}{k!}$$

$$= \sum_{j=0}^{\infty} x^j D_z^{-j} 1 = S_0(x,z).$$

We now look at $S_1(x,z)$. By definition we have

$$S_1(x,z) = \sum_{j=0}^{\infty} x^j D_z^{1-j} z = 1 + \sum_{j=1}^{\infty} x^j D_z^{1-j} z = 1 + \sum_{j=0}^{\infty} x^{j+1} D_z^{-j} z$$

$$= 1 + \sum_{j=0}^{\infty} x^{j+1} \left[ \frac{z^{j+1}}{(j+1)!} + \sum_{r=0}^{j} B_r z^r \right],$$

where the last equality makes use of the fact that a $j^{th}$ anti-derivative of $z$ is a polynomial in $z$ of degree $j + 1$ with highest order term $\frac{z^{j+1}}{(j+1)!}$.

Next write $\frac{x^2}{(e^x-1)^2} = \sum_{j=0}^{\infty} A_j x^j$ with $A_0 = 1$ since $\lim_{x \to 0} \frac{x^2}{(e^x-1)^2} = 1$. Then

$$f_1(x,z) = \left( \frac{x}{e^x - 1} \right)^2 e^{(z+1)x} = \sum_{k=0}^{\infty} A_k x^k \sum_{k=0}^{\infty} \frac{(z+1)^k}{k!} x^k$$

$$= \sum_{j=0}^{\infty} x^j \sum_{k=0}^{j} A_{j-k} \frac{(z+1)^k}{k!}$$

$$= 1 + \sum_{j=1}^{\infty} x^j \sum_{k=0}^{j} A_{j-k} \frac{(z+1)^k}{k!}$$

$$= 1 + \sum_{j=0}^{\infty} x^{j+1} \sum_{k=0}^{j+1} A_{j+1-k} \frac{(z+1)^k}{k!}.$$

Notice that $\sum_{k=0}^{j+1} A_{j+1-k} \frac{(z+1)^k}{k!}$ is a polynomial in $z$ of degree $j + 1$. Furthermore the coefficient of $z^{j+1}$ only comes from the term $A_0 \frac{(z+1)^{j+1}}{(j+1)!} =$

$\frac{(z+1)^{j+1}}{(j+1)!} = \frac{1}{(j+1)!} \sum_{i=0}^{j+1} \binom{j+1}{i} z^i = \frac{z^{j+1}}{(j+1)!} + \frac{1}{(j+1)!} \sum_{i=0}^{j} \binom{j+1}{i}$. Hence $\sum_{k=0}^{j+1} A_{j+1-k} \frac{(z+1)^k}{k!}$ is a $j^{th}$ anti-derivative of $z$ and

$$f_1(x, z) = 1 + \sum_{j=0}^{\infty} x^{j+1} \sum_{k=0}^{j+1} A_{j+1-k} \frac{(z+1)^k}{k!} = 1 + \sum_{j=0}^{\infty} x^{j+1} D_z^{-j} z = S_1(x, z).$$

# Chapter 14

# Number Theoretic Definitions of Stirling Numbers

We begin this chapter by defining two quantities which at first glance seem unrelated to Stirling numbers. Let $n$ and $j$ be nonnegative integers. Define $S_1(n, j)$ to be the sum of the products of the first $n$ integers taken $j$ at a time without repetitions. Since there are $\binom{n}{j}$ ways to choose $j$ integers from $\{1, 2, \ldots, n\}$, we may alternatively define $S_1(n, j)$ to be the sum of the $\binom{n}{j}$ products of the form $\alpha_1 \alpha_2 \ldots \alpha_j$ where $\alpha_i \in \{1, 2, \ldots, n\}$ and $\alpha_i \neq \alpha_j$ unless $i = j$. Schematically we have

$$S_1(n, k) = \sum_{1 \leq j_1 < j_2 < \cdots < j_n \leq n} j_1 j_2 \ldots j_k. \tag{14.1}$$

For example $S_1(4, 3) = 1 \cdot 2 \cdot 3 + 1 \cdot 2 \cdot 4 + 1 \cdot 3 \cdot 4 + 2 \cdot 3 \cdot 4 = 6 + 8 + 12 + 24 = 50$. Clearly $S_1(n, k) = 0$ if $k > n$ or if $k$ is negative. We make the convention that $S_1(n, 0) = 1$ for all nonnegative integers $n$. Equation (14.1) is equivalent to

$$S_1(n, k) = \sum_{j_k=1}^{n-k+1} j_k \sum_{j_{k-1}=j_k+1}^{n-k+2} j_{k-1} \cdots \sum_{j_2=j_3+1}^{n-1} j_2 \sum_{j_1=j_2+1}^{n} j_1$$

$$= \prod_{i=1}^{k} \sum_{j_i=j_{i+1}+1}^{n-i+1} j_i, \qquad j_{k+1} = 0. \tag{14.2}$$

Equation (14.2) has an advantage over Equation (14.1) in that it can be readily computed via a symbolic computational program and provides the following collection of explicit formulas for $S_1(n, k)$ with $1 \leq k \leq 6$:

$$S_1(n, 1) = \binom{n+1}{2}$$

$$S_1(n, 2) = \frac{3n+2}{4}\binom{n+1}{3}$$

$$S_1(n, 3) = \frac{n(n+1)}{2}\binom{n+1}{4}$$

$$S_1(n, 4) = \frac{15n^3 + 15n^2 - 10n - 8}{48}\binom{n+1}{5}$$

$$S_1(n, 5) = \frac{n(n+1)(3n^2 - n - 6)}{16}\binom{n+1}{6}$$

$$S_1(n, 6) = \frac{63n^5 - 315n^3 - 224n^2 + 140n + 96}{576}\binom{n+1}{7}.$$

By using these formulas we calculate the entries of Table 14.1.

| | $k = 0$ | $k = 1$ | $k = 2$ | $k = 3$ | $k = 4$ | $k = 5$ | $k = 6$ |
|---|---|---|---|---|---|---|---|
| $n = 0$ | 1 | | | | | | |
| $n = 1$ | 1 | 1 | | | | | |
| $n = 2$ | 1 | 3 | 2 | | | | |
| $n = 3$ | 1 | 6 | 11 | 6 | | | |
| $n = 4$ | 1 | 10 | 35 | 50 | 24 | | |
| $n = 5$ | 1 | 15 | 85 | 225 | 274 | 120 | |
| $n = 6$ | 1 | 21 | 175 | 735 | 1624 | 1764 | 720 |

Table 14.1: Values for $S_1(n, k)$ with $0 \leq k \leq n$ and $0 \leq n \leq 6$

There are many patterns present in Table 14.1. One of the easiest to discover is the Pascal-like recurrence

$$S_1(n, k) = S_1(n-1, k) + nS_1(n-1, k-1). \tag{14.3}$$

Here is a simple combinatorial proof of Equation (14.3). The left side of Equation (14.3) is the sum of the products $\alpha_1\alpha_2\ldots\alpha_k$, where each factor is distinct and an element of $\{1, 2, \ldots, n\}$. If such a product does not have $n$ as factor, $\alpha_i \in \{1, 2, \ldots, n-1\}$ for $1 \leq i \leq k$, and the sum of these products is $S_1(n-1, k)$. If the product has a factor of $n$, $\alpha_1\alpha_2\ldots\alpha_k = n\alpha_1\alpha_2\ldots\alpha_{k-1}$ with $\alpha_i \in \{1, 2, \ldots, n-1\}$ for $1 \leq i \leq k-1$. If we divide each such factor by $n$ and sum we have $S_1(n-1, k-1)$. Hence the sum of these products is $nS_1(n-1, k-1)$, and the right side of Equation (14.3) follows by the rule of sums.

Our next task is to derive a generating function for $S_1(n, k)$. We claim that

$$\prod_{k=0}^{n}(1 + kx) = \sum_{k=0}^{n} S_1(n, k)x^k. \tag{14.4}$$

To justify Equation (14.4) let us analyze the coefficient of $x^k$ in

$$\prod_{k=0}^{n}(1 + kx) = (1 + x)(1 + 2x)\ldots(1 + nx).$$

We obtain a monomial of $x^k$ by selecting $x$ from $k$ of the $n$ factors in $(1 + x)(1 + 2x)\ldots(1 + nx)$ and 1 from the remaining $n - k$ factors. The coefficient of a typical monomial is $j_1 j_2 \ldots j_k$ where $1 \le j_1 < j_2 < \cdots < j_k \le n$. Let $[x^k]f(x)$ denote the coefficient of $x^k$ in the formal power series expansion of $f(x)$. The rule of sums implies that

$$[x^k]\prod_{k=0}^{n}(1 + kx) = \sum_{1 \le j_1 < j_2 < \cdots < j_n \le n} j_1 j_2 \ldots j_k = S_1(n, k),$$

where the final equality follows from Equation (14.1).

Equation (14.4) provides a connection between $S_1(n, k)$ and Stirling numbers of the first kind. By Equation (12.26) we have $\prod_{k=0}^{n}(1 + kx) = \sum_{j=0}^{n+1}(-1)^j s(n + 1, n + 1 - j)x^j$. Combining this with Equation (14.4) implies that $\sum_{k=0}^{n+1} S_1(n, k)x^k = \sum_{k=0}^{n+1}(-1)^k s(n+1, n+1 - k)x^k$, or equivalently that

$$S_1(n, k) = (-1)^k s(n + 1, n + 1 - k). \tag{14.5}$$

Equation (14.5) often appears as

$$S_1(n - 1, k) = (-1)^k s(n, n - k). \tag{14.6}$$

We may combine Equation (14.6) with Equations (13.4) and (13.27) to obtain

$$S_1(n - 1, k) = (-1)^k \binom{n}{k} B_k^{(n+1)}(1) = (-1)^k \binom{n - 1}{k} B_k^{(n)}(0), \tag{14.7}$$

where $B_k^{(n)}(t)$ is the Nörlund Polynomial defined by Equation (13.1).

Equation (14.6) defines $S_1(n, k)$ in terms of a Stirling number of the first kind. By letting $k \to n - k$ we obtain

$$s(n, k) = (-1)^{n-k} S_1(n - 1, n - k), \tag{14.8}$$

a formula which defines a Stirling number of the first kind in terms of $S_1(n-1, n-k)$.

We now define our second number theoretic quantity $S_2(n, j)$. Let $n$ and $j$ be nonnegative integers. Define $S_2(n, j)$ to be the sum of the products of the first $n$ integers taken $j$ at a time *with* repetitions. Since there are $\binom{n+j-1}{j}$ possible ways to construct such products, we may define $S_2(n, j)$ as the sum of the $\binom{n+j-1}{j}$ possible products of the form $\alpha_1\alpha_2 \ldots \alpha_j$ where $\alpha_i \in \{1, 2, \ldots, n\}$. This is represented schematically as

$$S_2(n, k) = \sum_{1 \leq j_1 \leq j_2 \leq \cdots \leq j_n \leq n} j_1 j_2 \ldots j_k. \tag{14.9}$$

For example $S_2(2, 3) = 1 \cdot 1 \cdot 1 + 1 \cdot 1 \cdot 2 + 1 \cdot 2 \cdot 2 + 2 \cdot 2 \cdot 2 = 15$. By convention $S_2(n, 0) = 1$.

Equation (14.9) is equivalent to

$$S_2(n, k) = \sum_{j_k=1}^{n} j_k \sum_{j_{k-1}=j_k}^{n} j_{k-1} \cdots \sum_{j_2=j_3}^{n} j_2 \sum_{j_1=j_2}^{n} j_1$$

$$= \prod_{i=1}^{k} \sum_{j_i=j_{i+1}}^{n} j_i, \qquad j_{k+1} = 1. \tag{14.10}$$

By using a symbolic algebra program such as Maple to calculate the right side of Equation (14.10), we find that

$$S_2(n, 1) = \binom{n+1}{2}$$

$$S_2(n, 2) = \frac{(3n+1)}{4} \binom{n+2}{3}$$

$$S_2(n, 3) = \binom{n+1}{2} \binom{n+3}{4}$$

$$S_2(n, 4) = \frac{15n^3 + 30n^2 + 5n - 2}{48} \binom{n+4}{5}$$

$$S_2(n, 5) = \frac{3n^2 + 7n - 2}{8} \binom{n+1}{2} \binom{n+5}{6}$$

$$S_2(n, 6) = \frac{63n^5 + 315n^4 + 315n^3 - 91n^2 - 42n + 16}{576} \binom{n+6}{7}.$$

We use these formulas to compute Table 14.2.

|  | $k = 0$ | $k = 1$ | $k = 2$ | $k = 3$ | $k = 4$ | $k = 5$ | $k = 6$ |
|---|---|---|---|---|---|---|---|
| $n = 0$ | 1 | | | | | | |
| $n = 1$ | 1 | 1 | 1 | 1 | 1 | 1 | 1 |
| $n = 2$ | 1 | 3 | 7 | 15 | 31 | 63 | 127 |
| $n = 3$ | 1 | 6 | 25 | 90 | 301 | 966 | 3025 |
| $n = 4$ | 1 | 10 | 65 | 350 | 1701 | 7770 | 34105 |
| $n = 5$ | 1 | 15 | 140 | 1050 | 6951 | 42525 | 246730 |
| $n = 6$ | 1 | 21 | 266 | 2646 | 22827 | 179487 | 1323652 |

Table 14.2: Values for $S_2(n,k)$ with $0 \le n \le 6$

The Pascal-like recurrence present throughout Table 14.2 is
$$S_2(n,k) = S_2(n-1,k) + nS_2(n,k-1). \qquad (14.11)$$
The proof of Equation (14.11) is similar in nature to that of Equation (14.3) and is omitted.

There is a connection between Stirling numbers of the second kind and $S_2(n,k)$ which is discovered by comparing rows of Table 9.1 with the diagonals of Table 14.2 which begin and end in one. In particular
$$S_2(n,k) = S(n+k,n) = \binom{n+k}{k} B_k^{(-n)}(0), \qquad (14.12)$$
where the last equality follows from Equation (13.36).

We verify Equation (14.12) by deriving the generating function for $S_2(n,k)$. Observe that
$$\prod_{k=0}^{n} \frac{1}{1-kx} = \prod_{k=0}^{n}(1 + kx + k^2x^2 + k^3x^3 + k^3x^4 + \dots)$$
$$= \sum_{k=0}^{\infty} x^k \sum_{1 \le j_1 \le j_2 \le \dots \le j_n \le n} j_1 j_2 \dots j_k = \sum_{k=0}^{\infty} S_2(n,k)x^k.$$
On the other hand Equation (12.32) implies that
$$\prod_{k=0}^{n} \frac{1}{1-kx} = \prod_{k=1}^{n} \frac{1}{1-kx} = \sum_{k=0}^{\infty} S(n+k,n)x^k.$$
Comparing the coefficients of $x^k$ in the previous two lines gives Equation (14.12).

Equation (14.12) corresponds to Equation (14.6). The dual of Equation (14.12), which corresponds to Equation (14.8), is
$$S(n,k) = S_2(k, n-k). \qquad (14.13)$$

## 14.1 Relationships Between $S_1(n,k)$ and $S_2(n,k)$

Equation (14.6) relates $S_1(n,k)$ to Stirling numbers of the first kind while Equation (14.12) relates $S_2(n,k)$ to Stirling numbers of the second kind. We can apply these relationships to the various identities of Chapter 13 and discover new relationships between $S_1(n,k)$ and $S_2(n,k)$. As a case in point take Equations (13.28) and (13.29) and apply Equations (14.6) and (14.12) to find that

$$(-1)^k S_1(n-1,k) = \binom{n-1}{k} \sum_{j=0}^{k} (-1)^j \binom{k+1}{j+1} \frac{S_2(jn,k)}{\binom{k+jn}{k}}, \qquad (14.14)$$

and that

$$(-1)^k S_2(n,k) = \binom{k+n}{k} \sum_{j=0}^{k} (-1)^j \binom{k+1}{j+1} \frac{S_1(jn-1,k)}{\binom{jn-1}{k}}. \qquad (14.15)$$

For another example take Equation (13.42) and use Equations (14.6) and (14.12) to rewrite it as

$$S_2(n-k,k) = \sum_{j=0}^{k} \binom{k-n}{k+j} \binom{k+n}{k-j} S_1(k+j-1,k). \qquad (14.16)$$

We may also take Equation (13.31), substitute Equations (14.6) and (14.12), and transform it as

$$S_1(n-1,k) = \sum_{j=0}^{k} \binom{k-n}{k+j} \binom{k+n}{k-j} S_2(j,k). \qquad (14.17)$$

Equations (14.14)-(14.17) appear in Professor Gould's paper *Stirling Number Representation Problems* [Gould, 1960b]. Some further results regarding Equations (14.14) through (14.17) may be found in Professor Gould's paper *The Lagrange interpolation formula and Stirling numbers* [Gould, 1960a]. The reader may also refer to the Carlitz's paper entitled *Note on Nörlund's polynomial $B_n^{(z)}$* [Carlitz, 1960].

Equation (14.17) allows us to define $S_1(z,k)$ whenever $z$ is an arbitrary complex number since the right side is a polynomial $n$ of degree $2k$. Therefore we may interpret Equation (14.17) as a polynomial identity of degree $2k$ which is true for all nonnegative integers $n$. The fundamental theorem of algebra implies that such a polynomial identity must be true for all complex numbers and that we may replace $n$ with $z$ to obtain

$$S_1(z-1,k) = \sum_{j=0}^{k} \binom{k-z}{k+j} \binom{k+z}{k-j} S_2(j,k), \qquad z \text{ complex.} \qquad (14.18)$$

We may also define $S_2(z, k)$ for all complex $z$. In order to do so we start with Equation (13.30) and use Equation (14.12) to rewrite it as

$$D_x^k \left( \frac{e^x - 1}{x} \right)^n \Bigg|_{x=0} = \frac{S(n+k, k)}{\binom{n+k}{k}} = (-1)^k \frac{S_2(n, k)}{\binom{-n-1}{k}}. \tag{14.19}$$

Next take Equation (13.26) and apply Equation (14.6) to obtain

$$D_x^k \left( \frac{e^x - 1}{x} \right)^{-n} \Bigg|_{x=0} = D_x^k \left( \frac{x}{e^x - 1} \right)^n \Bigg|_{x=0} = \frac{s(n, n-k)}{\binom{n-1}{k}}$$

$$= (-1)^k \frac{S_1(n-1, k)}{\binom{n-1}{k}}. \tag{14.20}$$

Take Equation (14.20) and let $n \to -n$ to obtain

$$D_x^k \left( \frac{e^x - 1}{x} \right)^n \Bigg|_{x=0} = (-1)^k \frac{S_1(-n-1, k)}{\binom{-n-1}{k}}, \tag{14.21}$$

where $S_1(-n-1, k)$ is defined via Equation (14.18). The left side of Equation (14.21) is precisely the left side of Equation (14.19). Hence

$$S_2(n, k) = S_1(-n-1, k), \tag{14.22}$$

or equivalently

$$S(n+k, n) = (-1)^k s(-n, -n-k). \tag{14.23}$$

Take the previous equation with $n \to j$ and $k \to m - j$ to obtain

$$S(m, j) = (-1)^{m-j} s(-j, -m). \tag{14.24}$$

If we take Equation (14.22) and let $n \to -n-1$, we discover that

$$S_1(n, k) = S_2(-n-1, k), \tag{14.25}$$

or equivalently that

$$(-1)^k s(n+1, n+1-k) = S(-n-1+k, -n-1). \tag{14.26}$$

Take the preceding line with $n \to m-1$ and $k \to m-j$ to obtain

$$(-1)^{m-j} s(m, j) = S(-j, -m). \tag{14.27}$$

In regards to Equations (14.24) and (14.27), Knuth [2003] noted that H. W. Gould was the first twentieth-century mathematician to observe that we can use the polynomials $s(n, k)$ and $S(n, n-k)$ to extend the domain of Stirling numbers to negative values of $n$.

Equation (14.25) defines $S_2(-n, k)$ for all nonnegative integers $n$, namely as $S_2(-n, k) = S_1(n - 1, k)$. Take this relationship and substitute it into the left side of Equation (14.17) to obtain

$$S_2(-n, k) = \sum_{j=0}^{k} \binom{k-n}{k+j}\binom{k+n}{k-j} S_2(j, k). \qquad (14.28)$$

Since Equation (14.28) may be interpreted as a polynomial identity of degree $2k$ which is true for all nonnegative integers $n$, the fundamental theorem of algebra allows us to replace $n$ in Equation (14.28) with an arbitrary complex number $z$ and define

$$S_2(z, k) = \sum_{j=0}^{k} \binom{k-z}{k+j}\binom{k+z}{k-j} S_2(j, k), \qquad z \text{ complex.} \qquad (14.29)$$

Equations (14.16) and (14.17) when combined with Equation (14.25) are a fertile source of identities. For example take Equation (14.16) with $m \geq 1$, and let $k = m - 1$ and $n = -m$ to obtain

$$S_1(2m - 2, m - 1) = \sum_{j=0}^{m-1} (-1)^{m-1-j}\binom{2m-1}{m-1+j} S_1(m + j - 2, m - 1),$$

$$(14.30)$$

or equivalently

$$s(2m - 1, m) = \sum_{j=0}^{m-1} (-1)^{m-1-j}\binom{2m-1}{m-1+j} s(m + j - 1, j), \qquad m \geq 1.$$

$$(14.31)$$

A similar calculation applied to Equation (14.17) implies that

$$S_2(m, m - 1) = \sum_{j=0}^{m-1} (-1)^{m-1-j}\binom{2m-1}{m-1+j} S_2(j, m - 1), \qquad m \geq 1,$$

$$(14.32)$$

which is equivalent to

$$S(2m - 1, m) = \sum_{j=0}^{m-1} (-1)^{m-1-j}\binom{2m-1}{m-1+j} S(m + j - 1, j), \qquad m \geq 1.$$

$$(14.33)$$

Taking Equations (14.16) and (14.17) and setting $k = m$ and $n = m + 1$ implies that

$$m! = S_1(m, m) = \sum_{j=0} m(-1)^{m+j}\binom{2m+1}{m-j} S_2(j, m) \qquad (14.34)$$

$$1 = S_2(1, m)$$

$$= \sum_{j=0} m(-1)^{m+j}\binom{2m+1}{m-j} S_1(m + j - 1, m), \qquad (14.35)$$

which is equivalent to

$$s(m+1,1) = (-1)^m m!$$

$$= \sum_{j=0}^{m} (-1)^{m+j} \binom{2m+1}{m-j} S(m+j,j) \tag{14.36}$$

$$S(m+1,1) = 1 = \sum_{j=0}^{m} (-1)^j \binom{2m+1}{m-j} s(m+j,j). \tag{14.37}$$

## 14.2 Hagen Recurrences for $S_1(n,k)$ and $S_2(n,k)$

We end this chapter by deriving the elegant recursive formulas for $S_1(n,k)$ and $S_2(n,k)$ attributed to the nineteenth century German astronomer J. G. Hagen [1891, p. 60], namely

$$S_1(n,k) = \frac{1}{k} \sum_{j=0}^{k-1} S_1(n,j) \binom{n+1-j}{k+1-j}, \qquad k \geq 1, \tag{14.38}$$

and

$$S_2(n,k) = \frac{1}{k} \sum_{j=0}^{k-1} S_2(n,j) \binom{-n-j}{k+1-j}, \qquad k \geq 1. \tag{14.39}$$

We begin by proving Equation (14.38). We first use Equation (14.6) to transform Equation (14.38) into

$$s(n,n-k) = \frac{(-1)^k}{k} \sum_{j=0}^{k-1} (-1)^j \binom{n-j}{k+1-j} s(n,n-j). \tag{14.40}$$

Since Equation (14.40) is equivalent to

$$s(n,n-k) = \frac{(-1)^k}{k} \sum_{j=0}^{k} (-1)^j \binom{n-j}{k+1-j} s(n,n-j) - \frac{n-k}{k} s(n,n-k),$$

we may rewrite Equation (14.40) as

$$ns(n,n-k) = (-1)^k \sum_{j=0}^{k} (-1)^j \binom{n-j}{k+1-j} s(n,n-j). \tag{14.41}$$

If we can verify Equation (14.41) we will have verified Equation (14.40) and Equation (14.38). We verify Equation (14.41) by showing each side is the coefficient of $x^{n-k}$ in the expansion of $n\binom{x}{n}n!$. By Equation (12.1) we have

$$n\binom{x}{n}n! = n \sum_{k=0}^{n} s(n,k)x^k = n \sum_{k=0}^{n} s(n,n-k)x^{n-k}. \tag{14.42}$$

On the other hand

$$\sum_{k=0}^{n} x^{n-k}(-1)^k \sum_{j=0}^{k}(-1)^j \binom{n-j}{k+1-j} s(n, n-j)$$

$$= \sum_{j=0}^{n}(-1)^j s(n, n-j) \sum_{k=j}^{n}(-1)^k x^{n-k} \binom{n-j}{k+1-j}$$

$$= \sum_{j=0}^{n}(-1)^j s(n, n-j) \sum_{k=0}^{n-j}(-1)^{k+j} x^{n-k-j} \binom{n-j}{k+1}$$

$$= -x \sum_{j=0}^{n} s(n, n-j) \sum_{k=1}^{n-j+1}(-1)^k x^{n-j-k} \binom{n-j}{k}$$

$$= -x \sum_{j=0}^{n} s(n, n-j) \left[ \sum_{k=0}^{n-j}(-1)^k x^{n-j-k} \binom{n-j}{k} - x^{n-j} \right]$$

$$= -x \sum_{j=0}^{n} s(n, n-j)(x-1)^{n-j} + x \sum_{j=0}^{n} s(n, n-j) x^{n-j}$$

$$= -x \binom{x-1}{n} n! + x \binom{x}{n} n! = -(x-n) \binom{x}{n} n! + x \binom{x}{n} n! = n \binom{x}{n} n!.$$

Since

$$n \binom{x}{n} n! = \sum_{k=0}^{n} x^{n-k}(-1)^k \sum_{j=0}^{k}(-1)^j \binom{n-j}{k+1-j} s(n, n-j), \qquad (14.43)$$

we may compare the coefficient of $x^{n-k}$ in Equation (14.42) to that of Equation (14.43) and obtain Equation (14.41).

We next prove Equation (14.39). This time we apply Equation (14.12) and rewrite it as

$$S(n+k, n) = \frac{1}{k} \sum_{j=0}^{k-1} \binom{-n-j}{k+1-j} S(n+j, n). \qquad (14.44)$$

Equation (14.44) is equivalent to

$$-n S(n+k, n) = \sum_{j=0}^{k} \binom{-n-j}{k+1-j} S(n+j, n). \qquad (14.45)$$

If we can verify Equation (14.45) we will have verified Equation (14.44) and Equation (14.39). We once again turn to a generating function. By

Equation (13.7) we have

$$-n\sum_{k=0}^{\infty}\frac{x^k}{(n+k)!}S(n+k,n) = -\frac{1}{(n-1)!}\left(\frac{e^x-1}{x}\right)^n. \qquad (14.46)$$

We now take the right side of Equation (14.45), multiply by $\frac{x^k}{(n+k)!}$, and sum over $k$ to find that

$$\sum_{k=0}^{\infty}\frac{x^k}{(n+k)!}\sum_{j=0}^{k}\binom{-n-j}{k+1-j}S(n+j,n)$$

$$=\sum_{k=0}^{\infty}\frac{x^k}{(n+k)!}\sum_{j=0}^{k}(-1)^{k+1-j}\binom{n+k}{k+1-j}S(n+j,n)$$

$$=\sum_{j=0}^{\infty}\frac{S(n+j,n)}{(n+j-1)!}\sum_{k=j}^{\infty}(-1)^{k+1-j}\frac{x^k}{(k+1-j)!}$$

$$=\sum_{j=0}^{\infty}\frac{S(n+j,n)}{(n+j-1)!}x^j\sum_{k=0}^{\infty}(-1)^{k+1}\frac{x^k}{(k+1)!}$$

$$=\sum_{j=0}^{\infty}\frac{S(n+j,n)}{(n+j-1)!}x^{j-1}\sum_{k=1}^{\infty}(-1)^{k}\frac{x^k}{k!}$$

$$=(e^{-x}-1)\sum_{j=0}^{\infty}\frac{S(n+j,n)}{(n+j-1)!}x^{j-1}$$

$$=n\left(\frac{1-e^x}{xe^x}\right)\sum_{j=0}^{\infty}\frac{S(n+j,n)}{(n+j)!}x^j+\left(\frac{1-e^x}{e^x}\right)\sum_{j=0}^{\infty}j\frac{S(n+j,n)}{(n+j)!}x^{j-1}$$

$$=\frac{1}{(n-1)!}\left(\frac{1-e^x}{xe^x}\right)\left(\frac{e^x-1}{x}\right)^n+\left(\frac{1-e^x}{e^x}\right)\frac{1}{n!}D_x\left(\frac{e^x-1}{x}\right)^n$$

$$=-\frac{1}{(n-1)!}\left(\frac{e^x-1}{x}\right)^n.$$

Since these calculations show that

$$\sum_{k=0}^{\infty}\frac{x^k}{(n+k)!}\sum_{j=0}^{k}\binom{-n-j}{k+1-j}S(n+j,n) = -\frac{1}{(n-1)!}\left(\frac{e^x-1}{x}\right)^n, \quad (14.47)$$

we may compare the coefficient of $x^k$ in Equation (14.46) to the coefficient of $x^k$ in Equation (14.47) and obtain Equation (14.45).

Professor Gould considers Equation (14.39) to be a corollary of Equation

(14.38) since Equation (14.38) with $n \to -n - 1$ becomes

$$S_1(-n-1,k) = \frac{1}{k} \sum_{j=0}^{k-1} S_1(-n-1,j) \binom{-n-j}{k+1-j}.$$

By Equation (14.22) we may rewrite the preceding line as

$$S_2(n,k) = \frac{1}{k} \sum_{j=0}^{k-1} S_2(n,j) \binom{-n-j}{k+1-j},$$

which is precisely Equation (14.39).

# Chapter 15

# Bernoulli Numbers

Jacques Bernoulli was a seventeenth century mathematician known for his work on sums of powers of numbers. His investigations led to the discovery of a rational sequence of numbers known as the Bernoulli numbers. Let $n$ be a nonnegative integer. We define $B_n$ to be the $n^{th}$ Bernoulli number where

$$\sum_{n=0}^{\infty} \frac{x^n}{n!} B_n = \frac{x}{e^x - 1} =$$

$$1 - \frac{1}{2}x + \frac{1}{6}\frac{x^2}{2!} - \frac{1}{30}\frac{x^4}{4!} + \frac{1}{42}\frac{x^6}{6!} - \frac{1}{30}\frac{x^8}{8!} + \frac{5}{66}\frac{x^{10}}{10!} - \frac{691}{2730}\frac{x^{12}}{12!} + \dots \quad (15.1)$$

Notice that $B_n = B_n^{(1)}(0)$ where $B_n^{(k)}(t)$ is the Nörlund polynomial defined by Equation (13.1). Bernoulli numbers are intimately connected with Stirling numbers of the second kind. To discover one such connection we expand $\frac{x}{e^x-1}$ via the Taylor series for $\ln(1+x) = \sum_{k=1}^{\infty} \frac{(-1)^{k-1}}{k} x^k$ [Carlitz, 1953]. In particular

$$x = \ln e^x = \ln(1 + (e^x - 1)) = \sum_{n=1}^{\infty} \frac{(-1)^{n-1}}{n} (e^x - 1)^n = \sum_{n=0}^{\infty} \frac{(-1)^n}{n+1} (e^x - 1)^{n+1}.$$

Hence

$$\frac{x}{e^x - 1} = \sum_{n=0}^{\infty} \frac{(-1)^n}{n+1} (e^x - 1)^n$$

$$= \sum_{n=0}^{\infty} \frac{(-1)^n}{n+1} \sum_{k=0}^{n} (-1)^{n-k} \binom{n}{k} e^{kx}$$

$$= \sum_{n=0}^{\infty} \frac{(-1)^n}{n+1} \sum_{k=0}^{n} (-1)^{n-k} \binom{n}{k} \sum_{j=0}^{\infty} \frac{(kx)^j}{j!}$$

$$= \sum_{j=0}^{\infty} \frac{x^j}{j!} \sum_{n=0}^{\infty} \frac{(-1)^n}{n+1} \sum_{k=0}^{n} (-1)^{n-k} \binom{n}{k} k^j$$

$$= \sum_{j=0}^{\infty} \frac{x^j}{j!} \sum_{n=0}^{\infty} \frac{(-1)^n}{n+1} n! S(j,n).$$

This expansion when combined with Equation (15.1) implies that

$$B_n = \sum_{j=0}^{n} \frac{(-1)^j}{j+1} \sum_{k=0}^{j} (-1)^{j-k} \binom{j}{k} k^n = \sum_{j=0}^{n} \frac{(-1)^j}{j+1} j! S(n,j). \qquad (15.2)$$

Equation (15.2) is an explicit formula for the Bernoulli numbers in terms of Stirling numbers of the second kind. By using Equation (10.2) we can also describe $B_n$ in terms of Eulerian numbers,

$$B_n = \sum_{\alpha=0}^{n} \frac{(-1)^\alpha}{\alpha+1} \sum_{j=0}^{n} \binom{j-1}{n-\alpha} A_{n,j}, \qquad (15.3)$$

where

$$A_{n,j} = \sum_{k=0}^{j} (-1)^k \binom{n+1}{k} (j-k)^n.$$

In order to simplify Equation (15.3) assume that $n \geq 1$ and observe that

$$B_n = \sum_{\alpha=0}^{n} \frac{(-1)^\alpha}{\alpha+1} \sum_{j=0}^{n} \binom{j-1}{n-\alpha} A_{n,j}$$

$$= \sum_{\alpha=0}^{n} \frac{(-1)^\alpha}{\alpha+1} \sum_{j=n-\alpha}^{n} \binom{j-1}{n-\alpha} A_{n,j}$$

$$= \sum_{\alpha=0}^{n} \frac{(-1)^{n-\alpha}}{n-\alpha+1} \sum_{j=\alpha}^{n} \binom{j-1}{\alpha} A_{n,j}$$

$$= (-1)^n \sum_{j=1}^{n} A_{n,j} \sum_{\alpha=0}^{j-1} (-1)^\alpha \binom{j-1}{\alpha} \frac{1}{n-\alpha+1}.$$

An application of Melzak's theorem, Equation (7.12), implies that

$$\sum_{k=0}^{n} (-1)^k \binom{n}{k} \frac{1}{m-k} = \frac{(-1)^n}{(m-n)\binom{m}{n}}. \qquad (15.4)$$

We may rewrite the double summation as

$$B_n = \sum_{j=1}^{n} A_{n,j} \frac{(-1)^{n+j-1}}{(n+2-j)\binom{n+1}{j-1}} = \sum_{j=1}^{n} A_{n,j} \frac{(-1)^{n+j-1}}{(n+1)\binom{n}{j-1}}, \qquad n \geq 1.$$

(15.5)

Take Equation (15.5) with $j \to n-j+1$ to get

$$B_n = \sum_{j=1}^{n} A_{n,n-j+1} \frac{(-1)^j}{(n+1)\binom{n}{j}} = \sum_{j=1}^{n} A_{n,j} \frac{(-1)^j}{(n+1)\binom{n}{j}},$$

(15.6)

where the last equality follows from Equation (10.5). Since $A_{n,0} = 1$ for $n = 0$ we can extend the range of summation in Equation (15.6) to include the case of $n = 0$. In other words

$$B_n = \sum_{j=0}^{n} (-1)^j \frac{A_{n,j}}{(n+1)\binom{n}{j}}, \qquad n \geq 0.$$

(15.7)

To derive yet another explicit formula for $B_n$ in terms of Stirling numbers of the second kind, start with Equation (13.9) and let $n \to n-1$ to obtain

$$\left(\frac{x}{e^x - 1}\right)^n = \sum_{j=0}^{\infty} x^j \sum_{\beta=0}^{j} (-1)^\beta \binom{j+1}{\beta+1} \frac{(\beta n)! S(j+\beta n, \beta n)}{(j+\beta n)!}.$$

(15.8)

Take Equation (15.8) and let $n = 1$. This gives us

$$\frac{x}{e^x - 1} = \sum_{j=0}^{\infty} x^j \sum_{\beta=0}^{j} (-1)^\beta \binom{j+1}{\beta+1} \frac{\beta! S(j+\beta, \beta)}{(j+\beta)!}.$$

(15.9)

If we compare the coefficient of $x^j$ in Equation (15.1) with the coefficient of $x^j$ in Equation (15.9) we discover that

$$B_n = \sum_{k=0}^{n} (-1)^k \binom{n+1}{k+1} \frac{S(n+k, k)}{\binom{n+k}{n}}.$$

(15.10)

Equation (15.10) is attributed to Charles Jordan [1957] and should be compared to Equation (15.2).

## 15.1 Sum of Powers of Numbers

We have defined the Bernoulli numbers via the Taylor series expansion of $\frac{x}{e^x-1}$. There is another way to generate $\{B_n\}_{n=0}^{\infty}$ and this is through $S_p(n) = \sum_{k=0}^{n} k^p$ where $n$ and $p$ are nonnegative integers. Jacques Bernoulli in his seminal work *Ars Conjectandi* [1713] discovered that

$$S_p(n) = \sum_{k=0}^{n} k^p = \frac{n^{p+1}}{p+1} + \frac{n^p}{2} + \sum_{k=1}^{p-1} \binom{p}{k} \frac{n^{p-k}}{k+1} B_{k+1}, \qquad p \geq 1, \quad (15.11)$$

where the sum on the right side of Equation (15.11) is zero if $p < 2$. We will prove Equation (15.11) via induction on $n$. Before we get to the proof we will need some facts about $\{B_n\}_{n=0}^{\infty}$. First we show that

$$\sum_{k=0}^{n} \binom{n}{k} B_k = (-1)^n B_n. \tag{15.12}$$

To prove Equation (15.12) rewrite it as

$$\sum_{k=0}^{n} \frac{B_k}{k!(n-k)!} = (-1)^n \frac{B_n}{n!}. \tag{15.13}$$

Start with the left side of Equation (15.13), multiply by $x^n$, and sum over $n$ to discover that

$$\sum_{n=0}^{\infty} x^n \sum_{k=0}^{n} \frac{B_k}{k!(n-k)!} = \sum_{k=0}^{\infty} \frac{x^k}{k!} B_k \sum_{n=k}^{\infty} \frac{x^{n-k}}{(n-k)!} = \left(\frac{x}{e^x - 1}\right) e^x$$

$$= \frac{xe^x}{e^x - 1} \cdot \frac{e^{-x}}{e^{-x}} = \frac{-x}{1 - e^{-x}} = \sum_{n=0}^{\infty} \frac{B_n}{n!} (-x)^n$$

$$= \sum_{n=0}^{\infty} \frac{(-1)^n B_n}{n!} x^n.$$

Equation (15.13) follows from coefficient comparison.

Next we prove the identity

$$\sum_{k=0}^{n} \binom{n+1}{k} B_k = 0, \qquad n \geq 1. \tag{15.14}$$

To prove Equation (15.14) observe that Equation (15.1) is equivalent to

$$x = (e^x - 1) \sum_{n=0}^{\infty} \frac{x^n}{n!} B_n = x \sum_{n=0}^{\infty} \frac{x^n}{n!} B_n \sum_{j=0}^{\infty} \frac{x^j}{(j+1)!}. \tag{15.15}$$

Take Equation (15.15), divide by $x$, and use the Cauchy convolution to find that

$$1 = \sum_{n=0}^{\infty} x^n \sum_{k=0}^{n} \frac{B_k}{k!(n-k+1)!}. \tag{15.16}$$

If we compare the coefficient of $x^n$ on both sides of Equation (15.16) we find that $\sum_{k=0}^{n} \frac{B_k}{k!(n-k+1)!} = 0$ whenever $n \geq 1$, a result equivalent to Equation (15.14).

Equation (15.14) implies that $B_{2n+1} = 0$ whenever $n \geq 1$ since

$$(-1)^{2n+1} B_{2n+1} = \sum_{k=0}^{2n+1} \binom{2n+1}{k} B_k$$

$$= B_{2n+1} + \sum_{k=0}^{2n} \binom{2n+1}{k} B_k = B_{2n+1}.$$

Since $B_{2n+1} = 0$ whenever $n \geq 1$, Equation (15.12) is equivalent to

$$\sum_{k=0}^{n} \binom{n}{k} B_k = B_n, \qquad n \geq 2. \tag{15.17}$$

This recurrence for the Bernoulli numbers should be compared to recurrence for the Bell numbers given by Equation (9.66).

Back to our proof of Equation (15.11). Throughout this induction proof we assume $p$ is a fixed positive integer. It is easy to show that $S_p(0) = 0$ for $p \geq 1$ and $S_0(0) = 1$. With a little work we see that $\frac{0^{p+1}}{p+1} + \frac{0^p}{2} + \sum_{k=1}^{p-1} \binom{p}{k} \frac{0^{p-k}}{k+1} B_{k+1} = 0$ whenever $p \geq 1$. Thus Equation (15.11) is true for $n = 0$. Assume that Equation (15.11) is true for all nonnegative integers less than or equal to $n$. By definition

$$S_p(n+1) = \sum_{k=0}^{n+1} k^p = (n+1)^p + \sum_{k=0}^{n} k^p$$

$$= (n+1)^p + \frac{n^{p+1}}{p+1} + \frac{n^p}{2} + \sum_{j=1}^{p-1} \binom{p}{j} \frac{n^{p-j}}{j+1} B_{j+1}$$

$$= \sum_{j=0}^{p} \binom{p}{j} n^j + \frac{n^{p+1}}{p+1} + \frac{n^p}{2} + \sum_{j=1}^{p-1} \binom{p}{j} \frac{n^{p-j}}{j+1} B_{j+1}$$

$$= \sum_{j=0}^{p} \binom{p}{j} n^j + \frac{n^{p+1}}{p+1} + \frac{n^p}{2} + \sum_{j=1}^{p-1} \binom{p}{j} \frac{n^j}{p-j+1} B_{p-j+1}$$

$$= \frac{n^{p+1}}{p+1} + \frac{3}{2} n^p + \sum_{j=0}^{p-1} \binom{p}{j} n^j \left[ 1 + \frac{B_{p-j+1}}{p-j+1} \right].$$

To complete our proof we must show that

$$\frac{n^{p+1}}{p+1} + \frac{3}{2}n^p + \sum_{j=0}^{p-1}\binom{p}{j}n^j\left[1 + \frac{B_{p-j+1}}{p-j+1}\right]$$

$$= \frac{(n+1)^{p+1}}{p+1} + \frac{(n+1)^p}{2} + \sum_{k=1}^{p-1}\binom{p}{k}\frac{(n+1)^{p-k}}{k+1}B_{k+1}. \qquad (15.18)$$

The binomial theorem and Equation (15.14) imply that

$$\frac{(n+1)^{p+1}}{p+1} + \frac{(n+1)^p}{2} + \sum_{k=1}^{p-1}\binom{p}{k}\frac{(n+1)^{p-k}}{k+1}B_{k+1}$$

$$= \frac{1}{p+1}\sum_{j=0}^{p+1}\binom{p+1}{j}n^j + \frac{1}{2}\sum_{j=0}^{p}\binom{p}{j}n^j + \sum_{k=1}^{p-1}\binom{p}{k}\frac{B_{k+1}}{k+1}\sum_{j=0}^{p-k}\binom{p-k}{j}n^j$$

$$= \frac{n^{p+1}}{p+1} + \frac{1}{p+1}\sum_{j=0}^{p}\binom{p+1}{j}n^j + \frac{1}{2}\sum_{j=0}^{p}\binom{p}{j}n^j$$

$$+ \sum_{j=0}^{p-1}n^j\sum_{k=1}^{p-j}\binom{p}{p-k}\binom{p-k}{j}\frac{B_{k+1}}{k+1}$$

$$= \frac{n^{p+1}}{p+1} + \sum_{j=0}^{p}\frac{1}{p+1-j}\binom{p}{j}n^j + \frac{1}{2}\sum_{j=0}^{p}\binom{p}{j}n^j$$

$$+ \sum_{j=0}^{p-1}n^j\sum_{k=1}^{p-j}\binom{p}{p-k}\binom{p-k}{j}\frac{B_{k+1}}{k+1}$$

$$= \frac{n^{p+1}}{p+1} + \sum_{j=0}^{p}\frac{1}{p+1-j}\binom{p}{j}n^j + \frac{1}{2}\sum_{j=0}^{p}\binom{p}{j}n^j$$

$$+ \sum_{j=0}^{p-1}\binom{p}{j}n^j\sum_{k=1}^{p-j}\binom{p-j}{k}\frac{B_{k+1}}{k+1}$$

$$= \frac{n^{p+1}}{p+1} + \frac{3}{2}n^p + \sum_{j=0}^{p-1}\binom{p}{j}n^j\left[\frac{1}{p+1-j} + \frac{1}{2} + \sum_{k=1}^{p-j}\binom{p-j}{k}\frac{B_{k+1}}{k+1}\right]$$

$$= \frac{n^{p+1}}{p+1} + \frac{3}{2}n^p$$

$$+ \sum_{j=0}^{p-1}\binom{p}{j}n^j\left[\frac{1}{p+1-j} + \frac{1}{2} + \frac{1}{p-j+1}\sum_{k=1}^{p-j}\binom{p-j+1}{k+1}B_{k+1}\right]$$

$$= \frac{n^{p+1}}{p+1} + \frac{3}{2}n^p$$

$$+ \sum_{j=0}^{p-1} \binom{p}{j} n^j \left[ 1 + \frac{1}{p+1-j} + \frac{1}{p-j+1} \sum_{k=0}^{p-j} \binom{p-j+1}{k+1} B_{k+1} \right]$$

$$= \frac{n^{p+1}}{p+1} + \frac{3}{2}n^p$$

$$+ \sum_{j=0}^{p-1} \binom{p}{j} n^j \left[ 1 + \frac{1}{p+1-j} + \frac{1}{p-j+1} \sum_{k=1}^{p-j+1} \binom{p-j+1}{k} B_k \right]$$

$$= \frac{n^{p+1}}{p+1} + \frac{3}{2}n^p + \sum_{j=0}^{p-1} \binom{p}{j} n^j \left[ 1 + \frac{1}{p-j+1} \sum_{k=0}^{p-j+1} \binom{p-j+1}{k} B_k \right]$$

$$= \frac{n^{p+1}}{p+1} + \frac{3}{2}n^p + \sum_{j=0}^{p-1} \binom{p}{j} n^j \left[ 1 + \frac{B_{p-j+1}}{p-j+1} + \sum_{k=0}^{p-j} \binom{p-j+1}{k} B_k \right]$$

$$= \frac{n^{p+1}}{p+1} + \frac{3}{2}n^p + \sum_{j=0}^{p-1} \binom{p}{j} n^j \left[ 1 + \frac{B_{p-j+1}}{p-j+1} \right].$$

There are other ways to represent $S_p(n)$ using Bernoulli numbers. To develop these representations we introduce the Bernoulli polynomials $\{B_n(x)\}_{n=0}^{\infty}$ where

$$\sum_{n=0}^{\infty} \frac{t^n}{n!} B_n(x) = \frac{te^{tx}}{e^t - 1}, \qquad |t| < 2\pi. \tag{15.19}$$

Notice that $B_n(0) = B_n$ and $B_n(x) = B_n^{(1)}(x)$. Below we list $\{B_n(x)\}_{n=0}^{10}$:

$$B_0(x) = 1 \qquad B_1(x) = x - \frac{1}{2} \qquad B_3(x) = x^3 - \frac{3}{2}x^2 + \frac{1}{2}x$$

$$B_4(x) = x^4 - 2x^3 + x^2 - \frac{1}{30} \qquad B_5(x) = x^5 - \frac{5}{2}x^4 + \frac{5}{3}x^3 - \frac{1}{6}x$$

$$B_6(x) = x^6 - 3x^5 + \frac{5}{2}x^4 - \frac{1}{2}x^2 + \frac{1}{42}$$

$$B_7(x) = x^7 - \frac{7}{2}x^6 + \frac{7}{2}x^5 - \frac{7}{6}x^3 + \frac{x}{6}$$

$$B_8(x) = x^8 - 4x^7 + \frac{14}{3}x^6 - \frac{7}{3}x^4 + \frac{2}{3}x^2 - \frac{1}{30}$$

$$B_9(x) = x^9 - \frac{9}{2}x^8 + 6x^7 - \frac{21}{5}x^5 + 2x^3 - \frac{3}{10}x$$

$$B_{10}(x) = x^{10} - 5x^9 + \frac{15}{2}x^8 - 7x^6 + 5x^4 - \frac{3}{2}x^2 + \frac{5}{66}.$$

The Bernoulli polynomials satisfy

$$B_n(x+1) - B_n(x) = \Delta_{x,1}B_n(x) = nx^{n-1}. \tag{15.20}$$

To prove Equation (15.20) take the left side, multiple by $\frac{t^n}{n!}$, and sum over $n$. Equation (15.19) implies that

$$\sum_{n=0}^{\infty} [B_n(x+1) - B_n(x)] \frac{t^n}{n!} = \frac{te^{t(x+1)}}{e^t - 1} - \frac{te^{tx}}{e^t - 1}$$

$$= \frac{te^{tx}(e^t - 1)}{e^t - 1} = te^{tx}$$

$$= \sum_{n=0}^{\infty} \frac{t^{n+1}x^n}{n!} = \sum_{n=0}^{\infty} nx^{n-1}\frac{t^n}{n!},$$

and Equation (15.20) follows from coefficient comparison.

To relate Equation (15.20) with $S_p(n)$, take Equation (15.20), replace $p$ with $n$, let $x \to k$, and sum over $k$ where $0 \le k \le n$. We find that

$$\sum_{k=0}^{n} [B_p(k+1) - B_p(k)] = \sum_{k=0}^{n} pk^{p-1} = B_p(n+1) - B_p(0),$$

or equivalently that

$$S_p(n) = \sum_{k=0}^{n} k^p = \frac{B_{p+1}(n+1) - B_{p+1}}{p+1}. \tag{15.21}$$

To do more with Equation (15.21) we need to prove

$$B_n(x) = \sum_{k=0}^{n} \binom{n}{k} x^{n-k} B_k. \tag{15.22}$$

Equation (15.22) shows that $B_n(x)$ is a polynomial in $x$ of degree $n$. It readily follows from Equation (15.19) since

$$\sum_{n=0}^{\infty} \frac{t^n}{n!} B_n(x) = e^{tx} \cdot \frac{t}{e^t - 1} = \sum_{n=0}^{\infty} \frac{(tx)^n}{n!} \sum_{k=0}^{\infty} \frac{t^k}{k!} B_k$$

$$= \sum_{n=0}^{\infty} t^n \sum_{k=0}^{n} \frac{x^{n-k}}{(n-k)!} \frac{B_k}{k!} = \sum_{n=0}^{\infty} \frac{t^n}{n!} \sum_{k=0}^{n} \binom{n}{k} x^{n-k} B_k.$$

**Remark 15.1.** Equation (15.22) implies that

$$B_n(x) = \sum_{j=0}^{n} \frac{1}{j+1} \sum_{k=0}^{j} (-1)^k \binom{j}{k} (x+k)^n, \tag{15.23}$$

since

$$B_n(x) = \sum_{i=0}^{n} \binom{n}{i} x^{n-i} B_i = \sum_{i=0}^{n} \binom{n}{i} x^{n-i} \sum_{j=0}^{i} \frac{1}{j+1} \sum_{k=0}^{j} (-1)^k \binom{j}{k} k^i$$

$$= \sum_{i=0}^{n} \binom{n}{i} x^{n-i} \sum_{j=0}^{n} \frac{1}{j+1} \sum_{k=0}^{j} (-1)^k \binom{j}{k} k^i, \qquad (*)$$

$$= \sum_{j=0}^{n} \frac{1}{j+1} \sum_{k=0}^{j} (-1)^k \binom{j}{k} \sum_{i=0}^{n} \binom{n}{i} x^{n-i} k^i$$

$$= \sum_{j=0}^{n} \frac{1}{j+1} \sum_{k=0}^{j} (-1)^k \binom{j}{k} (x+k)^n.$$

Line $(*)$ is the $j^{th}$ difference of the monomial $k^i$. This difference is zero if $j > i$. Hence we can extend the range of $j$ to $n$. If $x = 0$, Equation (15.23) becomes Equation (15.2).

By combining Equations (15.21) and (15.22) we conclude that

$$S_p(n) = \sum_{k=0}^{n} k^p = \frac{1}{p+1} \sum_{k=0}^{p} \binom{p+1}{k} (n+1)^{p+1-k} B_k. \qquad (15.24)$$

A slight variation of Equation (15.24) is

$$S_p(n) = \sum_{k=0}^{n} k^p = \frac{1}{p+1} \sum_{k=0}^{p} (-1)^k \binom{p+1}{k} n^{p+1-k} B_k, \qquad p \geq 1. \quad (15.25)$$

Equation (15.25) is known as Bernoulli's formula.[1]  To verify Equation (15.25) we turn to Equation (15.12) and discover that

$$\frac{1}{p+1} \sum_{k=0}^{p} (-1)^k \binom{p+1}{k} n^{p+1-k} B_k$$

$$= \frac{1}{p+1} \sum_{k=0}^{p} \binom{p+1}{k} n^{p+1-k} \sum_{j=0}^{k} \binom{k}{j} B_j$$

$$= \frac{1}{p+1} \sum_{j=0}^{p} B_j \sum_{k=j}^{p} \binom{p+1}{k} \binom{k}{j} n^{p+1-k}$$

---

[1] Wikipedia, *Bernoulli number*, http://en.wikipedia.org/wiki/Bernoulli_number

$$= \frac{1}{p+1} \sum_{j=0}^{p} \binom{p+1}{j} B_j \sum_{k=j}^{p} \binom{p+1-j}{k-j} n^{p+1-k}$$

$$= \frac{1}{p+1} \sum_{j=0}^{p} \binom{p+1}{j} B_j \sum_{k=0}^{p-j} \binom{p+1-j}{k} n^{p+1-k-j}$$

$$= \frac{1}{p+1} \sum_{j=0}^{p} \binom{p+1}{j} B_j \left[ \sum_{k=0}^{p+1-j} \binom{p+1-j}{k} n^{p+1-k-j} - 1 \right]$$

$$= \frac{1}{p+1} \sum_{j=0}^{p} \binom{p+1}{j} B_j \left[ (n+1)^{p+1-j} - 1 \right]$$

$$= \frac{1}{p+1} \sum_{j=0}^{p} \binom{p+1}{j} (n+1)^{p+1-j} B_j - \frac{1}{p+1} \sum_{j=0}^{p} \binom{p+1}{j} B_j$$

$$= \frac{1}{p+1} \sum_{j=0}^{p} \binom{p+1}{j} (n+1)^{p+1-j} B_j = S_p(n).$$

## 15.2    Other Representations of $S_p(n)$

So far we have concentrated on using Bernoulli numbers to expand $S_p(n)$. But we may also use Stirling numbers of the second kind as witnessed by Equation (9.50) which says

$$S_p(n) = \sum_{k=0}^{n} k^p = \sum_{j=0}^{p} \binom{n+1}{j+1} j! S(p, j). \tag{15.26}$$

Equations (15.26) and (15.25) show that $S_p(n)$ is a polynomial in $n$ of degree $p+1$. Equation (15.25) writes this polynomial in terms of the basis $\{n^k\}_{k=0}^{p+1}$, while Equation (15.26) uses the basis $\left\{ \binom{n+1}{k+1} \right\}_{k=-1}^{p}$. If we use yet another basis, namely $\left\{ \binom{n+k}{k+1} \right\}_{k=-1}^{p}$, we get the formula

$$S_p(n) = \sum_{j=0}^{n} k^p = \sum_{j=0}^{p} (-1)^{p-j} \binom{n+j}{j+1} j! S(p, j), \qquad p \geq 1, \tag{15.27}$$

which is Equation (9.51). Equations (15.26) and (15.27) are restatements of previous results. To obtain a new formula for $S_p(n)$, we transform Equation

(15.26) through Pascal's identity to find that

$$S_p(n) = \sum_{k=0}^{n} k^p = \sum_{j=0}^{p} \binom{n+1}{j+1} j! S(p,j)$$

$$= \sum_{j=0}^{p} \left[ \binom{n}{j+1} + \binom{n}{j} \right] j! S(p,j)$$

$$= \sum_{j=0}^{p} \binom{n}{j+1} j! S(p,j) + \sum_{j=1}^{p} \binom{n}{j} j! S(p,j), \qquad \text{assuming } p \geq 1$$

$$= \sum_{j=0}^{p} \binom{n}{j+1} j! S(p,j) + \sum_{j=0}^{p-1} \binom{n}{j+1} (j+1)! S(p,j+1)$$

$$= \sum_{j=0}^{p} \binom{n}{j+1} j! \left[ S(p,j) + (j+1) S(p,j+1) \right]$$

$$= \sum_{j=0}^{p} \binom{n}{j+1} j! S(p+1,j+1).$$

These calculations imply that

$$S_p(n) = \sum_{k=0}^{p} \binom{n}{k+1} k! S(p+1,k+1), \qquad p \geq 1. \tag{15.28}$$

Equation (15.28) often appears in the literature as

$$S_p(n) = \sum_{k=0}^{p} (-1)^k \binom{n}{k+1} \sum_{j=0}^{k} (-1)^j \binom{k}{j} (j+1)^p, \qquad p \geq 1, \tag{15.29}$$

since

$$\sum_{j=0}^{k} (-1)^{k-j} \binom{k}{j} (j+1)^p = k! S(p+1,k+1). \tag{15.30}$$

To prove Equation (15.30) take the left side and expand via the binomial theorem. This gives us

$$\sum_{j=0}^{k} (-1)^{k-j} \binom{k}{j} (j+1)^p = \sum_{r=0}^{p} \binom{p}{r} \sum_{j=0}^{k} (-1)^{k-j} \binom{k}{j} j^r = k! \sum_{r=0}^{p} \binom{p}{r} S(r,k).$$

If we can show that

$$\sum_{r=0}^{p} \binom{p}{r} S(r,k) = S(p+1,k+1), \tag{15.31}$$

we will have verified Equation (15.30). By definition $S(p+1, k+1)$ is the number of set partitions of $\{1, 2, \ldots, p+1\}$ which have precisely $k+1$ subsets. We claim that the left side of Equation (15.31) also counts these set partitions. Rewrite the left side of Equation (15.31) as

$$\sum_{r=0}^{p} \binom{p}{r} S(r, k) = \sum_{r=0}^{p} \binom{p}{p-r} S(p-r, k) = \sum_{r=0}^{p} \binom{p}{r} S(p-r, k).$$

The $r$ counts the number of digits in $\{1, 2, \ldots p\}$ that are also elements of the subset containing $p+1$. The remaining $p-r$ digits form the remaining $k$ subsets of the set partition. With this interpretation of $r$, the rule of products and the rule of sums ensure that $\sum_{r=0}^{p} \binom{p}{r} S(p-r, k)$ counts each set partition of $\{1, 2, \ldots, p+1\}$ with $k+1$ subsets exactly once.

By carefully rewriting Equation (15.28) and comparing this result with Equation (15.11), we discover yet another explicit formula for $B_n$. Start with Equation (15.28) and observe that

$$S_p(n) = \sum_{k=0}^{n} k^p = \sum_{k=0}^{p} \binom{n}{k+1} k! S(p+1, k+1)$$

$$= \sum_{k=1}^{p+1} \binom{n}{k} (k-1)! S(p+1, k)$$

$$= \sum_{k=1}^{p+1} \frac{1}{k} \binom{n}{k} k! S(p+1, k)$$

$$= \sum_{k=1}^{p+1} \frac{1}{k} S(p+1, k) \sum_{j=0}^{k} s(k, j) n^j$$

$$= \sum_{j=1}^{p+1} n^j \sum_{k=j}^{p+1} \frac{1}{k} S(p+1, k) s(k, j).$$

These calculations show that

$$S_p(n) = \sum_{k=0}^{n} k^p = \sum_{j=1}^{p+1} n^j \sum_{k=j}^{p+1} \frac{1}{k} S(p+1, k) s(k, j), \qquad p \geq 1. \qquad (15.32)$$

Take Equation (15.11) and let $j = p - k$ to obtain

$$S_p(n) = \frac{n^{p+1}}{p+1} + \frac{n^p}{2} + \sum_{j=1}^{p-1} \binom{p}{j} \frac{n^j}{p-j+1} B_{p-j+1}, \qquad p \geq 1. \qquad (15.33)$$

Since Equations (15.32) and (15.33) are identical polynomials the coefficients of $n^j$ must agree. Recall that $S(p,p) = s(p,p) = 1$, $S(p+1,p) = \binom{p+1}{2}$, $s(p+1,p) = -\binom{p+1}{2}$ and rewrite Equation (15.32) as

$$S_p(n) = \frac{n^{p+1}}{p+1} + \frac{n^p}{2} + \sum_{j=1}^{p-1} n^j \sum_{k=j}^{p+1} \frac{1}{k} S(p+1,k)s(k,j), \qquad p \geq 2. \quad (15.34)$$

We then compare the coefficients of $n^j$ in Equations (15.33) and (15.34) to deduce that

$$\sum_{k=j}^{p+1} \frac{1}{k} S(p+1,k)s(k,j) = \binom{p}{j} \frac{B_{p-j+1}}{p-j+1}, \qquad 1 \leq j \leq p-1, \; p \geq 2. \quad (15.35)$$

Take Equation (15.35) and set $j = 1$. Since $s(k,1) = (-1)^{k-1}(k-1)!$, we discover that

$$B_p = \sum_{k=1}^{p+1} \frac{(-1)^{k-1}}{k}(k-1)!S(p+1,k) = \sum_{k=0}^{p} \frac{(-1)^k}{k+1} k!S(p+1,k+1), \qquad p \geq 2.$$
$$(15.36)$$

Equation (15.36) should be contrasted with Equation (15.2). Equation (15.2) says

$$B_n = \sum_{k=0}^{n} \frac{(-1)^k}{k+1} k!S(n,k) = \sum_{k=0}^{n} \frac{(-1)^k}{k+1} \sum_{j=0}^{k}(-1)^{k-j} \binom{k}{j} j^n$$
$$= \sum_{k=0}^{n} \frac{(-1)^k}{k+1} \Delta_{x,1}^k x^n \big|_{x=0}, \quad (15.37)$$

while Equation (15.36), when combined with Equation (15.30), says

$$B_n = \sum_{k=0}^{n} \frac{(-1)^k}{k+1} k!S(n+1,k+1) = \sum_{k=0}^{n} \frac{(-1)^k}{k+1} \sum_{j=0}^{k}(-1)^{k-j} \binom{k}{j}(j+1)^n$$
$$= \sum_{k=0}^{n} \frac{(-1)^k}{k+1} \Delta_{x,1}^k x^n \big|_{x=1}, \qquad n \geq 2. \quad (15.38)$$

To obtain yet another formula for $B_{n-j}$ in terms of Stirling numbers, take Equation (15.21), interchange the roles of $p$ and $n$, and combine with Equation (9.50) to obtain

$$S_n(p-1) = \sum_{k=0}^{p-1} k^n = \frac{B_{n+1}(p) - B_{n+1}}{n+1} = \sum_{j=0}^{n} \binom{p}{j+1} j!S(n,j).$$

If we take the previous line and solve for $B_{n+1}$, we find that

$$B_{n+1}(p) = (n+1) \sum_{j=0}^{n} \binom{p}{j+1} j! S(n,j) + B_{n+1}. \qquad (15.39)$$

for all nonnegative integers $n$ and $p$.

Since $B_{n+1}(p)$ is a polynomial in $p$ of degree $n+1$, Equation (15.39) may be interpreted as a polynomial identity which is true for all nonnegative integers $p$. The fundamental theorem of algebra allows for the replacement of $p$ with an arbitrary complex number $x$. In other words

$$B_{n+1}(x) = (n+1) \sum_{j=0}^{n} \binom{x}{j+1} j! S(n,j) + B_{n+1}, \qquad x \text{ complex.} \quad (15.40)$$

If $n \geq 1$, Equation (15.40) implies that

$$B_n(x) = n \sum_{\alpha=0}^{n-1} \binom{x}{\alpha+1} \alpha! S(n-1,\alpha) + B_n$$

$$= n \sum_{\alpha=0}^{n-1} \alpha! S(n-1,\alpha) \sum_{j=0}^{\alpha+1} \frac{s(\alpha+1,j)}{(\alpha+1)!} x^j + B_n$$

$$= n \sum_{\alpha=0}^{n-1} \frac{1}{\alpha+1} \sum_{j=0}^{\alpha} S(n-1,\alpha) s(\alpha+1,j) x^j$$

$$\quad + n \sum_{\alpha=0}^{n-1} \frac{S(n-1,\alpha) s(\alpha+1,\alpha+1)}{\alpha+1} x^{\alpha+1} + B_n$$

$$= n \sum_{j=0}^{n-1} x^j \sum_{\alpha=j}^{n-1} \frac{S(n-1,\alpha) s(\alpha+1,j)}{\alpha+1}$$

$$\quad + n \sum_{\alpha=1}^{n} \frac{S(n-1,\alpha-1)}{\alpha} x^\alpha + B_n$$

$$= n \sum_{j=0}^{n-1} x^j \sum_{\alpha=j}^{n-1} \frac{S(n-1,\alpha) s(\alpha+1,j)}{\alpha+1}$$

$$\quad + n \sum_{\alpha=1}^{n-1} \frac{S(n-1,\alpha-1)}{\alpha} x^\alpha + x^n + B_n$$

$$= n \sum_{j=1}^{n-1} x^j \sum_{\alpha=j}^{n-1} \frac{S(n-1,\alpha)s(\alpha+1,j)}{\alpha+1}$$

$$+ n \sum_{\alpha=1}^{n-1} \frac{S(n-1,\alpha-1)}{\alpha} x^\alpha + x^n + B_n$$

$$= n \sum_{j=1}^{n-1} x^j \sum_{\alpha=j+1}^{n} \frac{S(n-1,\alpha-1)s(\alpha,j)}{\alpha}$$

$$+ n \sum_{\alpha=1}^{n-1} \frac{S(n-1,\alpha-1)}{\alpha} x^\alpha + x^n + B_n$$

$$= n \sum_{j=1}^{n} x^j \sum_{\alpha=j}^{n} \frac{S(n-1,\alpha-1)s(\alpha,j)}{\alpha} + B_n.$$

If we compare the coefficient of $x^j$ in the previous line with that of $x^j$ in Equation (15.22), we discover that

$$\binom{n}{j} B_{n-j} = n \sum_{\alpha=j}^{n} \frac{1}{\alpha} S(n-1,\alpha-1)s(\alpha,j), \qquad 1 \le j \le n. \quad (15.41)$$

It is instructive to compare Equation (15.41) with Equation (15.35). Equation (15.41) says

$$\binom{n}{j} \frac{B_{n-j}}{n} = \sum_{k=j-1}^{n-1} \frac{1}{k+1} S(n-1,k)s(k+1,j), \qquad 1 \le j \le n,$$

while Equation (15.35) with $p \to n-1$ says

$$\binom{n-1}{j} \frac{B_{n-j}}{n-j} = \sum_{k=j}^{n} \frac{1}{k} S(n,k)s(k,j), \qquad 1 \le j \le n.$$

Since $\binom{n-1}{j} \frac{B_{n-j}}{n-j} = \binom{n}{j} \frac{B_{n-j}}{n}$, we discover the transformation

$$\sum_{k=j}^{n} \frac{1}{k} S(n,k)s(k,j) = \sum_{k=j-1}^{n-1} \frac{1}{k+1} S(n-1,k)s(k+1,j), \qquad n,j \ge 1.$$

## 15.3 Euler Polynomials and Euler Numbers

We now investigate a collection of polynomials which complement the Bernoulli polynomials. These polynomials are the Euler polynomials, $\{E_n(x)\}_{n=0}^{\infty}$, and are defined by

$$\sum_{n=0}^{\infty} E_n(x) \frac{t^n}{n!} = \frac{2e^{xt}}{e^t + 1}, \qquad |t| < \pi. \quad (15.42)$$

We use this section to develop connections between Euler polynomials and Stirling numbers of the second kind. To discover these connections we need to develop two explicit formulas for $E_n(x)$. The first formula follows from the generating function representations of $E_n(x)$ and $B_n(x)$. In particular

$$\sum_{k=0}^{\infty} \left[ B_k\left(\frac{x+1}{2}\right) - B_k\left(\frac{x}{2}\right) \right] \frac{2^k t^{k-1}}{k!} = \frac{2e^{xt}}{e^t + 1}$$

$$= \sum_{k=0}^{\infty} \frac{E_k(x)t^k}{k!} = \sum_{k=1}^{\infty} \frac{E_{k-1}(x)t^{k-1}}{(k-1)!},$$

or equivalently that

$$E_{n-1}(x) = \frac{2^n}{n} \left[ B_n\left(\frac{x+1}{2}\right) - B_n\left(\frac{x}{2}\right) \right], \qquad n \geq 1. \qquad (15.43)$$

We claim that

$$B_n\left(\frac{x+1}{2}\right) = \frac{B_n(x)}{2^{n-1}} - B_n\left(\frac{x}{2}\right). \qquad (15.44)$$

The proof of Equation (15.44) is presented in Section 15.4. By applying Equation (15.44) to the right side of Equation (15.43) we discover that

$$E_{n-1}(x) = \frac{2}{n} \left[ B_n(x) - 2^n B_n\left(\frac{x}{2}\right) \right], \qquad n \geq 1. \qquad (15.45)$$

Below we calculate $\{E_n\}_{n=0}^{10}$:

$$E_0(x) = 1 \qquad E_1(x) = x - \frac{1}{2} \qquad E_2(x) = x^2 - x$$

$$E_3(x) = x^3 - \frac{3}{2}x^2 + \frac{1}{4} \qquad E_4(x) = x^4 - 2x^3 + x$$

$$E_5(x) = x^5 - \frac{5}{2}x^4 + \frac{5}{2}x^2 - \frac{1}{2} \qquad E_6(x) = x^6 - 3x^5 + 5x^3 - 3x$$

$$E_7(x) = x^7 - \frac{7}{2}x^6 + \frac{35}{4} - \frac{21}{2}x^2 + \frac{17}{8}$$

$$E_8(x) = x^8 - 4x^7 + 14x^5 - 28x^3 + 17x$$

$$E_9(x) = x^9 - \frac{9}{2}x^8 + 21x^6 - 63x^4 + \frac{153}{2}x^2 - \frac{31}{2}$$

$$E_{10}(x) = x^{10} - 5x^9 + 30x^7 - 126x^5 + 255x^3 - 155x.$$

These examples imply that $E_n(x)$ is a polynomial in $x$ of degree $n$. This is indeed the case since an adaptation of the proof of Equation (15.22) shows that

$$E_n(x) = \sum_{j=0}^{n} \binom{n}{j} x^j E_{n-j}(0) = \sum_{j=0}^{n} \binom{n}{j} x^{n-j} E_j(0). \qquad (15.46)$$

Equation (15.45) when combined with Equation (15.44) provides a representation for $E_n(0)$ since

$$
\begin{aligned}
E_n(0) &= \frac{2^{n+1}}{n+1} \left[ B_{n+1}\left(\frac{1}{2}\right) - B_{n+1}(0) \right] \\
&= \frac{2^{n+1}}{n+1} \left[ \left(\frac{1}{2^n} - 1\right) B_{n+1} - B_{n+1} \right] \\
&= \frac{2\left(1 - 2^{n+1}\right)}{n+1} B_{n+1}.
\end{aligned} \tag{15.47}
$$

Equations (15.45) and (15.47) showcase the relationships between Bernoulli and Euler polynomials. Since Bernoulli numbers may be expressed in terms of Stirling numbers of the second kind, and since Euler polynomials are related to Bernoulli polynomials, it is not surprising to discover the $E_n(0)$ may also be expressed in terms of Stirling numbers. In particular we will show that

$$
E_n(0) = \sum_{k=0}^{n} \frac{(-1)^k}{2^k} k! S(n, k). \tag{15.48}
$$

Equation (15.48) is trivially true if $n = 0$. To prove Equation (15.48) for $n \geq 1$ we need some preliminary results. First is

$$
E_n(x) + E_n(x + 1) = 2x^n. \tag{15.49}
$$

The proof of Equation (15.49) follows from Equations (15.43) and (15.20) since

$$
\begin{aligned}
E_n(x) + E_n(x + 1) &= \frac{2^{n+1}}{n+1} \left[ B_{n+1}\left(\frac{x+2}{2}\right) - B_{n+1}\left(\frac{x}{2}\right) \right] \\
&\quad - \frac{2^{n+1}}{n+1} \left[ B_{n+1}\left(1 + \frac{x}{2}\right) - B_{n+1}\left(\frac{x}{2}\right) \right] \\
&= \frac{2^{n+1}}{n+1} \cdot \frac{n+1}{2^n} x^n = 2x^n.
\end{aligned}
$$

Equation (15.49) implies that

$$
2(-1)^k k^p = (-1)^k E_p(k) - (-1)^{k+1} E_p(k+1).
$$

Take this equation and sum over $k$ to discover that

$$
\sum_{k=0}^{n} (-1)^k k^p = \frac{(-1)^n E_p(n+1) + E_p(0)}{2}. \tag{15.50}
$$

Equation (15.50) complements Equation (15.21).

We are now ready to demonstrate Professor Gould's clever proof of Equation (15.48). Since $E_n(n-k+1)$ is a polynomial in $k$ of degree $n$, Euler's finite difference theorem implies that $\sum_{k=0}^{n+1}(-1)^k\binom{n+1}{k}E_n(n-k+1) = 0$. Professor Gould rewrote this equation as

$$E_n(0) = \frac{(-1)^n}{2^{n+1}}\sum_{k=0}^{n+1}(-1)^k\binom{n+1}{k}E_n(n-k+1)$$

$$+\frac{1}{2^{n+1}}E_n(0) + E_n(0) - \frac{1}{2^{n+1}}E_n(0)$$

$$= \frac{(-1)^n}{2^{n+1}}\sum_{k=0}^{n+1}(-1)^k\binom{n+1}{k}E_n(n-k+1)$$

$$-(-1)^n\frac{(-1)^{n+1}E_n(0)}{2^{n+1}} + \frac{E_n(0)}{2^{n+1}}\left[2^{n+1}-1\right]$$

$$= \frac{(-1)^n}{2^{n+1}}\left[\sum_{k=0}^{n+1}(-1)^k\binom{n+1}{k}E_n(n-k+1) - (-1)^{n+1}E_n(0)\right]$$

$$+\frac{E_n(0)}{2^{n+1}}\left[\sum_{k=0}^{n+1}\binom{n+1}{k} - 1\right]$$

$$= \frac{1}{2^{n+1}}\sum_{k=0}^{n}(-1)^{n-k}\binom{n+1}{k}E_n(n-k+1) + \frac{E_n(0)}{2^{n+1}}\sum_{k=0}^{n}\binom{n+1}{k}$$

$$= \frac{1}{2^{n+1}}\sum_{k=0}^{n}(-1)^k\binom{n+1}{n-k}E_n(k+1) + \frac{E_n(0)}{2^{n+1}}\sum_{k=0}^{n}\binom{n+1}{n-k}$$

$$= \frac{1}{2^n}\sum_{k=0}^{n}\binom{n+1}{n-k}\frac{(-1)^kE_n(k+1) + E_n(0)}{2}$$

$$= \frac{1}{2^n}\sum_{k=0}^{n}\binom{n+1}{n-k}\sum_{j=0}^{k}(-1)^jj^n$$

$$= \frac{1}{2^n}\sum_{k=0}^{n}\binom{n+1}{k}\sum_{j=0}^{n-k}(-1)^jj^n$$

$$= \frac{1}{2^n}\sum_{k=0}^{n}(-1)^k\binom{n+1}{k}\sum_{j=0}^{n-k}(-1)^{j+k}j^n$$

$$= \frac{1}{2^n} \sum_{k=0}^{n} (-1)^k \binom{n+1}{k} \sum_{j=k}^{n} (-1)^j (j-k)^n$$

$$= \frac{1}{2^n} \sum_{j=0}^{n} (-1)^j \sum_{k=0}^{j} (-1)^k \binom{n+1}{k} (j-k)^n.$$

These calculations imply that

$$E_n(0) = \frac{1}{2^n} \sum_{j=0}^{n} (-1)^j A_{n,j}, \qquad A_{n,j} = \sum_{k=0}^{j} (-1)^k \binom{n+1}{k} (j-k)^n. \quad (15.51)$$

If $n \geq 1$, $A_{n,0} = 0$, and Equation (15.51) becomes

$$E_n(0) = \frac{1}{2^n} \sum_{j=1}^{n} (-1)^j A_{n,j}, \qquad n \geq 1. \quad (15.52)$$

Since $A_{n,n-j+1} = A_{n,j}$ whenever $n \geq 1$, Equation (15.52) is equivalent to

$$E_n(0) = \frac{1}{2^n} \sum_{j=1}^{n} (-1)^j A_{n,j} = \frac{1}{2^n} \sum_{j=1}^{n} (-1)^{n-j+1} A_{n,n-j+1}$$

$$= \frac{(-1)^{n+1}}{2^n} \sum_{j=1}^{n} (-1)^j A_{n,j}. \quad (15.53)$$

Assume $n \geq 1$. Equation (15.52) implies that

$$E_n(0) = \frac{(-1)^n}{2^n} \sum_{j=0}^{n} (-1)^{j-1} A_{n,j} = \frac{(-1)^n}{2^n} \sum_{j=0}^{n} A_{n,j} (1-2)^{j-1}$$

$$- \frac{(-1)^n}{2^n} \sum_{j=0}^{n} A_{n,j} \sum_{k=0}^{j-1} (-1)^k \binom{j-1}{k} 2^k$$

$$= \sum_{k=0}^{n} \frac{(-1)^{n-k}}{2^{n-k}} \sum_{j=k}^{n} \binom{j-1}{k} A_{n,j}$$

$$= \sum_{k=0}^{n} \frac{(-1)^k}{2^k} \sum_{j=n-k}^{n} \binom{j-1}{n-k} A_{n,j}$$

$$= \sum_{k=0}^{n} \frac{(-1)^k}{2^k} \sum_{j=0}^{n} \binom{j-1}{n-k} A_{n,j}$$

$$= \sum_{k=0}^{n} \frac{(-1)^k}{2^k} k! S(n,k), \qquad \text{by Eq. (10.2).}$$

By combining Equation (15.47) with Equations (15.48) and (15.51), we discover that

$$B_{n+1} = \frac{n+1}{(2^{n+1}-1)} \sum_{j=0}^{n} \frac{(-1)^{j+1}}{2^{j+1}} j! S(n,j), \qquad (15.54)$$

and that

$$B_{n+1} = \frac{n+1}{2^{n+1}(2^{n+1}-1)} \sum_{j=0}^{n} (-1)^{j+1} A_{n,j}. \qquad (15.55)$$

To discover two other representations for $B_{n+1}$ attributed to L. Carlitz [1961], we combine Equation (15.48) with Equation (15.46) to obtain

$$E_n(x) = \sum_{k=0}^{n} \binom{n}{k} x^{n-k} \sum_{j=0}^{k} \frac{(-1)^j}{2^j} j! S(k,j)$$

$$= \sum_{k=0}^{n} \binom{n}{k} x^{n-k} \sum_{j=0}^{k} \frac{1}{2^j} \sum_{i=0}^{j} (-1)^i \binom{j}{i} i^k$$

$$= \sum_{k=0}^{n} \binom{n}{k} x^{n-k} \sum_{j=0}^{n} \frac{1}{2^j} \sum_{i=0}^{j} (-1)^i \binom{j}{i} i^k$$

$$= \sum_{j=0}^{n} \frac{1}{2^j} \sum_{i=0}^{j} (-1)^i \binom{j}{i} \sum_{k=0}^{n} \binom{n}{k} x^{n-k} i^k$$

$$= \sum_{j=0}^{n} \frac{1}{2^j} \sum_{i=0}^{j} (-1)^i \binom{j}{i} (x+i)^n$$

$$= \sum_{j=0}^{n} (-1)^j 2^{-j} \Delta_{x,1}^j x^n. \qquad (15.56)$$

Equation (15.56) implies that $E_n(0) = \sum_{k=0}^{n} (-1)^{-k} 2^k \left. \Delta_{x,1}^k x^n \right|_{x=0}$. Substitute Equation (15.47) in for $E_n(0)$ and solve for $B_{n+1}$ to obtain

$$B_{n+1} = \frac{(-1)^n(n+1)}{1-2^{n+1}} \sum_{k=0}^{n} (-1)^k 2^{-k} \left. \Delta_{x,1}^k x^n \right|_{x=0}. \qquad (15.57)$$

Equation (15.57) is the first of Carlitz's formulas for Bernoulli numbers. To obtain the second formula we observe that $E_n(1) = \sum_{k=0}^{n} (-1)^k 2^{-k} \left. \Delta_{x,1}^k x^n \right|_{x=1}$. Equation (15.42) with $x \to 1-x$ and $t \to -t$ implies that $E_n(x) = (-1)^n E_n(1-x)$. Hence $E_n(1) = (-1)^n E_n(0) = (-1)^n \frac{2(1-2^{n+1})}{n+1} B_{n+1}$ and

$$B_{n+1} = -\frac{n+1}{2^{n+1}-1} \sum_{k=0}^{n} (-1)^k 2^{-k-1} \left. \Delta_{x,1}^k x^n \right|_{x=1}. \qquad (15.58)$$

## 15.4 Polynomial Expansions Involving $B_n(x)$

In this section we will use polynomial expansions to provide a proof of Equation (15.44). To obtain the desired polynomial expansion, recall Equation (15.20), namely

$$B_{n+1}(x+1) - B_{n+1}(x) = (n+1)x^n. \tag{15.59}$$

Next observe that a simple modification of the proof of Equation (15.22) proves

$$B_n(x+y) = \sum_{j=0}^{n} \binom{n}{j} x^{n-j} B_j(y). \tag{15.60}$$

Take Equation (15.59) and use Equation (15.60) to expand $B_{n+1}(x+1)$.

$$x^n = \frac{1}{n+1} \left[ -B_{n+1}(x) + \sum_{k=0}^{n+1} \binom{n+1}{k} B_k(x) \right]$$

$$= \frac{1}{n+1} \sum_{k=0}^{n} \binom{n+1}{k} B_k(x). \tag{15.61}$$

We apply Equation (15.61) and expand an arbitrary polynomial in terms of the Bernoulli polynomials. Let $f(x) = \sum_{k=0}^{n} a_k x^k$. By Equation (15.61) we have

$$f(x) = \sum_{k=0}^{n} \frac{a_k}{k+1} \sum_{j=0}^{k} \binom{k+1}{j} B_j(x) = \sum_{j=0}^{n} B_j(x) \sum_{k=j}^{n} \binom{k+1}{j} \frac{a_k}{k+1}. \tag{15.62}$$

Since Taylor's theorem implies that $a_k = \frac{D_x^k f(0)}{k!}$, we may write Equation (15.62) as

$$f(x) = \sum_{j=0}^{n} B_j(x) \sum_{k=j}^{n} \binom{k+1}{j} \frac{D_x^k f(0)}{(k+1)!}, \qquad f(x) = \sum_{k=0}^{n} a_k x^k. \tag{15.63}$$

Equation (15.63) was know to Charles Jordan [1957, p. 248] albeit in a different formulation. To obtain Jordan's formulation first observe that

$$\int_0^1 f(x)\,dx = \sum_{k=0}^{n} a_k \int_0^1 x^k\,dx = \sum_{k=0}^{n} \frac{a_k}{k+1} = \sum_{k=0}^{n} \frac{D_x^k f(0)}{(k+1)!}. \tag{15.64}$$

We next calculate $\Delta_{x,1} D_x^{\alpha-1} f(x)$ where $\Delta_{x,1} F(x) = F(x+1) - F(x)$ and $\alpha \geq 1$. Since $\Delta_{x,1}$ is a linear operator we have

$$\Delta_{x,1} D_x^{\alpha-1} f(x) = \sum_{k=\alpha-1}^{n} a_k (\alpha - 1)! \binom{k}{\alpha - 1} \Delta_{x,1} x^{k-\alpha+1}$$

$$= \sum_{k=\alpha}^{n} a_k (\alpha - 1)! \binom{k}{\alpha - 1} \sum_{j=0}^{k-\alpha} \binom{k - \alpha + 1}{j} x^j. \quad (15.65)$$

If we set $x = 0$, the only nonzero term occurs when $j = 0$. In other words

$$\Delta_{x,1} D_x^{\alpha-1} f(x)\big|_{x=0} = \sum_{k=\alpha}^{n} a_k (\alpha - 1)! \binom{k}{\alpha - 1}$$

$$= \alpha! \sum_{k=\alpha}^{n} \binom{k+1}{\alpha} \frac{D_x^k f(0)}{(k+1)!}. \quad (15.66)$$

Take a look at the inner sum of Equation (15.63). If $j \geq 1$, Equation (15.66) shows it is precisely $\frac{1}{j!} \Delta_{x,1} D_x^{j-1} f(x)\big|_{x=0}$, while if $j = 0$ it becomes $\sum_{k=0}^{j} \frac{D_x^k f(0)}{(k+1)!} = \int_0^1 f(x)\,dx$. This analysis proves the following theorem found in [Jordan, 1957, Sec.48].

**Theorem 15.1.** *Let* $f(x) = \sum_{k=0}^{n} a_k x^k$. *Then*

$$f(x) = \sum_{j=0}^{n} B_j(x) \sum_{k=j}^{n} \binom{k+1}{j} \frac{D_x^k f(0)}{(k+1)!} = \sum_{j=0}^{n} C_j B_j(x), \quad (15.67)$$

*where* $C_0 = \int_0^1 f(x)\,dx$ *and* $C_j = \frac{1}{j!} \Delta_{x,1} D_x^{j-1} f(x)\big|_{x=0}$ *whenever* $1 \leq j \leq n$.

An important application of Theorem 15.1, known as Raabe's theorem [Jordan, 1957], expands $B_n(rx)$ whenever $r$ is a positive integer. Let $f(x) = r^{n-1} \sum_{k=0}^{r-1} B_n\left(\frac{x+k}{r}\right)$ where $n \geq 1$ and $r \geq 1$. Since $f(x)$ is a polynomial in $x$ of degree $n$ we may apply Theorem 15.1 and write

$$f(x) = \sum_{\alpha=0}^{n} C_\alpha B_\alpha(x),$$

$$C_0 = \int_0^1 f(x)\,dx, \qquad C_\alpha = \frac{1}{\alpha!} \Delta_{x,1} D_x^{\alpha-1} f(0), \quad \alpha \geq 1.$$

To simplify $C_0$ we need two properties of Bernoulli polynomials which readily follow from Equation (15.22). Term by term differentiation of Equation (15.22) shows that

$$B_n'(x) = \begin{cases} n B_{n-1}(x), & n \geq 1 \\ 0, & n = 0. \end{cases} \quad (15.68)$$

Equation (15.22) also implies that

$$B_n(1 - x) = (-1)^n B_n(x), \tag{15.69}$$

since

$$B_n(1 - x) = \sum_{j=0}^{n} \binom{n}{j}(1 - x)^{n-j} B_j = \sum_{k=0}^{n}(-1)^k x^k \sum_{j=0}^{n-k} \binom{n}{j}\binom{n-j}{k} B_j$$

$$= \sum_{k=0}^{n}(-1)^k x^k \binom{n}{k} \sum_{j=0}^{n-k} \binom{n-k}{j} B_j$$

$$= (-1)^n \sum_{k=0}^{n} \binom{n}{k} x^k B_{n-k} = (-1)^n B_n(x).$$

If $x = 0$, Equation (15.69) becomes

$$B_n(1) = (-1)^n B_n. \tag{15.70}$$

By applying Equations (15.68) and (15.70), and the fact that $B_{2n+1} = 0$ for $n \geq 1$, we simplify $C_0$ as follows:

$$C_0 = \int_0^1 f(x)\, dx = r^{n-1} \sum_{k=0}^{r-1} \int_0^1 B_n\left(\frac{x+k}{r}\right) dx$$

$$= r^n \sum_{k=0}^{r-1} \int_0^{\frac{1}{r}} B_n\left(z + \frac{k}{r}\right) dz$$

$$= r^n \sum_{k=0}^{r-1} \left[\frac{B_{n+1}\left(z + \frac{k}{r}\right)}{n+1}\right]_0^{\frac{1}{r}}$$

$$= r^n \sum_{k=0}^{r-1} \frac{B_{n+1}\left(\frac{k+1}{r}\right) - B_{n+1}\left(\frac{k}{r}\right)}{n+1}$$

$$= \frac{r^n}{n+1}\left[B_{n+1}(1) - B_{n+1}\right]$$

$$= \frac{r^n}{n+1}\left[(-1)^{n+1}B_{n+1} - B_{n+1}\right] = 0.$$

Next we calculate for $C_\alpha$ for $\alpha \geq 1$:

$$C_\alpha = \frac{1}{\alpha!}\Delta_{x,1} D_x^{\alpha-1} f(0) = \frac{1}{\alpha!}\Delta_{x,1} D_x^{\alpha-1}\left. r^{n-1} \sum_{k=0}^{r-1} B_n\left(\frac{x+k}{r}\right)\right|_{x=0}$$

$$= \frac{1}{\alpha!}\Delta_{x,1}\left. r^{n-1} \sum_{k=0}^{r-1}(\alpha - 1)!\binom{n}{\alpha - 1}\left(\frac{1}{r}\right)^{\alpha-1} B_{n-\alpha+1}\left(\frac{x+k}{r}\right)\right|_{x=0}$$

$$= \frac{r^{n-\alpha}}{\alpha}\binom{n}{\alpha - 1}\left. \sum_{k=0}^{r-1}\Delta_{x,1} B_{n-\alpha+1}\left(\frac{x+k}{r}\right)\right|_{x=0}$$

$$= \frac{r^{n-\alpha}}{\alpha} \binom{n}{\alpha-1} \sum_{k=0}^{r-1} B_{n-\alpha+1}\left(\frac{x+k+1}{r}\right) - B_{n-\alpha+1}\left(\frac{x+k}{r}\right)\bigg|_{x=0}$$

$$= \frac{r^{n-\alpha}}{\alpha} \binom{n}{\alpha-1} \sum_{k=0}^{r-1} \left[ B_{n-\alpha+1}\left(\frac{k+1}{r}\right) - B_{n-\alpha+1}\left(\frac{k}{r}\right) \right]$$

$$= \frac{r^{n-\alpha}}{\alpha} \binom{n}{\alpha-1} [B_{n-\alpha+1}(1) - B_{n-\alpha+1}]$$

$$= \frac{r^{n-\alpha}}{\alpha} \binom{n}{\alpha-1} [(-1)^{n-\alpha+1} B_{n-\alpha+1} - B_{n-\alpha+1}]$$

$$= \begin{cases} 0, & n > \alpha \\ -B_1 - B_1 = 1, & n = \alpha. \end{cases}$$

In summary we have shown that

$$r^{n-1} \sum_{k=0}^{r-1} B_n\left(\frac{x+k}{r}\right) = B_n(x), \qquad n \geq 1, \ r \geq 1. \tag{15.71}$$

We should note that Equation (15.71) is actually true for $n \geq 0$ and $r \geq 1$, the verification of which we leave for the reader. Equation (15.71) is known as Raabe's theorem and often appears as

$$B_n(rx) = r^{n-1} \sum_{k=0}^{r-1} B_n\left(x + \frac{k}{r}\right), \qquad n \geq 0, \ r \geq 1. \tag{15.72}$$

If $r = 2$, Equation (15.71) becomes

$$2^{n-1} \sum_{k=0}^{1} B_n\left(\frac{x+k}{2}\right) = 2^{n-1}\left[B_n\left(\frac{x}{2}\right) + B_n\left(\frac{x+1}{2}\right)\right] = B_n(x),$$

which is equivalent to Equation (15.44).

# Appendix A

# Newton-Gregory Expansions

At the beginning of Chapter 3 we discussed the Taylor series expansion for $f(x) = \sum_{i=0}^{n} a_i x^i$. In particular we showed that

$$f(x) = \sum_{k=0}^{n} \frac{f^{(k)}(a)}{k!}(x-a)^k, \qquad f^{(k)}(a) = \frac{d^k}{dx^k}f(x)\bigg|_{x=a} = D_x^k f(x)\big|_{x=a}.$$

$$(A.1)$$

See Equation (3.5). The Taylor series provide a way of expanding $f(x)$ in terms of its derivatives. Since $\lim_{h \to 0} \Delta_h^n f(x) = \frac{d^n}{dx^n}f(x)$, it is natural to wonder if we can expand $f(x)$ in terms of the difference operator. In order to answer this question we must transform the right side of Equation (A.1) into a series involving the difference operator. This is done by developing a formula which writes each derivative as a sum of difference operators and Stirling numbers of the first kind. Recall Equation (6.2) which states

$$\Delta_h^m f(x) = \Delta_{x,h}^m f(x) = \frac{(-1)^m}{h^m} \sum_{k=0}^{m} (-1)^k \binom{m}{k} f(x+kh).$$

Assume that $f(x) = x^\alpha$ where $\alpha$ is a nonnegative integer. Then

$$\Delta_h^m x^\alpha = \frac{(-1)^m}{h^m} \sum_{k=0}^{m} (-1)^k \binom{m}{k} (x+kh)^\alpha$$

$$= \frac{(-1)^m}{h^m} \sum_{k=0}^{m} (-1)^k \binom{m}{k} \sum_{j=0}^{\alpha} \binom{\alpha}{j} x^{\alpha-j} k^j h^j$$

$$= \frac{(-1)^m}{h^m} \sum_{j=0}^{\alpha} \binom{\alpha}{j} x^{\alpha-j} h^j \sum_{k=0}^{m} (-1)^k \binom{m}{k} k^j$$

$$= m! \sum_{j=0}^{\alpha} \binom{\alpha}{j} x^{\alpha-j} h^{j-m} S(j,m), \qquad \text{by Eq. (9.21)}$$

$$= m! \sum_{j=0}^{\alpha} \frac{1}{j!} \left[ j! \binom{\alpha}{j} x^{\alpha-j} \right] h^{j-m} S(j,m)$$

$$= m! \sum_{j=0}^{\alpha} \frac{1}{j!} h^{j-m} S(j,m) D_x^j x^{\alpha}.$$

In summary we have shown that

$$\Delta_h^m x^{\alpha} = m! \sum_{j=0}^{\alpha} \frac{1}{j!} h^{j-m} S(j,m) D_x^j x^{\alpha}, \qquad \alpha \text{ a nonnegative integer.} \quad \text{(A.2)}$$

If $\alpha$ is an arbitrary complex number the proof of Equation (A.2) is still valid as long as we extend the summation limit to infinity.

Since the difference operator is a linear operator we may extend Equation (A.2) to an arbitrary polynomial $f(x) = \sum_{\alpha=0}^{n} a_\alpha x^{\alpha}$ as follows:

$$\Delta_h^m f(x) = \sum_{\alpha=0}^{n} a_\alpha \Delta_h^m x^{\alpha} = m! \sum_{\alpha=0}^{n} a_\alpha \sum_{j=0}^{\alpha} \frac{h^{j-m}}{j!} S(j,m) D_x^j x^{\alpha}$$

$$= \sum_{j=0}^{n} \frac{h^{j-m}}{j!} S(j,m) \sum_{\alpha=j}^{n} a_\alpha D_x^j x^{\alpha} = \sum_{j=0}^{n} \frac{h^{j-m}}{j!} S(j,m) D_x^j f(x).$$

In summary we have proven the identity

$$\Delta_h^m f(x) = m! \sum_{j=0}^{n} \frac{1}{j!} h^{j-m} S(j,m) D_x^j f(x), \qquad f(x) = \sum_{\alpha=0}^{n} a_\alpha x^{\alpha}. \quad \text{(A.3)}$$

Equation (A.3) writes the difference operator as a sum of derivative operators and Stirling numbers of the second kind. We invert this identity and write the derivative as a sum of difference operators and Stirling numbers of the first kind. Assume $k$ is any positive integer with $k \geq n$. Multiply both sides of Equation (A.3) by $\frac{s(m,\alpha)}{m!} h^m$ and sum over $m$

$$\sum_{m=0}^{k} s(m,\alpha) \frac{h^m}{m!} \Delta_h^m f(x) = \sum_{m=0}^{k} s(m,\alpha) h^m \sum_{j=0}^{n} \frac{1}{j!} h^{j-m} S(j,m) D_x^j f(x)$$

$$= \sum_{j=0}^{n} \frac{1}{j!} h^j D_x^j f(x) \sum_{m=\alpha}^{k} S(j,m) s(m,\alpha)$$

$$= \begin{cases} 0, & j \neq \alpha \\ \frac{1}{\alpha!} h^{\alpha} D_x^{\alpha} f(x), & j = \alpha, \end{cases}$$

where the last equality follows by Equation (12.18). By solving for $D_x^\alpha f(x)$ we conclude that

$$D_x^\alpha f(x) = \alpha! \sum_{m=0}^{k} s(m,\alpha) \frac{h^{m-\alpha}}{m!} \Delta_h^m f(x), \qquad f(x) = \sum_{j=0}^{n} a_j x^j, \qquad k \geq n.$$
(A.4)

Equation (A.4) is the desired inversion of Equation (A.3). We now combine Equation (A.4) with Taylor series expansion of $f(x+y)$. First rewrite Equation (A.1) as

$$f(x+y) = \sum_{\alpha=0}^{n} \frac{x^\alpha}{\alpha!} f^{(\alpha)}(y), \qquad f(x) = \sum_{j=0}^{n} a_j x^j.$$
(A.5)

Then use Equation (A.4) to expand $f^{(\alpha)}(y)$ as follows:

$$f(x+y) = \sum_{\alpha=0}^{n} \frac{x^\alpha}{\alpha!} f^{(\alpha)}(y) = \sum_{\alpha=0}^{n} x^\alpha \sum_{j=0}^{n} s(j,\alpha) \frac{h^{j-\alpha}}{j!} \Delta_{y,h}^j f(y)$$

$$= \sum_{j=0}^{n} \frac{h^j}{j!} \Delta_{y,h}^j f(y) \sum_{\alpha=0}^{j} s(j,\alpha) \left(\frac{x}{h}\right)^\alpha$$

$$= \sum_{j=0}^{n} h^j \Delta_{y,h}^f(y) \binom{\frac{x}{h}}{j}, \qquad \text{by Eq. (12.1)}.$$

We have shown that if $f(x) = \sum_{j=0}^{n} a_j x^j$, then

$$f(x+y) = \sum_{\alpha=0}^{n} \frac{x^\alpha}{\alpha!} f^{(\alpha)}(y) = \sum_{\alpha=0}^{n} \binom{\frac{x}{h}}{\alpha} h^\alpha \Delta_{y,h}^\alpha f(y).$$
(A.6)

If $h \to 0$, the right sum of Equation (A.6) becomes the Taylor series.

Equation (A.6) answers the question of how to expand a polynomial as a series involving the difference operator. We remark that our proof of Equation (A.5) can be extended to $f(x) = \sum_{j=0}^{\infty} a_j x^j$. If we ignore questions of convergence and work in the context of formal power series we obtain

$$f(x+y) = \sum_{\alpha=0}^{\infty} \frac{x^\alpha}{\alpha!} f^{(\alpha)}(y) = \sum_{\alpha=0}^{\infty} \binom{\frac{x}{h}}{\alpha} h^\alpha \Delta_{y,h}^\alpha f(y).$$
(A.7)

If we let $h = 1$, Equation (A.7) becomes

$$f(x+y) = \sum_{\alpha=0}^{\infty} \binom{x}{\alpha} \Delta_{y,1}^\alpha f(y), \qquad f(x) = \sum_{j=0}^{\infty} a_j x^j.$$
(A.8)

Let $x \to x - a$ and $y \to a$ to obtain

$$f(x) = \sum_{k=0}^{\infty} \binom{x-a}{k} \Delta_1^k f(a), \qquad \Delta_1^k f(a) = \Delta_1^k f(x)|_{x=a}. \qquad (A.9)$$

Equation (A.9) is called the Newton-Gregory expansion for $f(x)$. [1]

For $f(x) = \sum_{j=0}^n a_j x^j$, Equation (A.8) often appears in the literature as

$$f(x+y) = \sum_{j=0}^n \binom{x}{j} f(y+j) \sum_{k=0}^{n-j} (-1)^k \binom{x-j}{k},$$

since

$$f(x+y) = \sum_{\alpha=0}^n \binom{x}{\alpha} \Delta_{y,1}^\alpha f(y)$$

$$= \sum_{\alpha=0}^n \binom{x}{\alpha} (-1)^\alpha \sum_{j=0}^\alpha (-1)^j \binom{\alpha}{j} f(y+j)$$

$$= \sum_{j=0}^n (-1)^j f(y+j) \sum_{\alpha=j}^n (-1)^\alpha \binom{x}{\alpha}\binom{\alpha}{j}$$

$$= \sum_{j=0}^n (-1)^j \binom{x}{j} f(y+j) \sum_{\alpha=j}^n (-1)^\alpha \binom{x-j}{\alpha-j}$$

$$= \sum_{j=0}^n \binom{x}{j} f(y+j) \sum_{\alpha=0}^{n-j} (-1)^\alpha \binom{x-j}{\alpha}.$$

We remark that Equation (A.8) appears in [Gould, 1972, Table Z] as

$$f(x+y) = \sum_{k=0}^n \binom{x}{k} \sum_{j=0}^k (-1)^j \binom{k}{j} f(y+k-j), \qquad f(x) = \sum_{j=0}^n a_j x^j, \quad (A.10)$$

since

$$f(x+y) = \sum_{\alpha=0}^n \binom{x}{\alpha} \Delta_{y,1}^\alpha f(y)$$

$$= \sum_{\alpha=0}^n \binom{x}{\alpha} (-1)^\alpha \sum_{j=0}^\alpha (-1)^j \binom{\alpha}{j} f(y+j)$$

$$= \sum_{\alpha=0}^n \binom{x}{\alpha} \sum_{j=0}^\alpha (-1)^j \binom{\alpha}{j} f(y+\alpha-j).$$

---

[1] Wikipedia, *Finite Difference*, http://en.wikipedia.org/wiki/Finite_difference

## Appendix B

# Generalized Bernoulli and Euler Polynomials

Let $a$ be a complex number. For a fixed $a$ we define two sets of polynomials $\left\{B_k^{(a)}(z)\right\}_{k=0}^{\infty}$ and $\left\{E_k^{(a)}(z)\right\}_{k=0}^{\infty}$ by

$$\sum_{k=0}^{\infty} B_k^{(a)}(z)\frac{x^k}{k!} = \frac{x^a e^{xz}}{(e^x - 1)^a}, \qquad |x| < 2\pi, \qquad \text{(B.1)}$$

$$\sum_{k=0}^{\infty} E_k^{(a)}(z)\frac{x^k}{k!} = \frac{2^a e^{xz}}{(e^x + 1)^a}, \qquad |x| < \pi. \qquad \text{(B.2)}$$

We say $B_k^{(a)}(z)$ is the generalized Bernoulli polynomial or Nörlund polynomial of degree $k$ and order $a$ while $E_k^{(a)}(z)$ is the generalized Euler polynomial of degree $k$ and order $a$. In Chapter 13 we discussed the intimate connection between Stirling numbers and Nörlund polynomials. In particular we showed that

$$\binom{n}{k}B_k^{(n+1)}(1) = s(n, n-k) \qquad \binom{n-1}{k}B_k^{(n)}(0) = s(n, n-k)$$

$$\binom{n+k}{k}B_k^{(-n)}(0) - S(n \mid k, n).$$

See Equations (13.4), (13.27), and (13.36) respectively. We also showed that

$$B_k^{(n+1)}(z+1) = k!D_z^{n-k}\binom{z}{n} \qquad B_k^{(n+1)}(1) = \frac{n-k}{n}B_k^{(n)}(0).$$

See Equations (13.2) and (13.24). The purpose of this appendix is to discuss algebraic properties shared by $\left\{B_k^{(a)}(z)\right\}_{k=0}^{\infty}$ and $\left\{E_k^{(a)}(z)\right\}_{k=0}^{\infty}$. In particular we show that $B_k^{(a)}(z)$ and $E_k^{(a)}(z)$ are Appell polynomials and exploit this Appell structure to derive formulas for expanding derivatives of $f(x) = \sum_{i=0}^{n} a_i x^i$ in terms of $\left\{B_k^{(a)}(z)\right\}_{k=0}^{\infty}$ and $\left\{E_k^{(a)}(z)\right\}_{k=0}^{\infty}$.[1]

---

[1]Wikipedia, *Appell Sequence*, http://en.wikipedia.org/wiki/Appell_sequence

# B.1   Basic Properties of $B_k^{(a)}(x)$ and $E_k^{(a)}(x)$

We begin our exploration of the basic properties by showing that $B_k^{(a)}(x)$ is an Appell polynomial. In Equation (B.1) interchange the role of $x$ and $z$. The derivative of the left side of (B.1) with respect to $x$ is

$$D_x \sum_{k=0}^{\infty} \frac{B_k^{(a)}(x)}{k!} z^k = \sum_{k=0}^{\infty} \frac{D_x B_k^{(a)}(x)}{k!} z^k \qquad (1),$$

while the derivative of the right side is

$$D_x \left[ \frac{z^a e^{xz}}{(e^z - 1)^a} \right] = \frac{z^{a+1}}{(e^z - 1)^a} e^{xz} = z \sum_{k=0}^{\infty} \frac{B_k^{(a)}(x)}{k!} z^k$$

$$= \sum_{k=1}^{\infty} \frac{B_{k-1}^{(a)}(x)}{(k-1)!} z^k = \sum_{k=0}^{\infty} \frac{k B_{k-1}^{(a)}(x)}{k!} z^k \qquad (2).$$

The equality of the coefficient of $z^k$ at Line (1) with that of Line (2) implies that

$$D_x B_k^{(a)}(x) = k B_{k-1}^{(a)}(x). \qquad (B.3)$$

Equation (B.3) shows that $\left\{ B_k^{(a)}(x) \right\}_{k=0}^{\infty}$ is a sequence of Appell polynomials.[2] If $a = 1$, Equation (B.3) becomes

$$D_x B_k(x) = k B_{k-1}(x). \qquad (B.4)$$

Iteration of Equation (B.3) provides

$$D_x^j B_k^{(a)}(x) = \prod_{i=0}^{j-1} (k-i) B_{k-j}^{(a)}(x) = j! \binom{k}{j} B_{k-j}^{(a)}(x). \qquad (B.5)$$

Equation (B.3), when substituted into the Taylor's series for $B_k^{(a)}(x+y)$, implies that

$$B_k^{(a)}(x+y) = \sum_{j=0}^{\infty} \frac{y^j}{j!} D_x^j B_k^{(a)}(x) = \sum_{j=0}^{k} \binom{k}{j} B_{k-j}^{(a)}(x) y^j$$

$$= \sum_{j=0}^{k} \binom{k}{j} B_j^{(a)}(x) y^{k-j}. \qquad (B.6)$$

---

[2]Wikipedia, *Appell Sequence*, http://en.wikipedia.org/wiki/Appell_sequence

Equation (B.6) is known as the addition property for $\left\{B_k^{(a)}(x)\right\}_{k=0}^{\infty}$. If $x = 0$, Equation (B.6) becomes

$$B_k^{(a)}(y) = \sum_{j=0}^{k} \binom{k}{j} B_{k-j}^{(a)}(0) y^j, \qquad (B.7)$$

which clearly shows that $B_k^{(a)}(x)$ is a polynomial in $x$ of degree $k$.

The generalized Euler polynomials also obey a similar set of properties. By adapting the proofs of Equations (B.3) through (B.6) we readily show that

$$D_x E_k^{(a)}(x) = k E_{k-1}^{(a)}(x)$$

$$D_x^j E_k^{(a)}(x) = \prod_{i=0}^{j-1} (k-i) E_{k-j}^{(a)}(x) = j! \binom{k}{j} E_{k-j}^{(a)}(x), \qquad (B.8)$$

and that

$$E_k^{(a)}(x + y) = \sum_{j=0}^{k} \binom{k}{j} E_j^{(a)}(x) y^{k-j}. \qquad (B.9)$$

Equation (B.8) shows that $\left\{E_k^{(a)}(x)\right\}_{k=0}^{\infty}$ is a collection of Appell Polynomials. If $a = 1$, Equation (B.8) becomes

$$D_x E_k(x) = k E_{k-1}(x). \qquad (B.10)$$

Equation (B.3) implies that

$$\int_0^a B_k^{(a)}(x)\, dx = \frac{B_{k+1}^{(n)}(a) - B_{k+1}^{(n)}(0)}{k + 1}, \qquad (B.11)$$

while Equation (B.8) implies that

$$\int_0^a E_k^{(a)}(x)\, dx = \frac{E_{k+1}^{(a)}(a) - E_{k+1}^{(a)}(0)}{k + 1}. \qquad (B.12)$$

To obtain Equation (B.3) we differentiated Equation (B.1) with respect to $z$. We now explore what happens when we differentiate with respect to $x$. Replace $z$ with $t$ and observe that

$$D_x \sum_{k=0}^{\infty} B_k^{(a)}(t) \frac{x^k}{k!} = \sum_{k=0}^{\infty} B_k^{(a)}(t) \frac{k x^{k-1}}{k!} = \frac{a x^{a-1} e^{tx}}{(e^x - 1)^a} + x^a D_x \frac{e^{tx}}{(e^x - 1)^a}.$$

Multiply both sides of this equation by $x$ to obtain

$$\sum_{k=0}^{\infty} B_k^{(a)}(t)\frac{kx^k}{k!} = \frac{ax^a e^{tx}}{(e^x-1)^a} + x^{a+1}\left[\frac{te^{tx}}{(e^x-1)^a} + \frac{(-a)e^{tx}e^x}{(e^x+1)^{a+1}}\right]$$

$$= \frac{ax^a e^{tx}}{(e^x-1)^a} + tx\frac{x^a e^{tx}}{(e^x-1)^a} - a\frac{x^{a+1}e^{(t+1)x}}{(e^x-1)^{a+1}}$$

$$= a\sum_{k=0}^{\infty} B_k^{(a)}(t)\frac{x^k}{k!} + tx\sum_{k=0}^{\infty} B_k^{(a)}(t)\frac{x^k}{k!}$$

$$-a\sum_{k=0}^{\infty} B_k^{(a+1)}(t+1)\frac{x^k}{k!}$$

$$= a\sum_{k=0}^{\infty} B_k^{(a)}(t)\frac{x^k}{k!} + t\sum_{k=0}^{\infty} B_{k-1}^{(a)}(t)\frac{kx^k}{k!}$$

$$-a\sum_{k=0}^{\infty} B_k^{(a+1)}(t+1)\frac{x^k}{k!}.$$

These calculations show that

$$\sum_{k=0}^{\infty}\frac{kB_k^{(a)}(t)}{k!}x^k = \sum_{k=0}^{\infty}\frac{aB_k^{(a)}(t) + tkB_{k-1}^{(a)}(t) - aB_k^{a+1}(t+1)}{k!}x^k,$$

which is equivalent to saying

$$kB_k^{(a)}(t) = aB_k^{(a)}(t) + tkB_{k-1}^{(a)}(t) - aB_k^{(a+1)}(t+1). \qquad \text{(B.13)}$$

An important case of Equation (B.13) occurs when $t = 0$ since $(k-a)B_k^{(a)}(0) = -aB_k^{(a+1)}(1)$, or equivalently

$$B_k^{(a+1)}(1) = \left(1 - \frac{k}{a}\right)B_k^{(a)}(0). \qquad \text{(B.14)}$$

Equation (B.13) provides a recurrence formula for $B_k^{(a+1)}(t+1)$ in terms of $B_{k-1}^{(a)}(t)$ and $B_k^{(a)}(t)$. Ideally we would like to have a recurrence for $B_k^{(a+1)}(t)$. In order to obtain the desired recurrence we observe that

$$\Delta_{x,1}B_k^{(a)}(x) = kB_{k-1}^{(a-1)}(x), \qquad \text{(B.15)}$$

since

$$\Delta_{x,1}\sum_{k=0}^{\infty} B_k^{(a)}(x)\frac{z^k}{k!} = \sum_{k=0}^{\infty}\frac{z^k}{k!}\Delta_{x,1}B_k^{(a)}(x) = \frac{z^a}{(e^z-1)^a}\Delta_{x,1}e^{zx}$$

$$= \frac{z^a}{(e^z-1)^a}\left[e^{z(x+1)} - e^{zx}\right] = z\cdot\frac{z^{a-1}e^{xz}}{(e^z-1)^{a-1}}$$

$$= \sum_{k=0}^{\infty} B_k^{(a-1)}(x)\frac{z^{k+1}}{k!} = \sum_{k=1}^{\infty} kB_{k-1}^{(a-1)}(x)\frac{z^k}{k!}.$$

If $a = 1$, Equation (B.15) becomes $\Delta_{x,1}B_k(x) = kB_{k-1}^{(0)}(x)$. Equation (B.1) with $a = 0$ implies that

$$\sum_{k=0}^{\infty} B_k^{(0)}(x)\frac{z^k}{k!} = e^{xz} = \sum_{k=0}^{\infty} x^k \frac{z^k}{k!},$$

Thus $B_k^{(0)}(x) = x^k$ and

$$\Delta_{x,1}B_k(x) = kB_{k-1}^{(0)} = kx^{k-1},$$

which is precisely Equation (15.20).

Equation (B.15) implies that

$$\Delta_{t,1}B_k^{(a)}(t) = B_k^{(a)}(t+1) - B_k^{(a)}(t) = kB_{k-1}^{(a-1)}(t).$$

Thus

$$B_k^{(a+1)}(t+1) = B_k^{(a+1)}(t) + kB_{k-1}^{(a)}(t).$$

We substitute this result into Equation (B.13) and simplify to obtain

$$B_k^{(a+1)}(t) = \left(1 - \frac{k}{a}\right) B_k^{(a)}(t) + \frac{k}{a}(t-a)B_{k-1}^{(a)}(t). \tag{B.16}$$

Equation (B.16) provides a Pascal-like recursion for $B_k^{(a+1)}(t)$ and provides an alternative verification of Equation (13.2). Take Equation (B.16), let $t \to x$, $a \to n$, and $k = n$ to obtain

$$B_n^{(n+1)}(x) = (x-n)B_{n-1}^{(n)}(x). \tag{B.17}$$

Take Equation (B.17) and iterate it $\alpha - 1$ times to find that

$$B_n^{(n+1)}(x) = \left[\prod_{j=0}^{\alpha}(x - n \quad j)\right] B_{n-\alpha-1}^{(n-\alpha)}(x). \tag{B.18}$$

Set $\alpha = n - 1$ and obtain

$$B_n^{(n+1)}(x) = B_0^{(1)}(x) \prod_{j=0}^{n-1}(x-n-j) = B_0(x) \prod_{j=0}^{n-1}(x-n-j)$$

$$= \prod_{j=0}^{n-1}(x-n-j) = \prod_{j=1}^{n}(x-n-j+1) = \prod_{j=1}^{n}(x-j).$$

In summary we have shown that

$$B_n^{(n+1)}(x) = \prod_{j=1}^{n}(x-j). \tag{B.19}$$

In Equation (B.5) let $a \to n + 1$, $k \to n$, and $j \to n - k$ to obtain

$$D_x^{n-k} B_n^{(n+1)}(x) = (n-k)! \binom{n}{n-k} B_k^{(n+1)}(x) = \frac{n!}{k!} B_k^{(n+1)}(x). \quad \text{(B.20)}$$

Combining Equations (B.19) and (B.20) gives us

$$D_x^{n-k} \prod_{j=1}^{n} (x - j) = \frac{n!}{k!} B_k^{(n+1)}(x).$$

Since $D_x x = 1 = D_x(x + 1)$ we may take the previous equation and let $x \to x + 1$. If we do so we find that

$$D_x^{n-k} \prod_{j=1}^{n} (x + 1 - j) = n! D_x^{n-k} \binom{x}{n} = \frac{n!}{k!} B_k^{(n+1)}(x + 1),$$

a result clearly equivalent to Equation (13.2).

The same procedure we used to derive Equation (B.16) may be applied to Equation (B.2). Take Equation (B.2), let $z \to t$, and differentiate with respect to $x$ to find that

$$D_x \sum_{k=0}^{\infty} E_k^{(a)}(t) \frac{x^k}{k!} = \sum_{k=0}^{\infty} E_k^{(a)}(t) \frac{k x^{k-1}}{k!}$$

$$= 2^a D_x \left[ \frac{e^{xt}}{(e^x + 1)^a} \right] = 2^a \left[ \frac{t e^{xt}}{(e^x + 1)^a} - \frac{a e^{xt} e^x}{(e^x + 1)^{a+1}} \right].$$

Multiply this equation by $x$ to discover that

$$\sum_{k=0}^{\infty} E_k^{(a)}(t) \frac{k x^k}{k!} = tx \frac{2^a e^{xt}}{(e^x + 1)^a} - \frac{ax}{2} \frac{2^{a+1} e^{(t+1)x}}{(e^x + 1)^{a+1}}$$

$$= tx \sum_{k=0}^{\infty} \frac{E_k^{(a)}(t)}{k!} x^k - \frac{ax}{2} \sum_{k=0}^{\infty} \frac{E_k^{(a+1)}(t + 1)}{k!} x^k$$

$$= t \sum_{k=0}^{\infty} \frac{k E_{k-1}^{(a)}(t)}{k!} x^k - \frac{a}{2} \sum_{k=0}^{\infty} \frac{k E_{k-1}^{(a+1)}(t + 1)}{k!} x^k.$$

In summary we have shown that

$$\sum_{k=0}^{\infty} \frac{k E_k^{(a)}(t)}{k!} x^k = \sum_{k=0}^{\infty} \frac{tk E_{k-1}^{(a)}(t) - \frac{a}{2} k E_{k-1}^{(a+1)}(t + 1)}{k!} x^k,$$

which is equivalent to saying

$$E_k^{(a)}(t) = t E_{k-1}^{(a)}(t) - \frac{a}{2} E_{k-1}^{(a+1)}(t + 1). \quad \text{(B.21)}$$

Equation (B.21) complements Equation (B.13). If $t = 0$, Equation (B.21) becomes

$$E_k^{(a)}(0) = -\frac{a}{2} E_{k-1}^{(a+1)}(1).$$ (B.22)

To find the complement to Equation (B.16) we introduce the Boole averaging operator $\nabla_{t,1} f(t) = \frac{f(t+1)+f(t)}{2}$. If we apply this linear operator to Equation (B.2) we find that

$$\nabla_{x,1} \sum_{k=0}^{\infty} E_k^{(a)}(x) \frac{z^k}{k!} = \sum_{k=0}^{\infty} \frac{z^k}{k!} \nabla_{x,1} E_k^{(a)}(x) = \frac{2^a}{(e^z + 1)^a} \nabla_{x,1} e^{xz}$$

$$= \frac{2^a}{(e^z + 1)^a} \frac{e^{z(x+1)} + e^{zx}}{2} = \frac{2^{a-1} e^{zx}}{(e^z + 1)^{a-1}}$$

$$= \sum_{k=0}^{\infty} E_k^{(a-1)}(x) \frac{z^k}{k!}.$$

By comparing the coefficient of $z^k$ in the second sum to that of $z^k$ in the final sum we discover that

$$\nabla_{x,1} E_k^{(a)}(x) = E_k^{(a-1)}(x).$$ (B.23)

Equation (B.23) is the complement to Equation (B.15). If $a = 1$, Equation (B.23) becomes

$$\nabla_{x,1} E_k(x) = E_k^{(0)}(x) = x^k,$$ (B.24)

where the last equality reflects the fact that $\sum_{k=0}^{\infty} E_k^{(0)}(x) \frac{z^k}{k!} = e^{xz}$.

Equation (B.23) implies that

$$E_{k-1}^{(a+1)}(t+1) = 2E_{k-1}^{(a)}(t) - E_{k-1}^{(a+1)}(t).$$ (B.25)

We place Equation (B.25) into Equation (B.21) and simplify to obtain

$$\frac{a}{2} E_{k-1}^{(a+1)}(t) = (a-t)E_{k-1}^{(a)}(t) + E_k^{(a)}(t),$$ (B.26)

which is the complement to Equation (B.16).

We end this section with two convolution identities. Equation (B.1) implies that

$$\sum_{k=0}^{\infty} \frac{B_k^{(z+w)}(x+y)}{k!} t^k = \frac{t^{z+w} e^{(x+y)t}}{(e^t - 1)^{w+z}} = \frac{t^z e^{xt}}{(e^t - 1)^z} \cdot \frac{t^w e^{yt}}{(e^t - 1)^w}$$

$$= \sum_{k=0}^{\infty} \frac{B_k^{(z)}(x)}{k!} t^k \sum_{j=0}^{\infty} \frac{B_k^{(w)}(y)}{k!} t^k$$

$$= \sum_{k=0}^{\infty} t^k \sum_{j=0}^{k} \frac{B_j^{(z)}(x) B_{k-j}^{(w)}(y)}{j!(k-j)!}.$$

Comparing the coefficients of $t^k$ shows that

$$B_k^{(z+w)}(x+y) = \sum_{j=0}^{k} \binom{k}{j} B_j^{(z)}(x) B_{k-j}^{(w)}(y). \tag{B.27}$$

An important case of Equation (B.27) is

$$B_k^{(0)}(x+y) = (x+y)^k = \sum_{j=0}^{k} \binom{k}{j} B_j^{(z)}(x) B_{k-j}^{(-z)}(y). \tag{B.28}$$

Take Equation (B.28), let $y = 0$ and $z = n$, where $n$ is a nonnegative integer. Equation (13.36) implies that

$$x^k = \sum_{j=0}^{k} \binom{k}{j} B_j^{(n)}(x) B_{k-j}^{(-n)}(0) = \sum_{j=0}^{k} \binom{k}{j} B_j^{(n)}(x) \frac{S(n+k-j,n)}{\binom{k-j+n}{k-j}}$$

$$= \sum_{j=0}^{k} \binom{k}{j} B_{k-j}^{(n)}(x) \frac{S(n+j,n)}{\binom{n+j}{n}}. \tag{B.29}$$

We may also take Equation (B.27), let $x = 0$ and $z = n$, with $n$ a nonnegative integer, to obtain

$$y^k = \sum_{j=0}^{k} \binom{k}{j} B_{k-j}^{(-n)}(y) B_j^{(n)}(0) = \sum_{j=0}^{k} \binom{k}{j} B_{k-j}^{(-n)}(y) \frac{s(n,n-k)}{\binom{n-1}{j}}, \tag{B.30}$$

where the last equality follows from Equation (13.27).

A similar argument shows that

$$E_k^{(z+w)}(x+y) = \sum_{j=0}^{k} \binom{k}{j} E_j^{(z)}(x) E_{k-j}^{(w)}(y). \tag{B.31}$$

A special case of Equation (B.31) is

$$E_k^{(0)}(x+y) = (x+y)^k = \sum_{j=0}^{k} \binom{k}{j} E_j^{(z)}(x) E_{k-j}^{(-z)}(y). \tag{B.32}$$

Equation (B.32) is the complement to Equation (B.28).

## B.2 Generalized Bernoulli and Euler Polynomial Derivative Expansions

We now explore how generalized Bernoulli and generalized Euler polynomials provide expansions for the derivatives of arbitrary polynomials. Recall Theorem 15.1 which states

$$f(x) = \int_0^1 f(t)\,dt + \sum_{k=1}^n \frac{B_k(x)}{k!}\Delta_{x,1}D_x^{k-1}f(0), \qquad f(x) = \sum_{i=0}^n a_i x^i. \quad \text{(B.33)}$$

We may generalize Theorem 15.1 as follows:

**Theorem B.1.** *Let* $f(x) = \sum_{k=0}^n a_k x^k$. *Then*

$$f(x+y) = \int_y^{y+1} f(t)\,dt + \sum_{\alpha=1}^n \frac{B_\alpha(x)}{\alpha!}\Delta_{y,1}D_y^{\alpha-1}f(y). \quad \text{(B.34)}$$

If $y = 0$, Equation (B.34) becomes Equation (B.33).

**Proof:** The binomial theorem implies that

$$f(x+y) = \sum_{k=0}^n a_k(x+y)^k = \sum_{k=0}^n a_k \sum_{j=0}^k \binom{k}{j} x^j y^{k-j}$$

$$= \sum_{k=0}^n a_k \sum_{j=0}^k \binom{k}{j} y^{k-j} \frac{1}{j+1} \sum_{\alpha=0}^j \binom{j+1}{\alpha} B_\alpha(x), \qquad \text{by Eq. (15.61)}$$

$$= \sum_{\alpha=0}^n B_\alpha(x) \sum_{k=\alpha}^n a_k \sum_{j=\alpha}^k \frac{1}{j+1} \binom{k}{j}\binom{j+1}{\alpha} y^{k-j}$$

$$- \sum_{\alpha=0}^n \frac{B_\alpha(x)}{\alpha!} \sum_{k=\alpha}^n k!a_k \sum_{j=0}^{k-\alpha} \frac{y^j}{j!(k-j+1-\alpha)!} \qquad (3).$$

The triple sum of Line (3) is the left side of Equation (B.34). If $\alpha = 0$, $B_0(x) = 1$ and Line (3) becomes

$$\sum_{k=0}^n k!a_k \sum_{j=0}^k \frac{y^j}{j!(k-j+1)!} = \sum_{k=0}^n \frac{a_k}{k+1} \sum_{j=0}^k \binom{k+1}{j} y^j$$

$$= \sum_{k=0}^n \frac{a_k}{k+1}\left[(y+1)^{k+1} - y^{k+1}\right]$$

$$= \sum_{k=0}^n a_k \int_y^{y+1} t^k\,dt = \int_y^{y+1} f(t)\,dt. \quad \text{(B.35)}$$

Now assume $1 \leq \alpha \leq n$. We claim that

$$\sum_{\alpha=1}^{n} \frac{B_\alpha(x)}{\alpha!} \sum_{k=\alpha}^{n} k! a_k \sum_{j=0}^{k-\alpha} \frac{y^j}{j!(k-j+1-\alpha)!} = \sum_{\alpha=1}^{n} \frac{B_\alpha(x)}{\alpha!} \Delta_{y,1} D_y^{\alpha-1} f(y),$$

(B.36)

since Equation (15.65) implies that

$$\sum_{\alpha=1}^{n} \frac{B_\alpha(x)}{\alpha!} \Delta_{y,1} D_y^{\alpha-1} f(y)$$

$$= \sum_{\alpha=1}^{n} \frac{B_\alpha(x)}{\alpha!} \sum_{k=\alpha}^{n} a_k(\alpha-1)! \binom{k}{\alpha-1} \sum_{j=0}^{k-\alpha} \binom{k-\alpha+1}{j} y^j$$

$$= \sum_{\alpha=1}^{n} \frac{B_\alpha(x)}{\alpha!} \sum_{k=\alpha}^{n} a_k k! \sum_{j=0}^{k-\alpha} \frac{y^j}{j!(k-j+1-\alpha)!}.$$

Combining Equations (B.35) and (B.36) with Line (3) proves Theorem B.1. $\qquad\qquad\square$

Because Equation (B.34) is symmetrical in $x$ and $y$, we may interchange the roles of $x$ and $y$ in the proof of Theorem B.1 to obtain

$$f(x+z) = \int_x^{x+1} f(t)\,dt + \sum_{k=1}^{n} \frac{B_k(z)}{k!} \Delta_{x,1} D_x^{k-1} f(x), \qquad f(x) = \sum_{i=0}^{n} a_i x^i.$$

(B.37)

Differentiate Equation (B.37) with respect to $x$ to obtain

$$D_x f(x+z) = D_x \left[ \int_x^{x+1} f(t)\,dt \right] + \sum_{k=1}^{n} \frac{B_k(z)}{k!} D_x \Delta_{x,1} D_x^{k-1} f(x)$$

$$= f(x+1) - f(x) + \sum_{k=1}^{n} \frac{B_k(z)}{k!} D_x \left[ D_x^{k-1} f(x+1) - D_x^{k-1} f(x) \right]$$

$$= \sum_{k=0}^{n} \frac{B_k(z)}{k!} \left[ D_x^k f(x+1) - D_x^k f(x) \right],$$

which we summarize as

$$D_x f(x+z) = \sum_{k=0}^{n} \frac{B_k(z)}{k!} \Delta_{x,1} D_x^k f(x), \qquad f(x) = \sum_{i=0}^{n} a_i x^i. \qquad (B.38)$$

Equation (B.38) writes the derivative of a polynomial in terms of the Bernoulli polynomials. Nörlund [1924] finds another quite remarkable formula which writes the derivative in terms of generalized Bernoulli polyno-mials. Nörlund claims

$$D_t^n f(t)|_{t=x+z} = f^{(n)}(x+z) = \sum_{k=0}^{m} \frac{B_k^{(n)}(z)}{k!} \Delta_{x,1}^n D_x^k f(x), \quad f(x) = \sum_{i=0}^{m} a_i x^i,$$

(B.39)

where $\Delta_{x,1}^n f(x) = \sum_{k=0}^{n} (-1)^{n-k} \binom{n}{k} f(x+k)$. If $n = 1$, Equation (B.39) becomes Equation (B.38).

Professor Gould derived a clever proof of Equation (B.39) based on Equation (B.28). By definition

$$D_t^n f(t)|_{t=x+z} = \sum_{k=0}^{m} a_k D_t^n t^k \Big|_{t=x+z} = \sum_{k=0}^{m} a_k n! \binom{k}{n} (x+z)^{k-n}$$

$$= \sum_{k=0}^{m} a_k n! \binom{k}{n} \sum_{j=0}^{k-n} \binom{k-n}{j} B_j^{(n)}(x) B_{k-n-j}^{(-n)}(z), \quad \text{by Eq. (B.28)}$$

$$= \sum_{j=0}^{m} B_j^{(n)}(x) \sum_{k=0}^{m} a_k n! \binom{k}{n} \binom{k-n}{j} B_{k-n-j}^{(-n)}(z)$$

$$= \sum_{j=0}^{m} \frac{B_j^{(n)}(x)}{j!} \sum_{k=0}^{m} \frac{a_k k!}{(k-n-j)!} \sum_{\gamma=0}^{k-n-j} \binom{k-n-j}{\gamma} B_\gamma^{(-n)}(0) z^{k-n-j-\gamma}$$

$$= \sum_{j=0}^{m} \frac{B_j^{(n)}(x)}{j!} \sum_{k=0}^{m} a_k k! \sum_{\gamma=0}^{m} \frac{B_\gamma^{(-n)}(0)}{\gamma!(k-n-j-\gamma)!} z^{k-n-j-\gamma}$$

$$= \sum_{j=0}^{m} \frac{B_j^{(n)}(x)}{j!} \sum_{\gamma=0}^{m} \frac{B_\gamma^{(-n)}(0)}{\gamma!} \sum_{k=0}^{m} a_k \frac{k!}{(k-n-j-\gamma)!} z^{k-n-j-\gamma}$$

$$= \sum_{j=0}^{m} \frac{B_j^{(n)}(x)}{j!} \sum_{\gamma=0}^{m} \frac{B_\gamma^{(-n)}(0)}{\gamma!} D_z^{n+j+\gamma} f(z)$$

$$= \sum_{j=0}^{m} \frac{B_j^{(n)}(x)}{j!} \sum_{\gamma=0}^{m} \frac{S(n+\gamma,n)}{\gamma!\binom{\gamma+n}{n}} D_z^{n+j+\gamma} f(z), \quad \text{by Eq. (13.36)}$$

$$= \sum_{j=0}^{m} \frac{B_j^{(n)}(x)}{j!} \sum_{\gamma=n}^{m} \frac{n! S(\gamma,n)}{\gamma!} D_z^\gamma \left[ D_z^j f(z) \right] \quad (4).$$

At this point Professor Gould refers to Equation (A.3) which states

$$\Delta_{z,1}^n G(z) = n! \sum_{\gamma=0}^{m} \frac{S(\gamma,n)}{\gamma!} D_z^\gamma G(z), \quad G(z) = \sum_{i=0}^{m} A_i z^i. \quad (B.40)$$

He then sets $G(z) = D_z^j f(z)$, which is itself a polynomial of degree $m - j$, and uses Equation (B.40) to rewrite the inner sum of Line (4) as

$$D_t^n f(t)|_{t=x+z} = \sum_{j=0}^{m} \frac{B_j^{(n)}(x)}{j!} \Delta_{z,1}^n D_z^j f(z).$$

If we interchange the roles of $x$ and $z$, the preceding line becomes Equation (B.39).

We next explore the role of generalized Euler polynomials in derivative expansions of polynomials. First we derive the complement of Theorem 15.1. Let $f(x) = \sum_{i=0}^{n} a_i x^i$. Since Equation (B.24) states that $x^i = \frac{1}{2}[E_i(x) + E_i(x+1)]$, we may rewrite $f(x)$ as

$$f(x) = \sum_{i=0}^{n} \frac{a_i}{2}[E_i(x) + E_i(x+1)] = \frac{1}{2}\sum_{i=0}^{n} a_i E_i(x) + \frac{1}{2}\sum_{i=0}^{n} a_i E_i(x+1)$$

$$= \frac{1}{2}\sum_{i=0}^{n} a_i E_i(x) + \frac{1}{2}\sum_{i=0}^{n} a_i \sum_{j=0}^{i} \binom{i}{j} E_j(x), \qquad \text{by Eq. (B.9)}$$

$$= \sum_{i=0}^{n} a_i E_i(x) + \frac{1}{2}\sum_{i=1}^{n} a_i \sum_{j=0}^{i-1} \binom{i}{j} E_j(x)$$

$$= \sum_{i=0}^{n} a_i E_i(x) + \frac{1}{2}\sum_{i=0}^{n-1} a_{i+1} \sum_{j=0}^{i} \binom{i+1}{j} E_j(x).$$

In summary we have shown that

$$f(x) = \sum_{i=0}^{n} a_i E_i(x) + \frac{1}{2}\sum_{i=0}^{n-1} a_{i+1} \sum_{j=0}^{i} \binom{i+1}{j} E_j(x), \qquad f(x) = \sum_{i=0}^{n} a_i x^i. \tag{B.41}$$

Equation (B.41) appears in Nörlund, albeit in a different formulation [Nörlund, 1924, pp. 20-34]. To arrive at Nörlund's formulation we need to define the forward averaging operator $\nabla_{x,w}\phi(x) = \frac{\phi(x+w)+\phi(x)}{2}$. We claim that Equation (B.41) is equivalent to

$$f(x) = \sum_{k=0}^{n} \frac{E_k(x)}{k!} \nabla_{x,1} D_x^k f(x)\Big|_{x=0}, \qquad f(x) = \sum_{i=0}^{n} a_i x^i. \tag{B.42}$$

Recall that $D_x^k f(x) = \sum_{i=k}^{n} a_i k! \binom{i}{k} x^{i-k}$. Since $\nabla_{x,w}$ is a linear operator we have

$$\nabla_{x,1} D_x^k f(x) = \sum_{i=k}^{n} a_i k! \binom{i}{k} \nabla_{x,1} x^{i-k} = \frac{1}{2}\sum_{i=k}^{n} a_i k! \binom{i}{k} \left[(x+1)^{i-k} + x^{i-k}\right].$$

Hence

$$\sum_{k=0}^{n} \frac{E_k(x)}{k!} \nabla_{x,1} D_x^k f(x)\Big|_{x=0}$$

$$= \sum_{k=0}^{n} \frac{E_k(x)}{k!} [a_k k! \nabla_{x,1} 1] + \sum_{k=0}^{n-1} \frac{E_k(x)}{k!} \sum_{i=k+1}^{n} a_i k! \binom{i}{k} \nabla_{x,1} x^{i-k}\Big|_{x=0}$$

$$= \sum_{k=0}^{n} a_k E_k(x) + \frac{1}{2} \sum_{k=0}^{n-1} E_k(x) \sum_{i=k+1}^{n} a_i \binom{i}{k} (x+1)^{i-k} + x^{i-k}\Big|_{x=0}$$

$$= \sum_{k=0}^{n} a_k E_k(x) + \frac{1}{2} \sum_{k=0}^{n-1} E_k(x) \sum_{i=k+1}^{n} a_i \binom{i}{k}$$

$$= \sum_{k=0}^{n} a_k E_k(x) + \frac{1}{2} \sum_{k=0}^{n-1} E_k(x) \sum_{i=k}^{n-1} a_{i+1} \binom{i+1}{k}$$

$$= \sum_{k=0}^{n} a_k E_k(x) + \frac{1}{2} \sum_{i=0}^{n-1} a_{i+1} \sum_{k=0}^{i} \binom{i+1}{k} E_k(x) \qquad (5).$$

Since Line (5) is the right side of Equation (B.41) the preceding calculations verify Equation (B.42) and prove the following theorem:

**Theorem B.2.** *Let* $f(x) = \sum_{i=0}^{n} a_i x^i$. *Define* $\nabla_{x,1}\phi(x) = \frac{\phi(x+1)+\phi(x)}{2}$. *Then*

$$f(x) = \sum_{k=0}^{n} \frac{E_k(x)}{k!} \nabla_{x,1} D_x^k f(0). \qquad (B.43)$$

Theorem B.2 complements Theorem 15.1. We may generalize Theorem B.2 as follows:

**Theorem B.3.** *Let* $f(x) = \sum_{i=0}^{n} a_i x^i$. *Define* $\nabla_{x,1}\phi(x) = \frac{\phi(x+1)+\phi(x)}{2}$. *Then*

$$f(x+z) = \sum_{k=0}^{n} \frac{E_k(z)}{k!} \nabla_{x,1} D_x^k f(x). \qquad (B.44)$$

If $x = 0$, Equation (B.44) becomes Equation (B.43). Theorem B.3 is the counterpart to Theorem B.1.

The proof of Theorem B.3 is not difficult. By definition

$$f(x+z) = \sum_{i=0}^{n} a_i(x+z)^i = \sum_{i=0}^{n} a_i \sum_{j=0}^{i} \binom{i}{j} x^i z^{i-j}$$

$$= \sum_{i=0}^{n} a_i \sum_{j=0}^{i} \binom{i}{j} z^{i-j} \left[ \frac{E_j(x) + E_j(x+1)}{2} \right]$$

$$= \frac{1}{2} \sum_{i=0}^{n} a_i \sum_{j=0}^{i} \binom{i}{j} z^{i-j} E_j(x) + \frac{1}{2} \sum_{i=0}^{n} a_i \sum_{j=0}^{i} \binom{i}{j} z^{i-j} E_j(x+1)$$

$$= \frac{1}{2} \sum_{i=0}^{n} a_i E_i(x+z) + \frac{1}{2} \sum_{i=0}^{n} a_i E_i(x+1+z) \qquad (6).$$

On the other hand

$$\sum_{k=0}^{n} \frac{E_k(z)}{k!} \, \nabla_{x,1} \, D_x^k f(x)$$

$$= \sum_{k=0}^{n} \frac{E_k(z)}{k!} \, \nabla_{x,1} \left[ \sum_{i=k}^{n} a_i k! \binom{i}{k} x^{i-k} \right]$$

$$= \sum_{k=0}^{n} E_k(z) \sum_{i=k}^{n} a_i \binom{i}{k} \nabla_{x,1} \, x^{i-k}$$

$$= \frac{1}{2} \sum_{k=0}^{n} E_k(z) \sum_{i=k}^{n} a_i \binom{i}{k} \left[ (x+1)^{i-k} + x^{i-k} \right]$$

$$= \frac{1}{2} \sum_{k=0}^{n} E_k(z) \sum_{i=k}^{n} a_i \binom{i}{k} (x+1)^{i-k} + \frac{1}{2} \sum_{k=0}^{n} E_k(z) \sum_{i=k}^{n} a_i \binom{i}{k} x^{i-k}$$

$$= \frac{1}{2} \sum_{i=0}^{n} a_i \sum_{k=0}^{i} \binom{i}{k} E_k(z)(x+1)^{i-k} + \frac{1}{2} \sum_{i=0}^{n} \sum_{k=0}^{i} \binom{i}{k} E_k(z) x^{i-j}$$

$$= \frac{1}{2} \sum_{i=0}^{n} a_i E_i(x+1+z) + \frac{1}{2} \sum_{i=0}^{n} E_i(x+z) \qquad (7).$$

Since Line (6) is equal to Line (7) we obtain Equation (B.44).

Take Equation (B.44) and differentiate with respect to $x$ to find that

$$D_x f(x+z) = \sum_{k=0}^{n} \frac{E_k(z)}{k!} D_x \nabla_{x,1} D_x^k f(x)$$

$$= \sum_{k=0}^{n} \frac{E_k(z)}{k!} D_x \left[ \frac{D_x^k f(x+1) + D_x^k f(x)}{2} \right]$$

$$= \sum_{k=0}^{n} \frac{E_k(z)}{k!} \frac{D_x^{k+1} f(x+1) + D_x^{k+1} f(x)}{2}$$

$$= \sum_{k=0}^{n} \frac{E_k(z)}{k!} \nabla_{x,1} D_x^{k+1} f(x)$$

$$= \sum_{k=0}^{n-1} \frac{E_k(z)}{k!} \nabla_{x,1} D_x^{k+1} f(x), \tag{B.45}$$

where the last equality reflects the fact that $f(x)$ is of degree $n$. Equation (B.45) is the analog of Equation (B.38) and writes the derivative of a polynomial in terms of the Euler polynomials.

There is also an analog to Equation (B.39) which we state as

$$D_t^n f(t)|_{t=x+z} = f^{(n)}(x+z) = \sum_{k=0}^{m} \frac{E_k^{(n)}(z)}{k!} \nabla_{x,1}^n D_x^{k+n} f(x), \qquad f(x) = \sum_{i=0}^{n+m} a_i x^i, \tag{B.46}$$

where $\nabla_{x,1}^n F(x) = \frac{1}{2^n} \sum_{k=0}^{n} \binom{n}{k} F(x+k)$. If $n = 1$, Equation (B.46) becomes Equation (B.45).

The proof of Equation (B.46) is similar to that of Equation (B.39). By definition we have

$$D_t^n f(t)|_{t=x+z} = f^{(n)}(x+z) = \sum_{k=n}^{n+m} a_k n! \binom{k}{n} (x+z)^{k-n}$$

$$= \sum_{k=n}^{n+m} a_k n! \binom{k}{n} \sum_{j=0}^{k-n} \binom{k-n}{j} E_j^{(n)}(x) E_{k-n-j}^{(-n)}(z), \qquad \text{Eq. (B.32)}$$

$$= \sum_{j=0}^{m} E_j^{(n)}(x) \sum_{k=n}^{n+m} a_k n! \binom{k}{n} \binom{k-n}{j} E_{k-n-j}^{(-n)}(z)$$

$$= \sum_{j=0}^{m} \frac{E_j^{(n)}(x)}{j!} \sum_{k=n}^{n+m} \frac{a_k k!}{(k-n-j)!} E_{k-n-j}^{(-n)}(z)$$

$$= \sum_{j=0}^{m} \frac{E_j^{(n)}(x)}{j!} \sum_{k=n}^{n+m} \frac{a_k k!}{(k-n-j)!} \sum_{p=0}^{k-n-j} \binom{k-n-j}{p} z^{k-n-j-p} E_p^{(-n)}(0) \quad (8).$$

Line (8) is the expansion of the left side of Equation (B.46). If we can show this expansion is equivalent to $\sum_{j=0}^{m} \frac{E_j^{(n)}(x)}{j!} \nabla_{z,1}^n D_z^{j+n} f(z)$, we will have proven Equation (B.46). It suffices to analyze $\nabla_{z,1}^n D_z^{j+n} f(z)$. The definition of $\nabla_{z,1}^n$ implies that

$$
\begin{aligned}
\nabla_{z,1}^n D_z^{j+n} f(z) &= \frac{1}{2^n} \sum_{k=0}^{n} \binom{n}{k} D_z^{j+n} f(z+k) \\
&= \frac{1}{2^n} \sum_{k=0}^{n} \binom{n}{k} \sum_{s=j+n}^{m+n} a_s (j+n)! \binom{s}{j+n} (z+k)^{s-j-n} \\
&= \frac{1}{2^n} \sum_{s=0}^{n+m} \frac{a_s s!}{(s-n-j)!} \sum_{k=0}^{n} \binom{n}{k} (z+k)^{s-j-n} \\
&= \frac{1}{2^n} \sum_{s=0}^{n+m} \frac{a_s s!}{(s-n-j)!} \sum_{k=0}^{n} \binom{n}{k} \sum_{p=0}^{s-j-n} \binom{s-j-n}{p} z^{s-j-n-p} k^p \\
&= \sum_{s=0}^{n+m} \frac{a_s s!}{(s-n-j)!} \sum_{p=0}^{s-j-n} \binom{s-j-n}{p} z^{s-j-n-p} \frac{1}{2^n} \sum_{k=0}^{n} \binom{n}{k} k^p \quad (9).
\end{aligned}
$$

We claim that

$$
\frac{1}{2^n} \sum_{k=0}^{n} \binom{n}{k} k^p = E_p^{(-n)}(0). \tag{B.47}
$$

Equation (B.47) follows from Equation (B.2) and coefficient comparison since

$$
\sum_{k=0}^{\infty} \frac{E_k^{(-n)}(0)}{k!} x^k = \frac{2^{-n}}{(e^x+1)^{-n}} = \frac{(e^x+1)^n}{2^n} = \frac{1}{2^n} \sum_{j=0}^{n} \binom{n}{j} e^{xj}
$$

$$
= \frac{1}{2^n} \sum_{j=0}^{n} \binom{n}{j} \sum_{k=0}^{\infty} \frac{(xj)^k}{k!} = \frac{1}{2^n} \sum_{k=0}^{\infty} \frac{x^k}{k!} \sum_{j=0}^{n} \binom{n}{j} j^k.
$$

Substitute Equation (B.47) into Line (9) to obtain

$$
\sum_{j=0}^{m} \frac{E_j^{(n)}(x)}{j!} \nabla_{x,1}^n D_z^{j+n} f(z) =
$$

$$
\sum_{j=0}^{m} \frac{E_j^{(n)}(x)}{j!} \sum_{s=0}^{n+m} \frac{a_s s!}{(s-n-j)!} \sum_{p=0}^{s-j-n} \binom{s-j-n}{p} z^{s-j-n-p} E_p^{(-n)}(0).
$$

If we let $s \to k$, we see that the preceding line is identical to Line (8). Equation (B.46) follows by interchanging the roles of $z$ and $x$.

## B.3 Additional Considerations Involving Newton Series

Equations (B.39) and (B.46) write the derivative of $f(x) = \sum_{i=0}^{n} a_i x^i$ in terms of generalized Bernoulli and generalized Euler polynomials respectively. We end this appendix with a brief discussion on deriving derivative expansions of polynomials via the Newton-Gregory series. Recall Equation (A.7) which states

$$f(x+y) = \sum_{j=0}^{n} \binom{\frac{x}{z}}{j} z^j \Delta_{y,z}^j f(y), \qquad f(x) = \sum_{i=0}^{n} a_i x^i. \qquad (B.48)$$

Simple algebra shows

$$\prod_{k=0}^{j-1} (x - kz) = z^j j! \binom{\frac{x}{z}}{j}, \qquad j \geq 1.$$

If we make the convention that $\prod_{k=0}^{-1}(x - kz) = 1$, we may use the previous identity to rewrite Equation (B.48) as

$$f(x+y) = \sum_{j=0}^{n} \frac{1}{j!} \prod_{k=0}^{j-1} (x - kz) \Delta_{y,z}^j f(y). \qquad (B.49)$$

Take Equation (B.49) and let $x \to y - z$ and $y \to x + z$. After simplification we obtain

$$f(x+y) = \sum_{j=0}^{n} \frac{1}{j!} \prod_{k=1}^{j} (y - kz) \Delta_{x,z}^j f(x+z), \qquad f(x) = \sum_{i=0}^{n} a_i x^i. \qquad (B.50)$$

Equation (B.50) appears as Relation (66) in [Nörlund, 1924, p.241]. Rewrite Equation (B.50) as

$$f(x+y) = \sum_{k=0}^{n} \binom{\frac{y-z}{z}}{k} z^k \Delta_{x,z}^k f(x+z). \qquad (B.51)$$

An application of the product rule shows that

$$D_y \left[ \binom{\frac{y-z}{z}}{k} z^k \right] = \binom{\frac{y-z}{z}}{k} z^k \sum_{j=1}^{k} \frac{1}{y - jz}. \qquad (B.52)$$

Equation (B.52) helps us differentiate Equation (B.51) with respect to $y$. In particular we find that

$$D_y f(x+y) = \sum_{k=1}^{n} \binom{\frac{y-z}{z}}{k} z^k \Delta_{x,z}^k f(x+z) \sum_{j=1}^{k} \frac{1}{y - jz}, \qquad n \geq 1. \qquad (B.53)$$

Take Equation (B.53) and let $y \to 0$ to obtain

$$D_y f(x+y)|_{y=0} = f'(x) = \sum_{k=1}^{n} (-1)^{k-1} z^{k-1} \Delta_{x,z}^{k} f(x+z) \sum_{j=1}^{k} \frac{1}{j}$$

$$= \sum_{k=0}^{n-1} (-1)^{k} z^{k} \Delta_{x,z}^{k+1} f(x+z) \sum_{j=0}^{k} \frac{1}{j+1}, \qquad \text{(B.54)}$$

whenenever $n \geq 1$. Since $B_k^{(n)}(0) = \frac{s(n,n-k)}{\binom{n-1}{k}}$, let $k \to n$ and $n \to n+2$ to obtain $B_n^{(n+2)}(0) = \frac{s(n+2,2)}{n+1}$. Because $\frac{n!}{k!} D_x^k \binom{x}{n}\big|_{x=0} = s(n,k)$, we can use the product rule to show

$$s(n,2) = (-1)^n (n-1)! \sum_{k=1}^{n-1} \frac{1}{k}. \qquad \text{(B.55)}$$

Take Equation (B.55), let $n \to n+2$ and obtain

$$B_n^{(n+2)}(0) = (-1)^n n! \sum_{k=0}^{n} \frac{1}{k+1}. \qquad \text{(B.56)}$$

Equation (B.56) evaluates the inner sum of Equation (B.54). In particular we find that

$$f'(x) = \sum_{k=0}^{n-1} \frac{z^k}{k!} B_k^{(k+2)}(0) \Delta_{x,z}^{k+1} f(x+z), \qquad f(x) = \sum_{i=0}^{n} a_i x^i. \qquad \text{(B.57)}$$

Equation (B.57) appears as Relation (67) in [Nörlund, 1924, p.241]. It should be contrasted with Equation (B.38) with $z = 0$, namely

$$f'(x) = \sum_{k=0}^{n} \frac{B_k}{k!} \Delta_{x,1} D_x^k f(x), \qquad f(x) = \sum_{i=0}^{n} a_i x^i.$$

To obtain yet another formula for $f'(x)$, we take Equation (B.51), differentiate with respect to $y$, and then set $y = z$.

$$D_y f(x+y)|_{y=z} = f'(x+z) = \sum_{k=1}^{n} D_y \binom{\frac{y-z}{z}}{k} \bigg|_{y=z} z^k \Delta_{x,z}^{k} f(x+z)$$

$$= \sum_{k=1}^{n} D_u \binom{\frac{u}{z}}{k} \bigg|_{u=0} z^k \Delta_{x,z}^{k} f(x+z)$$

$$= \sum_{k=1}^{n} (-1)^{k-1} \frac{z^{k-1}}{k} \Delta_{x,z}^{k} f(x+z)$$

$$= \sum_{k=0}^{n-1} (-1)^{k} \frac{z^k}{k+1} \Delta_{x,z}^{k+1} f(x+z).$$

Now replace $x \to x - z$ to discover that

$$f'(x) = \sum_{k=0}^{n-1}(-1)^k \frac{z^k}{k+1}\Delta_{x,z}^{k+1}f(x), \qquad f(x) = \sum_{i=0}^{n}a_i x^i, \qquad \text{(B.58)}$$

which is Relation (68) in [Nörlund, 1924].

Since $B_k^{(n+1)}(z+1) = k!D_z^{n-k}\binom{z}{n}$, we deduce that $B_n^{(n+1)}(z+1) = n!\binom{z}{n}$. Set $z = -1$ to obtain $B_n^{(n+1)}(0) = n!\binom{-1}{n} = (-1)^n n!$. This means we may rewrite Equation (B.58) as

$$f'(x) = \sum_{k=0}^{n-1}\frac{z^k}{(k+1)k!}B_k^{(k+1)}(0)\Delta_{x,z}^{k+1}f(x), \qquad f(x) = \sum_{i=0}^{n}a_i x^i. \qquad \text{(B.59)}$$

Generalizations of Equations (B.57) and (B.59) are found through the identity

$$D_y^\alpha \prod_{j=1}^{k}(y - jz) = z^{k-\alpha}\frac{k!}{(k-\alpha)!}B_{k-\alpha}^{k+1}\left(\frac{y}{z}\right). \qquad \text{(B.60)}$$

The proof of Equation (B.60) follows from the observation that

$$B_{k-\alpha}^{(k+1)}(x) = (k-\alpha)!D_x^\alpha\binom{x-1}{k} = \frac{(k-\alpha)!}{k!}D_x^\alpha\left[\prod_{j=1}^{k}(x-j)\right].$$

See Equations (13.2) and (B.5). Let $x = \frac{y}{z}$. Then

$$B_{k-\alpha}^{(k+1)}\left(\frac{y}{z}\right) = \frac{(k-\alpha)!}{k!}D_{\frac{y}{z}}^\alpha\prod_{j=1}^{k}\left(\frac{y}{z}-j\right) = \frac{(k-\alpha)!}{k!z^k}D_{\frac{y}{z}}^\alpha\prod_{j=1}^{k}(y-jz)$$

$$= \frac{(k-\alpha)!z^\alpha}{k!z^k}D_y^\alpha\prod_{j=1}^{k}(y-iz),$$

since $D_{\frac{y}{z}}f(x) = D_x f(x) = zD_y f\left(\frac{y}{z}\right)$.

Take Equation (B.50) and observe that

$$D_y^\alpha f(x+y) = \sum_{k=0}^{n}\frac{1}{k!}\Delta_{x,z}^k f(x+z)D_y^\alpha\prod_{j=1}^{k}(y-jz)$$

$$= \sum_{k=\alpha}^{n}\frac{z^{k-\alpha}}{(k-\alpha)!}B_{k-\alpha}^{(k+1)}\left(\frac{y}{z}\right)\Delta_{x,z}^k f(x+z)$$

$$= \sum_{k=0}^{n-\alpha}\frac{z^k}{k!}B_k^{(k+\alpha+1)}\left(\frac{y}{z}\right)\Delta_{x,z}^{k+\alpha}f(x+z).$$

Let $y = 0$ and obtain

$$D_y^\alpha f(x+y)\big|_{y=0} = f^{(\alpha)}(x) = \sum_{k=0}^{n-\alpha} \frac{z^k}{k!} B_k^{(k+\alpha+1)}(0)\Delta_{x,z}^{k+\alpha} f(x+z), \quad f(x) = \sum_{i=0}^{n} a_i x^i,$$

(B.61)

which is the desired generalization of Equation (B.57).

To obtain the generalization for Equation (B.59) let $y = z$. This gives us

$$D_z^\alpha f(x+z) = \sum_{k=0}^{n-\alpha} \frac{z^k}{k!} B_k^{(k+\alpha+1)}(1)\Delta_{x,z}^{k+\alpha} f(x+z)$$

$$= \alpha \sum_{k=0}^{n-\alpha} \frac{z^k}{(k+\alpha)k!} B_k^{(k+\alpha)}(0)\Delta_{x,z}^{k+\alpha} f(x+z), \qquad \text{by Eq. (B.14).}$$

It is now a matter of setting $x \to x - z$ in the preceding line to obtain

$$D_x^\alpha f(x) = f^{(\alpha)}(x) = \alpha \sum_{k=0}^{n-\alpha} \frac{z^k}{(k+\alpha)k!} B_k^{(k+\alpha)}(0)\Delta_{x,z}^{k+\alpha} f(x), \qquad (\text{B.62})$$

whenever $f(x) = \sum_{i=0}^{n} a_i x^i$. Equation (B.62) appears as Relation (71) in [Nörlund, 1924, p.242].

We mention that Equation (B.60) is valid for $\alpha = 0$, in which case we have

$$z^k B_k^{(k+1)}\left(\frac{y}{z}\right) = \prod_{j=1}^{k}(y - jz).$$

(B.63)

We may substitute Equation (B.63) into Equation (B.50) to obtain

$$f(x+y) = \sum_{k=0}^{n} \frac{z^k}{k!} B_k^{(k+1)}\left(\frac{y}{z}\right) \Delta_{x,z}^k f(x+z), \qquad f(x) = \sum_{i=0}^{n} a_i x^i. \quad (\text{B.64})$$

Equation (B.64) has an advantage over Equation (B.50) in that it can be readily integrated. First set $y = tz$ to obtain

$$f(x+tz) = \sum_{k=0}^{n} \frac{z^k}{k!} B_k^{(k+1)}(t)\Delta_{x,z}^k f(x+z).$$

Next integrate with respect to $t$ where $0 \le t \le 1$. Since

$$\int_x^{x+1} B_n^{(n+1)}(t)\, dt = \frac{B_{n+1}^{(n+1)}(x+1) - B_{n+1}^{(n+1)}(x)}{n+1}$$

$$= \frac{\Delta_{x,1} B_{n+1}^{(n+1)}(x)}{n+1} = B_n^{(n)}(x),$$

$$\int_0^1 f(x+tz)\,dt = \sum_{k=0}^n \frac{z^k}{k!} \Delta_{x,z}^k f(x+z) \int_0^1 B_k^{(k+1)}(t)\,dt$$

$$= \sum_{k=0}^n \frac{z^k}{k!} B_k^{(k)}(0) \Delta_{x,z}^k f(x+z). \qquad \text{(B.65)}$$

If $t = x + tz$, Equation (B.65) is equivalent to

$$\frac{1}{z}\int_x^{x+z} f(t)\,dt = \sum_{k=0}^n \frac{z^k}{k!} B_k^{(k)}(0) \Delta_{x,z}^k f(x+z), \qquad f(x) = \sum_{i=0}^n a_i x^i. \text{ (B.66)}$$

For readers interested in learning more about generalized Bernoulli and generalized Euler polynomials we refer them to [Nörlund, 1924] and in particular Chapter 9 where Nörlund extends Equations (B.61), (B.62), and (B.66) to encompass $f(x) = \sum_{i=0}^\infty a_i x^i$. When dealing with the infinite extensions Nörlund carefully discusses issues of convergence. He then uses these infinite extensions to derive series representations for $x^{-\alpha}$ and $\log\left(1 + \frac{1}{x}\right)$ in terms of $B_k^{(k)}(0)$.

# Bibliography

J. Bernoulli (1713). *Ars conjectandi, opus posthumum. Accedit Tractatus de seriebus infinitis, et epistola gallic scripta de ludo pilae reticularis*, Basel: Thurneysen Brothers, OCLC 7073795

R. A. Brualdi (2014). *Introductory Combinatorics*, Fourth Edition, Pearson Prentice Hall, New Jersey.

L. Carlitz (1952). Note on a formula of Szily, *Scripta Mathematica*, Vol. 18, pp. 249-253.

L. Carlitz (1953). Remark on a formula for the Bernoulli numbers, *Proc. Amer. Math. Soc.*, Vol. 4, pp. 400-401.

L. Carlitz (1959). Eulerian numbers and polynomials, *Math. Magazine*, Vol. 32, pp. 247-260.

L. Carlitz (1960). Note on Nörlund's polynomial $B_n^{(z)}$, *Proc. Amer. Math. Soc.*, Vol. 11, pp. 452-455.

L. Carlitz (1961). The Staudt-Clausen theoerm, *Math. Magazine*, Vol. 34, pp. 131-146.

J. B. Conway (1978). *Functions of One Complex Variable I*, Second Edition, Springer-Verlag, New York.

D. File and S. Miller (2003). *Fundamental Theorem of Algebra Lecture Notes from the Reading Classics (Euler) Working Group, Autumn 2003*, available online at http://web.williams.edu/Mathematics/sjmiller/public_html/OSUClasses/683L/FundThmAlg_DFile.pdf.

G. Dobinski (1877). *Archiv der Mathematik und Physik*, Vol. 61, pp. 333-336.

R. Frisch (1926). Sur les semi-invariants et moments employés dans l'étude des distributions. statistiques, Skrifter utgitt av Det Norske Videnskaps-Akademi i Oslo, II. Historisk-Filosofisk Klasse. No 3, 87 pp. Quoted by Th. Skolem, p. 337, in Netto's Lehrbuch.

H. W. Gould (1960a). The Lagrange interpolation formula and Stirling numbers, *Proc. Amer. Math. Soc.*, Vol. 11, pp. 421-425.

H. W. Gould (1960b). Stirling number representation problems, *Proc. Amer. Math. Soc.*, Vol. 11, pp. 447-451.

H. W. Gould (1972). *Combinatorial Identities: A Standardized Set of Tables*

*Listing 500 Binomial Coefficient Summations*, Revised Edition, Morgantown, WV.

H. W. Gould (circa 1974). *Sums of Powers of Numbers*, unpublished 52 page manuscript.

H. W. Gould (1987). *Topics in Combinatorics*, second edition, Published by the author, Morgantown, W.Va.

H. W. Gould (2002). The generalized chain rule of differentiation with historical notes, *Utilitas Mathematica*, Vol. 61, pp. 97-106.

H. W. Gould and T. Glatzer (1979). *Catalan and Bell Numbers: Research Bibliography of Two Special Number Sequences*, fifth edition, ( Mathematica Monongaliae No. 12.); on-line edition available at http://www.math.wvu.edu/~gould/.

J. G. Hagen (1891). *Synopsis der höeheren Mathematik*, Berlin, Vol. I, pp. 79-80.

G. H. Halphen (1879-80). Sur une formula d'analyse, *Bull. Soc. Math France*, Vol. 8, 62-64; also found in *Oeuvres*, Vol. 2, pp. 447-449.

C. Jordan (1957). *Calculus of Finite Differences*, New York.

D. E. Knuth (2003). *Selected Papers on Discrete Mathematics*, xvi+812pp. (CSLI Lecture Notes, no. 106.).

Z. A. Melzak (1953). Problem 4458, *Amer. Math. Monthly*, Vol. 58(1951), 636. Solutions. ibid. 60, pp. 53-54. (R. V. Parker was one of the solvers.)

L. Moser (1948). Problem E799, *Amer. Math. Monthly*, Vol. 55(1948), 30. Solutions, ibid. 55, pp. 502-504.

T. S. Nanjundiah (1958). Remark on a note of P. Turán, *American Math Monthly*, Vol. 65, pp. 354.

N. Nielsen (1923). *Traité élémentaire des nombres de Bernoulli*, Paris, pp. 26-30.

N. E. Nörlund (1924). *Vorlesungen über Differenzenrechnung*, Springer-Verlag, Berlin, Germany.

F. R. Olson (1963). Abstract on talk entitled Some special determinants, *Amer. Math. Monthly*, Vol. 63, pp. 612.

*The OnLine Encyclopedia of Integer Sequences*, published electronically at http://oeis.org, 2015.

R. V. Parker (1953). Solution to Problem 4458, *Amer. Math Monthly*, Vol. 60, pp. 54.

E. D. Rainville (1960). *Special Functions*, Macmillan Company, New York.

K. A. Ross (1980). *Elementary Analysis: The Theory of Calculus*, Springer-Verlag, New York.

L. Schläfli (1852). Sur les coefficients du développement du produit $1(1 + x)(1 + 2x) \ldots (1 + (n-1)x)$ suivant les puissance ascendantes de $x$, *Crelle's Journal für die reine und angewandte Mathematik*, Vol. 43, pp. 1-22.

M. R. Spiegel (1964). *Theory and Problems of Complex Variables*, Schaum's Outline Series, McGraw-Hill.

J. Stewart (2007). *Essential Calculus: Early Transcendentals*, Thomson Brooks/Cole, California.

V. R. Rao Uppuluri and J. A. Carpenter (1969). Numbers generated by the function $\exp(1 - e^x)$, *Fibonacci Quarterly*, Vol. 7, pp. 437-448.

H. S. Wilf (2014). *generatingfunctionology*, Academic Press, New York.

B. C. Wong (1930). Problem 3399, *Amer. Math. Monthly*, Vol. 36, 543. Solution, ibid. 37(1930), pp. 322-324.

B C. Wong (1931). Problem 3426, *Amer. Math. Monthly*, Vol. 37, 260. Solution, ibid. 38(1931), pp. 116.

J. Worpitzky (1883). Studien über die Bernoullischen und Eulerschen Zahlen, *Jour. reine u. angew. Math.*, Vol. 94, pp. 203-232.

D. Zeilberger (n.d.). Personal communication with Doron Zeilberger.

# Index